THERMODYNAMICS
OF NUCLEAR MATERIALS
1979
VOL. II

The following States are Members of the International Atomic Energy Agency:

AFGHANISTAN	HOLY SEE	PHILIPPINES
ALBANIA	HUNGARY	POLAND
ALGERIA	ICELAND	PORTUGAL
ARGENTINA	INDIA	QATAR
AUSTRALIA	INDONESIA	ROMANIA
AUSTRIA	IRAN	SAUDI ARABIA
BANGLADESH	IRAQ	SENEGAL
BELGIUM	IRELAND	SIERRA LEONE
BOLIVIA	ISRAEL	SINGAPORE
BRAZIL	ITALY	SOUTH AFRICA
BULGARIA	IVORY COAST	SPAIN
BURMA	JAMAICA	SRI LANKA
BYELORUSSIAN SOVIET	JAPAN	SUDAN
SOCIALIST REPUBLIC	JORDAN	SWEDEN
CANADA	KENYA	SWITZERLAND
CHILE	KOREA, REPUBLIC OF	SYRIAN ARAB REPUBLIC
COLOMBIA	KUWAIT	THAILAND
COSTA RICA	LEBANON	TUNISIA
CUBA	LIBERIA	TURKEY
CYPRUS	LIBYAN ARAB JAMAHIRIYA	UGANDA
CZECHOSLOVAKIA	LIECHTENSTEIN	UKRAINIAN SOVIET SOCIALIST
DEMOCRATIC KAMPUCHEA	LUXEMBOURG	REPUBLIC
DEMOCRATIC PEOPLE'S	MADAGASCAR	UNION OF SOVIET SOCIALIST
REPUBLIC OF KOREA	MALAYSIA	REPUBLICS
DENMARK	MALI	UNITED ARAB EMIRATES
DOMINICAN REPUBLIC	MAURITIUS	UNITED KINGDOM OF GREAT
ECUADOR	MEXICO	BRITAIN AND NORTHERN
EGYPT	MONACO	IRELAND
EL SALVADOR	MONGOLIA	UNITED REPUBLIC OF
ETHIOPIA	MOROCCO	CAMEROON
FINLAND	NETHERLANDS	UNITED REPUBLIC OF
FRANCE	NEW ZEALAND	TANZANIA
GABON	NICARAGUA	UNITED STATES OF AMERICA
GERMAN DEMOCRATIC REPUBLIC	NIGER	URUGUAY
GERMANY, FEDERAL REPUBLIC OF	NIGERIA	VENEZUELA
GHANA	NORWAY	VIET NAM
GREECE	PAKISTAN	YUGOSLAVIA
GUATEMALA	PANAMA	ZAIRE
HAITI	PARAGUAY	ZAMBIA
	PERU	

The Agency's Statute was approved on 23 October 1956 by the Conference on the Statute of the IAEA held at United Nations Headquarters, New York; it entered into force on 29 July 1957. The Headquarters of the Agency are situated in Vienna. Its principal objective is "to accelerate and enlarge the contribution of atomic energy to peace, health and prosperity throughout the world".

 © IAEA, 1980

Printed by the IAEA in Austria
August 1980

PROCEEDINGS SERIES

THERMODYNAMICS
OF NUCLEAR MATERIALS
1979

PROCEEDINGS OF AN INTERNATIONAL SYMPOSIUM ON
THERMODYNAMICS OF NUCLEAR MATERIALS
HELD BY THE
INTERNATIONAL ATOMIC ENERGY AGENCY
IN JÜLICH, FEDERAL REPUBLIC OF GERMANY,
FROM 29 JANUARY TO 2 FEBRUARY 1979

In two volumes

VOL. II

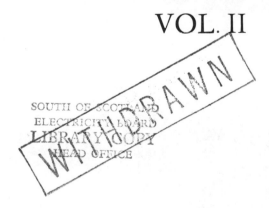

INTERNATIONAL ATOMIC ENERGY AGENCY
VIENNA, 1980

EDITORIAL NOTE

The papers and discussions have been edited by the editorial staff of the International Atomic Energy Agency to the extent considered necessary for the reader's assistance. The views expressed and the general style adopted remain, however, the responsibility of the named authors or participants. In addition, the views are not necessarily those of the governments of the nominating Member States or of the nominating organizations.

Where papers have been incorporated into these Proceedings without resetting by the Agency, this has been done with the knowledge of the authors and their government authorities, and their cooperation is gratefully acknowledged. The Proceedings have been printed by composition typing and photo-offset lithography. Within the limitations imposed by this method, every effort has been made to maintain a high editorial standard, in particular to achieve, wherever practicable, consistency of units and symbols and conformity to the standards recommended by competent international bodies.

The use in these Proceedings of particular designations of countries or territories does not imply any judgement by the publisher, the IAEA, as to the legal status of such countries or territories, of their authorities and institutions or of the delimitation of their boundaries.

The mention of specific companies or of their products or brand names does not imply any endorsement or recommendation on the part of the IAEA.

Authors are themselves responsible for obtaining the necessary permission to reproduce copyright material from other sources.

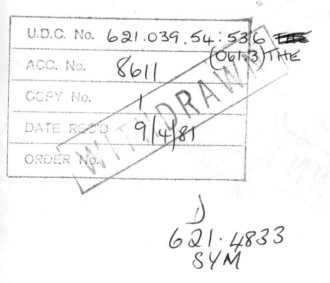
THERMODYNAMICS OF NUCLEAR MATERIALS 1979, VOL. II
IAEA, VIENNA, 1980
STI/PUB/520
ISBN 92−0−040180−5

FOREWORD

The design of nuclear fuels and their containment requires thermodynamic data on which to base estimates of fuel life, of cladding or coating performance and integrity, and of the behaviour of fuels under accident conditions involving high fuel temperatures and pressures. In the past 20 years, co-operation has been growing between the nuclear engineers and the scientists providing thermodynamic data. For example, nuclear engineers require such data on which to base their safety analyses and, because practical experience based on reactor accidents is almost totally lacking, they have to place considerable trust in the thermodynamicist's ability to provide them with accurate information. The cost of thermodynamics research in terms of materials, equipment, time and manpower is very high, and great experience is required in choosing the experiments or studies that will provide the necessary data. International meetings at which thermodynamicists and engineers meet to discuss data and to seek ways of obtaining information that is lacking serve to ensure that there is a minimum of unnecessary duplication of effort on a world-wide scale.

The International Atomic Energy Agency has played a major role in furthering the exchange of information on the thermodynamics of nuclear materials for over 17 years. In addition to the monographs being published on materials of importance in nuclear technology — for example the Atomic Energy Review Special Issues, and the series on The Chemical Thermodynamics of Actinide Elements and Compounds — and the holding of several small meetings on specific topics, the Agency has convened five symposia on the Thermodynamics of Nuclear Materials. The meetings of 1962, 1965 and 1967 covered the full range of basic thermodynamic studies and methodologies; the 1974 meeting continued the work of the earlier meetings and evidenced the growing interaction between thermodynamicists and engineers, with many studies initiated to obtain data that the engineers lacked. The 1979 Symposium, therefore, was aimed at 'applied thermodynamics' and at those basic studies of direct relevance to nuclear engineering. The very positive response engendered showed the value of supporting such co-operation between 'science' and 'technology'.

A significant portion of the symposium programme was devoted to oxide fuels, particular reference being made to fast breeder reactor applications. Oxygen diffusion and its effect on fuel/cladding interactions were discussed, while the behaviour of fission products, especially caesium, were considered in detail. Binary, ternary and even some quaternary phase diagrams were presented for fuel—fission product systems, these studies being based on both experimental and theoretical work; in one paper, the formation of two compounds was postulated that have yet to be experimentally identified. Sophisticated experiments had been undertaken to obtain experimental data at temperatures in excess of 7000 K, temperatures at which core components vaporize, and which could be expected to occur in core melt-down accidents. That such experiments were very necessary was supported by authors who showed that normal extrapolation of existing lower temperature data to temperatures above 2000 K could be seriously in error, if not completely valueless.

A new feature, one that will be of growing importance as moves are made towards setting up the first international fusion experiment (INTOR), is the presentation of work on the thermodynamics of materials required in fusion reactors — liquid lithium, solid lithium alloys and lithium-containing ceramics. The life of reactor blankets will depend on choosing suitable thermodynamic conditions for operation. Studies were also reported on diffusion of elements in glass matrices intended for fixing high-level radioactive waste. The stability of such vitreous storage media and the resistance of the glasses to leaching can be improved by considering the thermodynamic status of the major components.

The Proceedings of the 1979 Symposium is published in two volumes, and contains the texts of the 55 technical papers presented and of the resulting discussions. There is also a summary paper, and the publication includes a detailed subject index.

The Agency records with regret the death, shortly before the Symposium was held, of two eminent scientists who worked on the thermodynamics of nuclear materials, Dr. R.J. Ackermann (USA) and Professor O.S. Ivanov (USSR). They were both closely linked for many years with Agency activities on this subject, and their enthusiastic support will be sorely missed.

The Agency is indebted to the Government of the Federal Republic of Germany for the invitation to hold the meeting at the Jülich Nuclear Research Centre. The excellent facilities provided contributed in no small measure to the success of the meeting.

CONTENTS OF VOL. II

Section J: PHASE DIAGRAMS

Section K: STUDIES RELATED TO CLAD PERFORMANCE

Section L: MELT/CONCRETE INTERACTIONS

Section M: SUMMARY

Section H

FUSION THERMODYNAMICS

THE CURRENT STATUS OF
FUSION REACTOR BLANKET THERMODYNAMICS*

E. VELECKIS, R.M. YONCO, V.A. MARONI
Chemical Engineering Division,
Argonne National Laboratory,
Argonne, Illinois,
United States of America

Abstract

THE CURRENT STATUS OF FUSION REACTOR BLANKET THERMODYNAMICS.
The available thermodynamic information is reviewed for three categories of materials
that meet essential criteria for use as breeding blankets in D-T fuelled fusion reactors: liquid
lithium, solid lithium alloys, and lithium-containing ceramics. The leading candidate, liquid
lithium, which also has potential for use as a coolant, has been studied more extensively than
have the solid alloys or ceramics. Recent studies of liquid lithium have concentrated on its
sorption characteristics for hydrogen isotopes and its interaction with common impurity
elements. Hydrogen isotope sorption data (P-C-T relations, activity coefficients, Sieverts'
constants, plateau pressures, isotope effects, free energies of formation, phase boundaries, etc.)
are presented in a tabular form that can be conveniently used to extract thermodynamic
information for the α-phases of the Li-LiH, Li-LiD and Li-LiT systems and to construct complete
phase diagrams. Recent solubility data for Li_3N, Li_2O, and Li_2C_2 in liquid lithium are discussed
with emphasis on the prospects for removing these species by cold-trapping methods. Current
studies on the sorption of hydrogen in solid lithium alloys (e.g. Li-Al and Li-Pb), made using
a new technique (the hydrogen titration method), have shown that these alloys should lead to
smaller blanket-tritium inventories than are attainable with liquid lithium and that the P-C-T
relationships for hydrogen in Li-M alloys can be estimated from lithium activity data for these
alloys. There is essentially no refined thermodynamic information on the prospective ceramic
blanket materials. The kinetics of tritium release from these materials is briefly discussed.
Research areas are pointed out where additional thermodynamic information is needed for all
three material categories.

1. INTRODUCTION

Of all the potentially useful thermonuclear fuel cycles, the
deuterium-tritium reaction

$$D + T \rightarrow {}^4He(3.5 \text{ MeV}) + n(14.1 \text{ MeV}) \tag{1}$$

remains the leading candidate for near-term fusion reactor applications because of its relatively low ignition temperature ($\sim 10^8$K) and

---

* Work performed under the auspices of the United States Department of Energy. The
work originating at Argonne National Laboratory was supported by the Division of Basic
Energy Sciences, United States Department of Energy.

3

because most of the thermonuclear energy appears in the form of energetic neutrons which can be converted to thermal energy using the same principles that are employed in present-day fission reactors [1]. The deuterium required for fueling a D-T reactor is in sufficient natural abundance to last for many hundreds of years. The tritium, however, does not occur naturally in significant quantities and must be produced by some other means. While the prospects for supplying tritium by external manufacture appear to be severely constrained from an economic viewpoint [2], breeding tritium within the reactor is well recognized as a practically achievable solution [3]. This breeding can be accomplished via the $^6Li(n,\alpha)T$ and $^7Li(n,n'\alpha)T$ nuclear reactions, and in some designs it must be further augmented by the addition of materials with large $(n,2n)$ cross-sections ($e.g.$, Be,Pb).

The basic features of a D-T fusion reactor consist of a magnetically confined D-T plasma (or an inertially confined D-T pellet) contained in an evacuated chamber. The chamber is surrounded by a blanket material that is interspersed with some type of coolant/heat transfer medium. The tritium breeding and neutron thermalization occur in the blanket from which the thermal energy is carried off by the coolant in much the same way as in a fission reactor.

Although the technology required to effect the utilization of fusion energy must still be considered to be in the very early stages of development, a significant amount of encouraging information has been generated in recent years [1,4,5]. The availability of the thermodynamic data to support work in many areas of fusion technology will be pivotal to the development of workable concepts for fusion reactors. We have limited the scope of this paper to recent work on the thermodynamic characteristics of plausible blanket material options for D-T fueled fusion reactors. There are, of course, other areas of fusion technology where thermodynamic considerations are important, $e.g.$, cryogenics, fluid/structure interactions, fuel purification, isotopic enrichment, and surface chemistry. The future expansion of the supportive research base for fusion technology will, no doubt, spur the preparation of separate treatises in these areas.

This review will summarize the pertinent data on three basic classes of prospective blanket materials: liquid lithium, solid lithium alloys, and ceramics. Emphasis will be placed on dilute solutions of hydrogen isotopes and other non-metallic elements in these materials and on the contribution of such data to the development of blanket processing technology.

2.   STATUS REPORT

In developing generic criteria for the selection of potential breeding materials the following factors are important: (1) the lithium atom density must be high enough to yield a positive breeding gain in a reasonable thickness (<1 m) of material; (2) the constituents present with the lithium must not have parasitic effects on the neutron flux and energy distribution; and (3) the material must be chemically stable in the blanket environment. Other considerations [3] which relate to tritium recovery, hydraulic and mechanical performance of the blanket, availability of resources, and blanket system maintainability are highly concept-dependent and must be weighed against one another in evaluating alternative design choices.

2.1. Liquid lithium

Liquid lithium is regarded as a leading candidate for the breeding medium. Its major attributes are its high atomic density which results in a favorable tritium breeding gain, low melting point which makes it possible to circulate lithium into and out of the blanket for processing, and excellent heat transfer characteristics which make it potentially useful as a primary reactor coolant. The drawbacks of liquid lithium stem from its high reactivity with air and water, its corrosivity, and its large MHD interactions with the confining fields of magnetic fusion reactors [6].

2.2. Solid lithium alloys

A large number of the fusion reactor design concepts advanced to date have incorporated solid lithium-containing alloys as the breeding material [3]. A number of binary lithium alloy systems are known to have a least one Li-rich, high melting compound (e.g., LiAl, Li_7Pb_2, Li_4Si, Li_4Sn, and Li_3Bi) that has the capability of yielding an adequate breeding gain. Cursory thermal hydraulic analyses [3] have indicated that the nuclear heat generated in appropriately-sized pebbles of these alloys can be removed without running into meltdown problems or compromising overall reactor power cycle efficiency. Bench-scale studies [7,8] on selected alloys have shown that, under steady-state tritium production conditions, it is possible to reduce their tritium concentrations to acceptable levels for fusion reactor blankets (<1 ppm T by wt) by extracting tritium into a stream of helium.

One design study has appeared [9] that employs a liquid Li-Pb alloy (~38 at. % Pb) as the coolant and breeding medium. During the refueling phase of the reactor operating cycle, the alloy is allowed to freeze (F.P. = 450°C) in the heat exchangers so that the heat of fusion of the alloy can sustain the energy conversion cycle. This design tends to facilitate tritium removal but may create materials compatibility problems that are more severe than those with lithium alone.

2.3. Ceramic materials

Numerous fusion reactor conceptual designs have employed solid lithium-containing ceramics as the breeding material [3]. With the exception of Li_2O which has adequate lithium atom density, most of the materials considered have marginal breeding gains and would require addition of neutron multiplication schemes. Although tritium recovery from the ceramics appears to be straightforward [7,8,10], the adequacy of their thermal hydraulic performance, stability in

the radiation field, and their compatibility with structural materials
at elevated temperatures remains to be demonstrated.

3.   THERMODYNAMIC STUDIES RELATED TO LIQUID LITHIUM BLANKETS

Of all the candidate blanket materials discussed in the preceding
section, liquid lithium has been subjected to the most extensive
study from a thermodynamic viewpoint. This section contains a comprehensive review of recent work on the thermodynamics of the systems
comprised of lithium with selected non-metallic elements and the
implications of the results to fusion reactor technology.

3.1. The Li-LiH, Li-LiD, and Li-LiT systems

Until recently, information on the thermodynamics of the systems
of lithium with hydrogen isotopes has been sparse. Early work on the
Li-LiH system was summarized in a 1960 review article by Messer [11]
and in a 1972 book "Lithium Hydride" by Shpilrain and Yakimovich [12].
The 1968 book "Metal Hydrides" also contains a chapter on saline hydrides [13]. In this section we will discuss the more recent work on
the Li-LiH, Li-LiD, and Li-LiT systems that was not covered by the above
reviews.

3.1.1. Isothermal studies

The measurement of the pressure-composition-temperature (P-C-T)
relationships under isothermal conditions constitutes perhaps the
most widely accepted technique used in investigating the metal-hydro-
gen systems. The data produced by this technique form a family of $\sqrt{P}$
vs. C isotherms whose shapes reflect the phase relationships in the
system under investigation. In the lithium systems, each isotherm
is comprised of two rising portions that are separated by a horizontal
plateau [14,15]. The composition ranges of the first and second
rising portions correspond to homogeneous α- and β-phases. The constant pressure plateau defines a two-phase (α + β) coexistence region.

The isothermal reaction between gaseous hydrogen[1] and liquid
lithium and the corresponding equilibrium constant (K) can be
written as

$$Li(soln) + \tfrac{1}{2} H_2(g) \rightleftarrows LiH(soln) \tag{2}$$

$$K = N_2\gamma_2/(N_1\gamma_1\sqrt{P}) \tag{3}$$

where the subscripts 1 and 2 refer to Li and LiH, respectively, N is
the mole fraction, γ, the activity coefficient, P is the equilibrium
hydrogen pressure, and the hydrogen is assumed to be dissolved in

---

[1] In order to simplify notation, the hydrogen symbol serves as a stand-in for all three
hydrogen isotopes.

lithium as a monohydride species [13]. It is customary to express γ
as a power series in N according to Margules' equations [16], *i.e.*,

$$\ln \gamma_1 = \alpha N_2{}^2 + \beta N_2{}^3 \tag{4}$$

and

$$\ln \gamma_2 = (\alpha + {}^3\!/_2\beta) N_1{}^2 - \beta N_1{}^3 \tag{5}$$

Substitution of eqs. 4 and 5 in eq. 3 and rearrangement gives a second-order power series in N_2

$$\ln(\sqrt{P} \cdot N_1/N_2) = -\ln K + \alpha(1 - 2N_2) + \frac{1}{2}\beta(1 - 3N_2{}^2) \tag{6}$$

in which experimentally measured quantities appear on the left-hand
side and the unknown, temperature-dependent constants (K, α, and β)
are on the right-hand side.

Experimental $\sqrt{P}$ *vs.* N_2 data can be subjected to iterative least-squares procedures to give numerical expressions for the parameters
ln K, α, and β as linear functions of $1/T$. These data fitting procedures
have been described in detail elsewhere [14,15]. The resulting
thermodynamic expressions for the lithium-rich (α) phases of the Li-LiH
and Li-LiD systems are given in TABLE I. They are based on work carried
out at the Argonne National Laboratory (ANL) [14,15,17]. Sieverts' law
constants reported at the Oak Ridge National Laboratory (ORNL) [18,19]
are also listed in TABLE I.

In liquid lithium blankets hydrogen isotopes are expected to
be present at very low concentrations. Thus, for the condition $N_2 \rightarrow 0$,
eq. 6 becomes

$$\ln K_S = \ln K - \alpha - \frac{1}{2}\beta \tag{7}$$

where $K_S = N_2/\sqrt{P}$ is the Sieverts' law constant corresponding, in this
case, to the change in state $RT \ln K_S = \mu^\circ_{Li}(\ell,T,P) + \mu^\circ_{H_2}(g,T,1 \text{ Atm}) - \mu^*_{LiH}(T,P)$, where μ^*_{LiH}, μ°_{Li}, and $\mu^\circ_{H_2}$ are the standard chemical potentials for the species indicated defined in terms of an infinitely
dilute solution, a pure liquid and a pure ideal gas, respectively.

The analytical form of the results in TABLE I permits an extrapolation into dilute solution regions where the data are important
for fusion reactor applications. The reliability of such an extrapolation was tested by comparisons with the data that had been obtained
experimentally for the Li-LiD system by Smith *et al.* [18], by Goodall
and McCracken [15,20], and by Ihle and Wu [21]. The latter two studies
were made at very low deuterium concentrations using mass-spectrometric
techniques. As shown in Fig. 1, the agreement among the four studies
is very good thus verifying the accuracy of the equations in TABLE I.

3.1.2. Solubility studies

Above the monotectic temperatures the solubility of hydrides in
the α-phases can be deduced from the points of intersection of the first

TABLE I. Thermodynamic Data for the α-Fields of the Li-LiH, Li-LiD, and Li-LiT Systems.

P-C-T Relations [14,15]

$$\ln(P_{H_2}/atm)^{\frac{1}{2}} = \ln(N_{LiH}/N_{Li}) + 6.498 + 0.4589\ N_{LiH} + 0.1558\ N_{LiH}^2$$
$$- (1/T)(6182 + 1973\ N_{LiH} + 3462\ N_{LiH}^2)$$

$$\ln(P_{D_2}/atm)^{\frac{1}{2}} = \ln(N_{LiD}/N_{Li}) + 6.138 + 1.8516\ N_{LiD} - 1.3010\ N_{LiD}^2$$
$$- (1/T)(5599 + 3492\ N_{LiD} + 1951\ N_{LiD}^2)$$

Activity Coefficients of the Solute Species [17]

$$\ln\ \gamma_{LiH} = 1.135\ N_{Li}^2 - 2.690\ N_{Li}^3 + (1/T)(3118\ N_{Li}^2 + 211.4\ N_{Li}^3)$$

$$\ln\ \gamma_{LiD} = 1.124\ N_{Li}^2 - 2.776\ N_{Li}^3 + (1/T)(3285\ N_{Li}^2 + 143.8\ N_{Li}^3)$$

Sieverts' Law Constants[a] Reported at ANL [14,15]

Li-LiH: $\ln(K_S/atm^{-\frac{1}{2}}) = -6.498 + 6182/T$

Li-LiD: $\ln(K_S/atm^{-\frac{1}{2}}) = -6.138 + 5599/T$

Sieverts' Law Constants[a] Reported at ORNL [18,19]

Li-LiH: $\ln(K_S/atm^{-\frac{1}{2}}) = -6.525 + 6242/T$

Li-LiD: $\ln(K_S/atm^{-\frac{1}{2}}) = -6.198 + 5644/T$

Li-LiT: $\ln(K_S/atm^{-\frac{1}{2}}) = -5.909 + 5085/T$

---

[a]The Sieverts' law constant is defined by $K_S = N_2/\sqrt{P}$, where N_2 is the mole fraction of LiH, LiD, or LiT in lithium and P is the pressure of H_2, D_2, or T_2 in atmospheres.

rising portion and plateau portion of the isotherms. Below the mono-tectic temperature, where the precipitating phases are solid hydrides, this technique is ineffective [14] and more direct experimental methods have been sought. At the University of Nottingham the solubilities of LiH [22] and LiD [23] in lithium have been determined using a resisto-metric technique; at ANL the solubility of LiD in lithium was measured by a direct melt-sampling technique [17]. The results of these investi-gations are shown in Fig. 2 on a plot of 1/T vs. the logarithm of the mole fraction of hydride. In this particular type of plot the data are presented as a conventional phase diagram but with an expanded scale towards low temperatures where the solubility data are more important for fusion reactor applications.

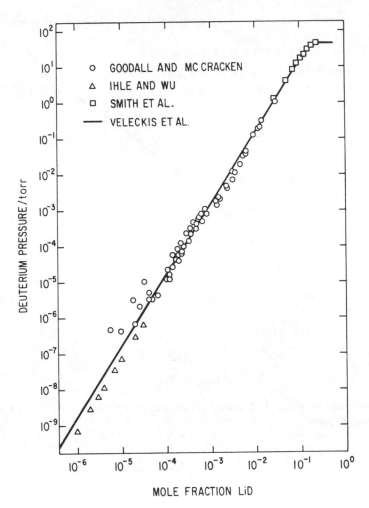

FIG.1. Comparison of the calculated P versus N isotherm with the literature data at 700°C for the Li-LiD system.

        The solubility data in Fig. 2 do not lie on straight lines. Smooth
liquidus curves, best representing the observed data, can be determined
as follows [17]. Consider a solution having a composition that corresponds to the liquidus curve at the monotectic temperature. At this
point the activity of hydride is near unity and, therefore, the monotectic temperature may be looked upon as the freezing point of pure
hydride (solvent). As more lithium (solute) is added, the freezing
point of the solution will be lowered as shown by the descending solubility curves in Fig. 2. The activity of saturated solvent, a_{LiH}, may

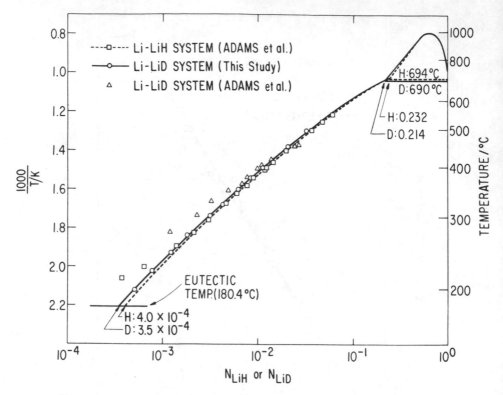

FIG.2. Phase diagrams for the Li-LiH and Li-LiD systems projected on the (1/T) versus log (composition) plane.

be related to the freezing point lowering [24] by the equation

$$\ln a_{LiH} = \ln \gamma_{LiH} N_{LiH}$$

$$= \ln a^\circ_{LiH} - \frac{\Delta H_m}{R} \frac{\theta}{TT_m} + \frac{\Delta a}{R} \left[\frac{\theta}{T} + \ln(1 - \frac{\theta}{T_m}) \right] + \frac{\Delta b}{R} \frac{\theta^2}{2T} \quad (8)$$

where γ_{LiH} and N_{LiH} are the activity coefficient and mole fraction of LiH at the liquidus line, $\Delta H_m/cal \cdot mol^{-1} = 5307$ is the latent heat of fusion of LiH [25], T_m is the monotectic temperature, $\theta = (T_m - T)$ is the freezing point lowering, $\Delta C_p/cal \cdot mol^{-1} \cdot K^{-1} = \Delta a + \Delta bT = 20.61 - 0.0192 T$ is the difference in the heat capacities between liquid and solid LiH [25], and $a^\circ_{LiH} \cong 0.98$ is the activity of LiH at the monotectic temperature, serving here as the standard state for liquid LiH. Eq. 8 can be taken to be valid also for LiD, assuming identical values for ΔC_p and ΔH_m.

TABLE II. Miscibility Gap Boundaries of the Li-LiH and Li-LiD
Systems.[a]

Between the Eutectic and Monotectic Temperatures [17, 22, 23][b]

$\ln N'_{LiH} - 10.372 \ln T + 4.8314 \times 10^{-3} T + (68.024 + 487.77/T)$
$\quad\quad + (5.8000 - 6870.2/T) N'_{LiH} - (6.9350 - 3752.2/T) N'^{2}_{LiH}$
$\quad\quad + (2.6900 - 211.40/T) N'^{3}_{LiH} = 0$

$\ln N'_{LiD} - 10.372 \ln T + 4.8314 \times 10^{-3} T + (67.917 + 591.35/T)$
$\quad\quad + (6.0800 - 7001.4/T) N'_{LiD} - (7.2040 - 3716.4/T) N'^{2}_{LiD}$
$\quad\quad + (2.7760 - 143.80/T) N'^{3}_{LiD} = 0$

Above the Monotectic Temperature [14,15]

$\ln N'_{LiH} = 2.235 - 3576/T$[c]
$\ln N'_{LiD} = 2.604 - 3992/T$[c]

$N''_{LiH} = 0.984 \text{(at 710°C)}, \ 0.971 \text{(759°C)}, \ 0.957 \text{(803°C)},$
$\quad\quad 0.923 \text{(847°C)}, \ 0.888 \text{(878°C)}, \text{ and } 0.855 \text{(903°C)}$

$N''_{LiD} = 0.980 \text{(at 705°C)}, \ 0.972 \text{(756°C)}, \ 0.953 \text{(805°C)},$
$\quad\quad 0.931 \text{(840°C)}, \text{ and } 0.905 \text{(871°C)}$

---

[a]The prime and double prime refer to the lithium-rich and
hydride-rich limits of the miscibility gap, respectively.

[b]In the temperature range 200-600°C these expressions can
be approximated by the linear equations: $\ln N'_{LiH} =$
$3.769 - 5472/T$ and $\ln N'_{LiD} = 3.165 - 5117/T$.

[c]These equations are valid up to ∿900°C.

Substitution of equations given in TABLE I for γ_{LiH} and γ_{LiD} into
eq. 8 yields expressions for the solubility of LiH and LiD in lithium
between the eutectic and monotectic temperatures. These expressions
and the miscibility gap boundary data for temperatures above the monotectic are presented in TABLE II.

Extrapolation of the liquidus curves to the eutectic temperature
yields the eutectic compositions and the freezing point depressions for
Li-LiH and Li-LiD. Since the isotope effects along the liquidus lines
are small, approximately the same eutectic composition would apply also
to the Li-LiT system. The lowest tritium concentration achievable by
direct cold-trapping of lithium would be well in excess of 100 ppm by wt —
a level that would be intolerably high for a fusion reactor blanket
system.

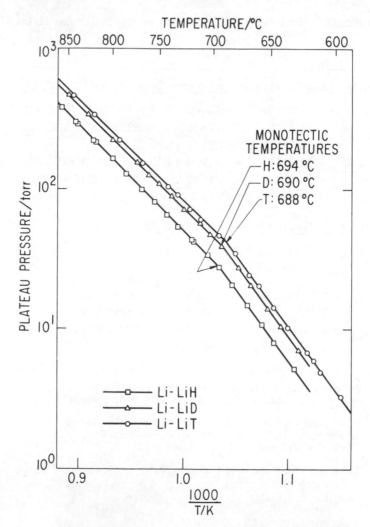

FIG.3.   *Plateau pressure data for the Li-LiH, Li-LiD, and Li-LiT systems.*

3.1.3. Decomposition pressures in the (α + β) fields

Within the (α + β) fields of the Li-LiH phase diagrams, the decomposition pressures do not vary with composition of the corresponding condensed phases. This property tends to simplify experimental procedures to allow the inclusion of tritium and the determination of precise H/D/T isotope effects.

Several recent studies on the plateau pressures have been reported. For temperatures above the monotectic, measurements have been made by

TABLE III. Decomposition Pressures in the $(\alpha + \beta)$ Fields of the
Li-LiH, Li-LiD, and Li-LiT Systems [31].

	ln(P/Atm) = A + B/T			
	Below Monotectic		Above Monotectic	
	A	B (10³K)	A	B (10³K)
Li-LiH	20.79	-23.31	14.71	-17.42
Li-LiD	21.19	-23.31	14.52	-16.87
Li-LiT	21.35	-23.31	14.29	-16.53

Heumann and Salmon [26], by Smith and Land [27], and by Shpilrain *et al.*
[28] using the customary equilibration techniques. Below the monotectic, measurements were made for the Li-LiH system by Ihle and Wu [29],
using a combined mass spectrometric-Knudsen effusion technique and by
Katsuta *et al.* [30], using gas chromatography. At ANL we used an
equilibration technique to make comprehensive plateau pressure measurements both below and above the monotectic temperatures for all three
hydrogen isotopes [31]. The results reported in this section are based
mainly on this latter study.

Figure 3 shows the plateau-pressure plots. For a given tempera-
ture the pressures are in the order $P_{T_2} > P_{D_2} > P_{H_2}$. Each plot has two
distinct linear segments. The point at which the segments intersect
defines the monotectic temperature. Linear ln P = A + B/T equations
derived for each segment are listed in TABLE III. Above the monotectic temperature these equations represent least-squares fits of the
data in Fig. 3. Below the monotectic the plateau pressure data were
collectively fitted to individual equations (shown in TABLE III) in
which the parameter B was constrained to a value common to all three
systems. These equations yield temperature-independent isotope
effects, $P_{D_2}/P_{H_2} = 1.492$ and $P_{T_2}/P_{H_2} = 1.751$, that are only slightly
different from $\sqrt{2}$ and $\sqrt{3}$ as predicted from the D/H and T/H mass ratios.

3.1.4. Species characterization in the gas phase

In addition to hydrogen molecules, the gas phase over very dilute
solutions of hydrogen in lithium is expected to include molecular
species containing lithium atoms. The complexity of the gas phase has
been recently revealed by Ihle and Wu [32]. Employing the combined mass
spectrometric-Knudsen effusion method, they observed the following
molecular species in the gas phase over dilute solutions of deuterium
in lithium: Li_2, Li_3, LiD, LiD_2, Li_2D, Li_2D_2, and D_2. From the data
they evaluated the partial pressures of each species as functions of
liquid phase composition ($10^{-7} < N_D < 10^{-4}$) and temperature (800 to 1600 K),

their atomization and binding energies, and the equilibrium constants for the exchange reactions among the species.

Although the existence of gaseous hydrides does not affect the Sieverts' law relationships [32], it leads to an increased density of hydrogen isotopes in the saturated vapors over the liquid lithium blanket. Ihle and Wu [33], have pointed out that, owing to this enrichment, it is feasible, in principle, to separate hydrogen isotopes from lithium by fractional distillation performed at temperatures above 1240 K.

3.1.5. Summary of thermodynamic data

The general features of the Li-LiH phase diagrams have been known for many years [11,12]. There is a wide monotectic horizontal along which two liquid phases, one rich in lithium (α) the other rich in hydride (β), are in equilibrium with essentially stoichiometric LiH. Above the monotectic temperature there is a miscibility gap [$\alpha(\ell)+\beta(\ell)$] that closes at the consolute point. Below the monotectic temperature there is a coexistence region [$\alpha(\ell)+\text{LiH}(s)$] that terminates at the eutectic temperature. The data presented in section 3.1 permit an accurate construction of the phase diagrams for the Li-LiH and Li-LiD systems and a good approximation for the Li-LiT system, owing to the small effect of isotopic substitution on the phase boundary compositions. The liquidus lines can be drawn by solving the equations in TABLE II. Other phase boundary data required for phase diagram construction are given in Figs. 2 and 3.

In addition to phase boundary information, the relations in TABLES I, II, and III can be used to derive various properties pertaining to the solution thermodynamics of the Li-LiH systems. For example, they have been used to estimate the standard free energies of formation of the liquid [14,15] and solid [31] hydrides (see TABLE IV).

3.2. Interaction of non-hydrogenous elements with liquid lithium

The binary systems of lithium with non-metals other than hydrogen isotopes are of interest to the fusion program primarily from the standpoint of materials compatibility. These systems have been reviewed by Messer [34], by Cairns et al. [35], by Adams et al. [36], and by Smith and Moser [37]. Because of the limited scope of this presentation, we will concentrate on published work related to the systems Li-N, Li-O, and Li-C which has appeared since the earlier reviews.

3.2.1. The Li-N system

Nitrogen gas reacts with lithium to form Li_3N which exists in liquid lithium solutions probably as an ionic species, since the phase in equilibrium with the saturated solutions is solid Li_3N and since ammonia is evolved upon hydrolysis of these solutions.

The system $Li-Li_3N$ was first investigated by Bolshakov et al. [38] who employed thermal analysis techniques to determine the liquidus line

TABLE IV. Standard Free Energy of Formation of Solid and
Liquid Hydrides.

Above the Monotectic Temperature[a] [14,15]

   LiH(ℓ): ΔG_f^o/kcal·mol^{-1} = 13.47 x 10^{-3}T - 16.55

   LiD(ℓ): ΔG_f^o/kcal·mol^{-1} = 13.17 x 10^{-3}T - 15.87

Below the Monotectic Temperature [31]

   LiH(s): ΔG_f^o/kcal·mol^{-1} = 19.76 x 10^{-3}T - 22.63

   LiD(s): ΔG_f^o/kcal·mol^{-1} = 20.30 x 10^{-3}T - 22.73

   LiT(s): ΔG_f^o/kcal·mol^{-1} = 20.64 x 10^{-3}T - 22.82

---

[a]Smith *et al.* [18b] have recently reported the following
expressions for liquid hydrides.

   LiH(ℓ): ΔG_f^o/kcal·mol^{-1} = 12.83 x 10^{-3}T - 15.71

   LiD(ℓ): ΔG_f^o/kcal·mol^{-1} = 11.86 x 10^{-3}T - 14.18

   LiT(ℓ): ΔG_f^o/kcal·mol^{-1} = 10.94 x 10^{-3}T - 12.72 .

between the eutectic temperature and the melting point of Li$_3$N. Adams
et al. [22] determined the solubility of Li$_3$N in lithium from 200 to
450°C using a resistometric technique in which the saturation concentrations are detected by corresponding breaks in the resistivity
curves. Yonco *et al.* [39] determined the solubility of Li$_3$N in lithium
from 195 to 441°C by taking filtered samples of saturated solutions and
assaying them for nitrogen using the micro-Kjeldahl method. They also
measured the equilibrium nitrogen pressure over solid Li$_3$N between 660
and 778°C and the melting point of Li$_3$N (81$_5$ ± 1°C).

Figure 4 shows a plot of log(mol % Li$_3$N) *vs.*1/T constructed from
the composite results of these three studies. Below 450°C the combined
data of the latter two studies were statistically fitted to the linear
equation

$$\ln N_{Li_3N} = 3.086 - 4878/T \tag{9}$$

with 4% uncertainty in the mean value of N_{Li_3N}. Above 450°C the
solubility can be represented by the curve that is formed by extrapolation of eq. 9 to a point where the line joins smoothly with the
data of Bolshakov *et al.*, although the exact shape of the curve at
the melting point of Li$_3$N is subject to question [39].

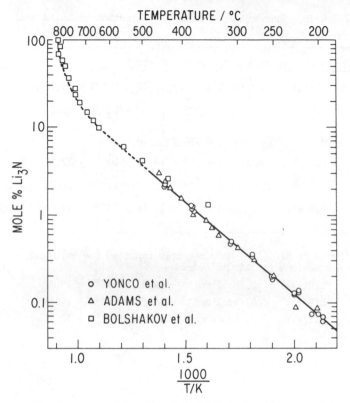

FIG.4.   *Comparison of data for the solubility of Li₃N in liquid lithium.*

A eutectic point of 0.047 mol % Li₃N (940 ppm N by wt) and 180.2°C, inferred from Eq. 9 and from the freezing point depression of lithium, is in good agreement with the values of 0.068 mol % Li₃N and 180.24°C reported by Hubberstey *et al.* [40] who determined the hypoeutectic liquidus by thermal analysis. The high eutectic concentration precludes cold-trapping as an effective way to remove nitrogen from lithium.

The solubility information in Fig. 4, when used in conjunction with the Li₃N decomposition pressures [39], permits the evaluation of the thermodynamic properties of the Li-Li₃N system. The equilibrium reaction between gaseous nitrogen and liquid lithium to form a liquid solution of Li₃N in lithium and the corresponding equilibrium constant (K) may be written as

$$\tfrac{1}{2}\ N_2\,(g) + 3\ Li\,(soln) \rightleftarrows Li_3N\,(soln) \tag{10}$$

$$K = N_2\gamma_2 / (N_1{}^3\gamma_1{}^3\sqrt{P}) \tag{11}$$

where N_1 and N_2 are the mole fractions of Li and Li_3N, γ_1 and γ_2 are their activity coefficients, P is the nitrogen pressure, and the standard states of Li and Li_3N are the pure liquid phases.

A reasonable estimate of the γ's can be made from the Margules' equations [16], truncated at the quadratic term. For the solvent lithium this would give

$$\ln \gamma_1 = (\omega/RT) \, N_2^2 \qquad\qquad (12)$$

where ω is a temperature-independent constant. By application of the Gibbs-Duhem relationship to eq. 12 one obtains for the activity coefficient of the solute Li_3N

$$\ln \gamma_2 = (\omega/RT)(N_1^2 - N_1'^2) - \ln N_2' \qquad\qquad (13)$$

where the primed symbols refer to the values at the liquidus boundary. The constant ω was evaluated [39] by substituting eq. 12 into eq. 11, selecting values of P and N_2 existing at the liquidus boundary, defining solid Li_3N as the standard state (*i.e.*, $N_2'\gamma_2' = 1$), and choosing a value for ω such that a plot of $\ln K$ *vs.* $1/T$ would give a statistically best straight line. The resulting linear relationship (at $\omega = 3.14$ kcal/mol) for $\ln K$ *vs.* $1/T$ yielded the following standard free energy of formation of solid Li_3N

$$\Delta G_f^o/\text{kcal·mol}^{-1} = 33.2 \times 10^{-3}T - 39.1 \qquad\qquad (14)$$

In the temperature range 600-1000 K, ΔG_f^o values calculated from eq. 14 are 1.5 to 2.1 kcal/mol less negative than those listed in the JANAF tables for solid Li_3N [41], based on the recent calorimetric work of O'Hare and Johnson [42] and of Osborne and Flotow [43]. The discrepancy can be considered to be small when one takes into account the approximations made by Yonco *et al.* [39] in deriving the expression for ΔG_f^o. The coefficients in eq. 14 also compare favorably with those reported in the recent emf studies by Bonomi *et al.* [44] ($\Delta G_f^o/\text{kcal·mol}^{-1} = 35.2 \times 10^{-3}T - 40.65$).

Dilute solutions of Li_3N in lithium are important in the use of lithium as the blanket material. For the conditions $N_2 \to 0$ and $N_1\gamma_1 \to 1$, eq. 11 may be rewritten to define the Sieverts' law constant, K_S. Thus,

$$\lim_{N_2 \to 0} N_2/\sqrt{P} = K_S = K/\gamma_2^* \qquad\qquad (15)$$

where γ_2^* is the activity coefficient of Li_3N at infinite dilution, calculated by applying the condition $N_1 \to 1$ to eq. 13. For the temperature range 661-778°C, the Sieverts' law constant can be represented by

$$\ln(K_S/\text{atm}^{-\frac{1}{2}}) = -13.80 - 14\,590/T \qquad\qquad (16)$$

3.2.2 The Li-O system

The compounds Li_2O and Li_2O_2 are known to exist in the lithium-oxygen system, although only the normal oxide is formed by the direct reaction of oxygen with lithium [34]. Available thermochemical information for Li_2O and Li_2O_2 has been reviewed elsewhere [37]. Yonco et al. [45] determined the solubility of Li_2O in lithium for temperatures between 195 and 734°C by taking equilibrated melt samples filtered through 2 μm pore size filters. The samples were assayed for oxygen by a fast neutron activation method. The data can be represented by

$$\ln N_{Li_2O} = 1.449 - 6669/T \qquad (17)$$

with a correlation coefficient of 0.997.

The solubilities calculated from eq. 17 are considerably lower than those reported previously by Hoffman [46] and by Konovalov et al. [47] who used similar experimental techniques but coarser filters (20μm [46] and 30-40μm [47]). Since undissolved Li_2O may pass through coarse pore filters [35], the lower solubility values reported in the ANL work are preferred over those of the earlier studies.

The solubility of oxygen at the melting point of lithium (180.49°C) is 1.76×10^{-4} mol % Li_2O (4 ppm O by wt). This concentration defines a theoretical limit for oxygen removal from liquid lithium that can be reached by cold trapping in conjunction with the use of fine pore size (<2 μm) filters.

3.2.3. The Li-C system

This system appears to have a single intermediate phase, Li_2C_2, which dissolves in lithium as an acetylide species. Carbon added to lithium in any form will ultimately be converted to acetylide if heated to 600°C [48].

The solubility of acetylide in liquid lithium is being measured at ANL by Yonco et al. [49] by taking filtered samples (2 μm pore size filters) of saturated solutions of Li_2C_2 in lithium and analyzing for carbon by the acetylene evolution method [48]. The data collected thus far vary between 9.0×10^{-5} mol % Li_2C_2 at 200°C and 2.7×10^{-2} mol % Li_2C_2 at 600°C. The indicated solubility at the melting point of lithium is 5×10^{-5} mol % Li_2C_2 (2 ppm C by wt) which strongly implies that carbon elimination by cold-trapping methods should be feasible in liquid lithium systems. Thermal analysis studies of Fedorov and Su [50] yield acetylide solubilities that are at least ten times higher than those of Yonco et al. and are believed to be in error [40]. The available thermochemical data for Li_2C_2 have been summarized elsewhere [37].

Recently Down *et al.* [51] reported the synthesis of dilithium cyanamide by reaction of lithium nitride with dilithium acetylide in excess lithium

$$4 \, Li_3N + Li_2C_2 \rightarrow 2 \, Li_2NCN + 10 \, Li \tag{18}$$

This is the first observation of a complex salt whose stability in liquid lithium is greater than that of the binary reactants, and could contribute significantly to an understanding of lithium corrosion phenomena and nitrogen/carbon chemistry in liquid lithium.

4.   THERMODYNAMIC STUDIES RELATED TO SOLID BLANKETS

The sorption characteristics of solid lithium-containing alloys towards the hydrogen isotopes constitute important considerations in evaluating their suitability as fusion reactor blanket materials. There is very little information available on this subject in the literature.   Early work, devoted mainly to the thermal decomposition of higher hydrides (*e.g.* LiAlH₄ [52]), has contributed little to the understanding of the dilute solutions of hydrogen in Li-M alloys. The dissolution of hydrogen in aluminum-rich solid Li-Al alloys has been analyzed by Owen and Randall [53].   The sorption of hydrogen in solid Li-Al and Li-Pb is currently under a thorough investigation at ANL using the hydrogen titration method (HTM) [54].   The latter studies have shown that hydrogen sorption characteristics in solid Li-M alloys can also be determined indirectly from the lithium activity data in binary Li-M alloys.   Such data are also available for a number of solid alloys from studies in which the traditional coulometric titration method (CTM) was employed.

4.1. Hydrogen titration studies

During the early stages of work at ANL it became apparent that lithium-based alloys react with hydrogen in a different manner than do the alloys of the transition metals.   Instead of forming stable ternary solutions, hydrogen reacts with the lithium in the alloy to form solid LiH, according to the following reversible reaction

$$Li \, (in \, alloy) + \tfrac{1}{2} \, H_2(g) \rightleftarrows LiH(s) \tag{19}$$

for which the equilibrium constant is given by

$$K = a_{LiH} / (a_{Li} \sqrt{P}) \tag{20}$$

where a is the activity of the subscripted component and P is the equilibrium hydrogen pressure in atmospheres.

According to eq. 19, when small quantities of hydrogen are added to an alloy LiM_x, the composition of the alloy will change from $x = n_M/n_{Li}$ to a new composition $x = n_M/(n_{Li} - n_H)$. Equilibrium hydrogen pressures, measured after each consecutive hydrogen addition, produce isothermal $\sqrt{P}$ $vs.$ x traces whose cascade-like shapes reflect the phase relationships in the binary Li-M systems-- the homogeneous intermediate phases reveal themselves as rising segments and the two-phase coexistence fields appear as constant pressure plateaus. The process leading to the equilibrium in eq. 19 is equivalent to a titration in which hydrogen is the titrant, LiH the titration product, and the equivalence points are identifiable from the shape of the isotherms. Hence, this method of investigating binary alloys is referred to as the hydrogen titration method (HTM).

Another important feature of HTM arises from the simple relationship between the measured decomposition pressures and the lithium activity in the binary Li-M alloy. Below the monotectic temperature of the Li-LiH system the LiH-rich phase is solid LiH. Therefore, the LiH precipitating out of LiM_x alloy may be taken to have unit activity and eq. 20 becomes

$$a_{Li} = 1/(K\sqrt{P}) \tag{21}$$

The proportionality factor in eq. 21 is the inverse of the equilibrium constant for the reaction in eq. 19 which can be evaluated from the standard free energy of formation of solid LiH, given in TABLE IV. HTM thus provides a powerful new technique for investigating both the phase relationships and the lithium activities in Li-M alloys. The technique has been applied to two alloy systems: Li-Al-H and Li-Pb-H [54].

4.1.1. The Li-Al-H system

According to a recent phase diagram [55], the Li-Al system has four homogeneous solid phases: α(dilute solution of lithium in aluminum), β("LiAl"), γ("Li$_3$Al$_2$") and δ("Li$_9$Al$_4$"). The latter two phases are believed to be line compounds, but the highest melting β-phase has a wide homogeneity range (46-52 at. % Al). Since the β-phase has been emphasized for fusion reactor applications, we have prepared two Li-Al alloys having compositions within this phase.

The results are shown in Fig. 5 as log P $vs.$ at % Al isotherms. The fact that both alloys have produced essentially the same Al-rich β-phase boundary ($\sim$52 at. % Al) serves as a verification for the reaction in eq. 19. On the left side of this boundary the rising isothermal segments correspond to the homogeneity ranges of the β-phase. Constant pressure plateaus on the right side indicate the onset of a wide ($\alpha + \beta$) two-phase region. Yao $et\ al.$ [56] have determined lithium activities from emf measurements on the 50 at. % Al alloy at three temperatures (314, 343, and 380°C). The corresponding hydrogen pressure values, extrapolated from their data to the temperatures of the HTM study, are in excellent agreement. Further agreement is indicated by comparisons in the ($\alpha + \beta$) region, where the data produced by either technique are more reliable because of their insensitivity to variations in alloy composition.

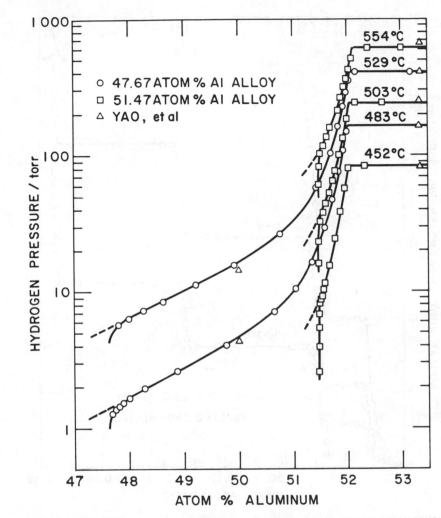

FIG.5. *Hydrogen pressure versus alloy composition isotherms for the system Li-Al.*

4.1.2. The Li-Pb-H system

The reliability of HIM is also being tested on the more complex Li-Pb alloy system. The reported phase diagram [57] reveals several intermediate phases: $Li_{22}Pb_5$, Li_7Pb_2, Li_3Pb, Li_8Pb_3, and LiPb. HIM results are shown in Fig. 6. Three of the isotherms (at 500, 549, and 598°C) were obtained with an alloy having an initial Pb/Li ratio of 0.2610. The fourth isotherm (also at 500°C) had an initial Pb/Li ratio of 0.2858.

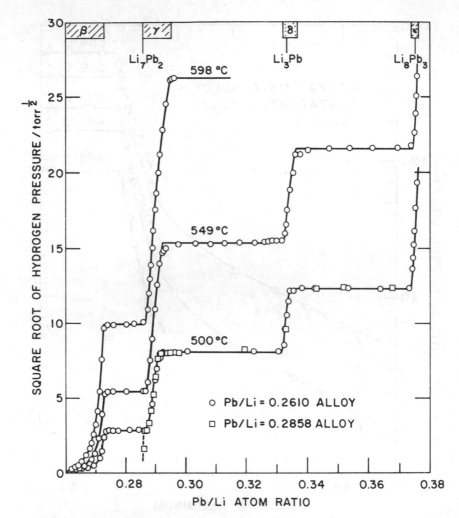

FIG.6.   *Square root of hydrogen pressure versus alloy composition isotherms for the Li-Pb system.*

      The isotherms have the expected cascade-like shapes with four rising portions separated by three pressure plateaus. Each of the rising portions was assigned a band (β, γ, δ, and ϵ) representing the maximum limits of homogeneity for that phase. The bands γ, δ, and ϵ include the stoichiometries of the phase Li_7Pb_2, Li_3Pb, and Li_8Pb_3, respectively. For concentrations below Pb/Li = 2/7, the shape of the isotherms suggest the existence of a wide homogeneous phase (β), whose lead-rich boundary may correspond to $Li_{11}Pb_3$.

Lithium activities, calculated for each two-phase plateau portion from the corresponding hydrogen pressure using eq. 21 show an excellent agreement with those recently determined by the coulometric titration technique [58]. This agreement and the superposition of the two isotherms at 500°C in Fig. 6, despite their different starting compositions, lends further credence to the applicability of HIM in investigating the thermodynamic properties and phase relations of binary lithium-containing alloys.

4.2. Coulometric titration studies

According to eq. 21, if isothermal a_{Li} vs. C data were available for a Li-M alloy, one could construct a corresponding set of $\sqrt{P}$ vs. C isotherms that represents hydrogen sorption characteristics of that alloy. Electromotive force measurements provide a convenient means of determining activities in binary alloys. For Li-M systems these measurements can be carried out in cells of the type [Li(ℓ)/Li-salt/Li-M(alloy)] in which the isothermal cell emfs are continuously recorded as functions of the alloy concentration which is varied by the coulometric titration method (CIM).

CIM studies on the Li-Al [56,59] and Li-Pb [58] systems and their agreement with the HIM data were mentioned above. Other solid alloy systems, for which CIM data are available, are: Li-Bi [60], Li-Sb [60], and Li-Si [61,62]. Lithium activities and the corresponding hydrogen decomposition pressures, calculated using data of these studies at 500°C, are given in TABLE V for selected intermetallic compounds that show promise in fusion reactor applications.

4.3. Solubility of hydrogen in solid lithium alloys

The slight inclination in the initial vertical portions of the isotherms in Fig. 5 implies that there is some dissolution of hydrogen in Li-M alloys. Only when the hydrogen solubility limit is exceeded, can the precipitation of LiH occur, according to eq. 19. This information is important in fusion reactor applications because it relates to the concentration of tritium present in the blanket medium under steady-state operating conditions.

Very little information is available on these solutions. Talbot et al. have determined the Sieverts' law constants ($K_S = N_H/\sqrt{P}$) for tritium in LiAl and LiBi$_{5.6}$ alloys [63] and for hydrogen in LiAl alloy [64]. The current work at ANL on the dissolution of hydrogen in Li-Al and Li-Pb alloys [54], however, indicates considerably lower K_S values.

Lacking exact hydrogen solubility data, one can attempt to estimate the magnitude of K_S from CIM data, such as those listed in TABLE V. The estimates are made by assuming that the same hydrogen solubility limit (e.g., $N_H = 10^{-4}$) exists for all alloys and that Sieverts' law is obeyed throughout the solution range, i.e., $K_S \propto 1/\sqrt{P_{max}}$.

TABLE V.   Maximum hydrogen pressures (or minimum lithium activities) at 500°C that can be reached in solutions of hydrogen in selected intermetallic compounds containing lithium.

Compound	M.P. (°C)	a_{Li}	P_{max} (atm)
Li_3Sb	~1200	2.5×10^{-6}	1.0×10^7
Li_3Bi	1145	2.5×10^{-5}	1.1×10^5
Li_8Pb_3	642[b]	4.8×10^{-3}	3.0
Li_3Pb	658[b]	2.0×10^{-2}	1.6×10^{-1}
Li_7Pb_2	726	3.2×10^{-2}	6.8×10^{-2}
β-LiAl[a]	700	7.5×10^{-2}	1.2×10^{-2}
Li_2Si	703	1.7×10^{-2}	2.4×10^{-1}
Li_4Si	635[b]	5.9×10^{-1}	1.9×10^{-4}
Lithium	liquid	1.0	5.0×10^{-10} [c]

[a]At 50 at. % Al.   [b]Incongruent melting.   [c]At $N_H = 10^{-4}$.

Materials possessing larger P_{max} values are more desirable for blanket applications because of the increased tendency of hydrogen isotopes to favor the gas phase, thereby facilitating tritium removal and reducing tritium inventory in the condensed blanket phase. The compounds listed in TABLE V show large variations in P_{max} which could be useful as a criterion in selecting an appropriate alloy. In all cases the P_{max} values are much larger than the equivalent value for liquid lithium, which is also shown in TABLE V for comparison.

5.   STUDIES RELATED TO CERAMIC BLANKETS

There is little thermodynamic information that pertains to the use of ceramics as breeder blanket materials. This section will be devoted to a brief discussion of available literature studies on the lithium-containing ceramics [7,8,10,53,65], describing the kinetics of tritium extraction from these materials. In these studies tritium is introduced into the ceramic specimens by neutron irradiation *via* the (n, ^6Li) and (n, ^7Li) transmutations. Since this procedure is analogous to the mode of tritium production in fusion reactor blankets, the experience gained should readily extrapolate to fusion reactor operating conditions.

Powell, *et al.* [7] and Wiswall and Wirsing [8] have reported that temperatures in excess of 600°C were necessary to recover >99 % of the tritium from Li_2O, $LiAlO_2$, and Li_2SiO_3 granules in one hour or less. Johnson *et al.* [10] have observed that, during the neutron irradiation itself, >80 % of helium and from 9 to 90 % of tritium

was released from the ceramic targets, the amount of tritium released
being inversely proportional to the bulk density of the target
material. The tritium appeared mostly in a noncondensible form (HT
or T_2) but the fraction of the condensible form (HTO or T_2O) increased
with increasing temperature of the target. Release of tritium from
$LiAlO_2$ targets was also investigated by Yunker [65]. His data have
shown that essentially all of the tritium can be recovered by vacuum
outgassing at 850°C with a large fraction of tritium appearing in a
condensible form.

It can be concluded from the results of these studies that
thermal extraction can be an effective method of removing tritium
from the oxide-type ceramics. A small fraction of tritium bred in
a ceramic blanket will be ejected *in situ* into the gas space by
direct recoil. The remaining tritium will diffuse to the surface
of the hot ceramic granules where it can be swept away by a purge
gas. Lower tritium inventory in the blanket can be effected by
increasing its steady-state operating temperature and/or by decreasing the mean size of the granules.

6.   RECOMMENDATIONS FOR FUTURE RESEARCH

The impetus of the fusion energy program has already stimulated
a significant amount of thermodynamic research on the candidate
blanket materials. Nevertheless, more work is required for full
evaluation of the in-reactor performance of these materials. In this
section we attempted to point out some areas where thermodynamic
data are either lacking or are in need of verification.

As was shown above, the existing thermodynamic data on lithium
is quite substantive. There still are, however, areas where additional information would be beneficial. The uncertainty remaining
in the Sieverts' law constants for tritium would be clarified by
additional experiments on very dilute solutions of tritium in lithium.
The stability and precipitation of oxide species in liquid lithium
and conditions for their removal by cold-trapping need further investigation. Identification and characterization of the carbon/
nitrogen species formed in dilute solutions in lithium should be
undertaken. Correlation of these latter studies with lithium
corrosion data could provide a clearer understanding of the compatibility of lithium with structural materials.

Considerably more work is needed to complete the thermodynamic
data base for solid lithium alloy systems. Accurate measurements
of Sieverts' law constants for tritium in these alloys would provide
a firmer basis for evaluating blanket inventories. Thermodynamic
properties of the systems formed between the alloys and non-metallic
impurity elements may be important in determining long-term tritium
release characteristics. Thermochemical data for many of the candidate
alloys are needed in evaluating thermal hydraulic response, safety
aspects related to accidental melt-down, and possible uses as thermal
storage sinks to sustain pulsed thermonuclear power production.
Pressure-composition-temperature data on solutions of hydrogen isotopes
in the alloys that are on the lithium-rich ends of the corresponding

phase diagrams would contribute insight on the potential utility of "slightly-alloyed" liquid lithium as a blanket material.

From the standpoint of thermodynamic data, ceramics are the least-studied category of the blanket materials. There is essentially no data available on the sorption of hydrogen isotopes in the prospective ceramic materials under anticipated fusion reactor operating conditions. The effect of impurity elements on the performance of these materials needs to be investigated. Thermochemical and sintering information are also needed for the same reasons as were mentioned in the preceding paragraph for the solid lithium alloys.

ACKNOWLEDGEMENTS

The work originating at the Argonne National Laboratory was supported by the Division of Basic Energy Sciences, U. S. Department of Energy. The interest and encouragement of R. P. Epple, F. A. Cafasso, and L. Burris, Jr. is gratefully acknowledged.

REFERENCES

[1] RIBE, F. L., Rev. Mod. Phys. 47 (1975) 7.

[2] JOHNSON, E. F., AIChE Journal 23 (1977) 617.

[3] Proc. Magnetic Fusion Energy Blanket and Shield Workshop (POWELL, J. R., FILLO, J. A., TWINING, B. G., DORNING, J. J., Eds.) U.S. Dept. of Energy Rep. CONF-760343 (1975).

[4] Proc. 2nd Topical Meeting, Techn. of CTR (KULCINSKI, J. L., Ed.) U.S. Dept. of Energy Rep. CONF-760935 (1976).

[5] Proc. 7th Symp. on Engineering Problems of Fusion Research (LUBELL, M. S., WHITMORE, C., Eds.) IEEE Publication No. 77CH1267-4-NPS (1977).

[6] CARLSON, G. A., "Magnetohydrodynamic Pressure Drop in Lithium Flowing in Conducting Wall Pipe in a Transverse Magnetic Field-- Theory and Experiment," Lawrence Livermore Laboratory Rep. UCRL-75307 (1974).

[7] POWELL, J. R., WISWALL, R. H., WIRSING, E., "Tritium Recovery from Fusion Blankets Using Solid Lithium Compounds," Brookhaven National Laboratory Rep. BNL-20563 (1975).

[8] WISWALL, R. H., WIRSING, E., "The Removal of Tritium from Fusion Reactor Blankets," Brookhaven National Laboratory Rep. BNL-50748 (1977).

[9] SZE, D. K., SCHLUDERBERG, D. C., Trans. Am. Nuc. Soc. 30 (1978)
68.

[10] JOHNSON, A. B., KABELE, T. J., GURWELL, W. E., "Tritium Production from Ceramic Targets: A Summary of the Hanford Coproduct
Program," Battelle Pacific Northwest Laboratories Rep. BNWL-2097
(1976).

[11] MESSER, C. E., "A Survey Report on Lithium Hydride," AEC Research and Development Rep. NYO-9470 (1960).

[12] SHPILRAIN, E. E., YAKIMOVICH, K. A., "Lithium Hydride. Physico-
chemical and Thermophysical Properties," State Service of Standard and Reference Data. Monograph Series No. 10, Moscow (1972).

[13] MAGEE, C. B., "Metal Hydrides," (MUELLER, W. M., BLACKLEDGE, J. P.,
LIBOWITZ, G. G., Eds.) Academic Press, New York (1968) Chap. 6.

[14] VELECKIS, E., VAN DEVENTER, E. H., BLANDER, M., J. Phys. Chem. 78
(1974) 1933.

[15] VELECKIS, E., J. Phys. Chem. 81 (1977) 526.

[16] HILDEBRAND, J. H. SCOTT, R. L., "The Solubility of Nonelectro-
lytes," 3rd ed., Reinhold, New York (1950) 34.

[17] VELECKIS, E., YONCO, R. M., MARONI, V. A., J. Less-Common
Metals 55 (1977) 85.

[18] SMITH, F. J., BATISTONI, A. M., BEGUN, G. M., LAND, J. F.: (a) Proc.
9th Symp. Fusion Technology, Pergamon Press, New York (1976) 325,
(b) J. Inorg. Nucl. Chem., submitted for publication.

[19] SMITH, F. J., REDMAN, J. D., STREHLOW, R. A., BELL, J. T., Proc.
Symp. Tritium Technology Related to Fusion Reactor Systems
(SMITH, W. H., WILKES, W. R., WITTENBERG, L. J., Eds) ERDA-50
(1975).

[20] McCRACKEN, G. M., GOODALL, D. H. J., LONG, G., "The Extraction of
Tritium from Liquid Lithium," IAEA Workshop on Fusion Reactor
Design Problems, Culham, U. K. (1974).

[21] IHLE, H. R., WU, C. H., Proc. 8th Symp. on Fusion Technology,
Noordwijkerhout, The Netherlands, EUR-5182C (1974) 787.

[22] ADAMS, P. F., DOWN, M. G., HUBBERSTEY, P., PULHAM, R. J., J. Less-
Common Metals 42 (1975) 325.

[23] ADAMS, P. F., HUBBERSTEY, P., PULHAM, R. J., THUNDER, A. E., J.
Less-Common Metals 46 (1976) 285.

[24] LEWIS, G. N., RANDALL, M., "Thermodynamics" (revised by PITZER,
K. S., BREWER, L.) 2nd ed., McGraw-Hill, New York (1961) 404.

[25] SHPILRAIN, E. E., YAKIMOVICH, K. A., KAGAN, D. N., SHVALB, V. G.,
      Fluid Mech.-Sov. Res. 3 (1974) 3.

[26] HEUMANN, F. K., SALMON, O. N., "The Lithium Hydride, Deuteride,
      and Tritide Systems," USAEC Rep. KAPL-1667 (1956).

[27] SMITH, F. J., LAND, J. F., Trans. Am. Nucl. Soc., Annual Meeting
      (1975) 167.

[28] SHPILRAIN, E. E., YAKIMOVICH, K. A., SHERESHEVSKII, V. A.,
      Teplofiz. Vys. Temp. 15 (1977) 661.

[29] IHLE, H. R., WU, C. H., J. Inorg. Nucl. Chem. 36 (1974) 2167.

[30] KATSUTA, H., ISHIGAI, T., FURUKAWA, K., Nucl. Technol. 32
      (1977) 297.

[31] VELECKIS, E., J. Nucl. Mater. 79 (1979) 20.

[32] IHLE, H. R., WU. C. H., Proc. 2nd Int. Congress on Hydrogen in
      Metals, Pergamon Press, New York (1977) paper 1D8.

[33] IHLE, H. R., WU, C. H., J. Phys. Chem. 79 (1975) 2386.

[34] MESSER, C. E., "The Alkali Metals," The Chem. Soc. (London)
      Spec. Bull. No. 22 (1966) 183.

[35] CAIRNS, E. J., CAFASSO, F. A., MARONI, V. A., "The Chemistry of
      Fusion Technology" (GRUEN, D. M., Ed.) Plenum Press, New York
      (1972).

[36] ADAMS, P. F., HUBBERSTEY, P., PULHAM, R. J., J. Less-Common
      Metals 42 (1975) 1.

[37] SMITH, J. F., MOSER, Z., J. Nucl. Mater. 59 (1976) 158.

[38] BOLSHAKOV, K. A., FEDOROV, P. I., STEPINA, F. A., Izvest. Vysshikh.
      Ucheb. Zavedenii, Tsvetnaya Met. 2 (1959) 52.

[39] YONCO, R. M., VELECKIS, E., MARONI, V. A., J. Nucl. Mater.
      57 (1975) 317.

[40] HUBBERSTEY, P., PULHAM, R. J., THUNDER, A. E., J. Chem. Soc.,
      Faraday Trans. 1 (1976) 431.

[41] JANAF Thermochemical Tables, Dow Chemical Co., Midland, MI (1978).

[42] O'HARE, P. A. G., JOHNSON, G. K., J. Chem. Thermod. 7 (1975) 13.

[43] OSBORNE, D. W., FLOTOW, H. E., J. Chem. Thermod. 10 (1978) 675.

[44] BONOMI, A., HADATE, M., GENTAZ, C., Proc. Int. Symp. on Molten
      Salts (Pemsler, J. P., Braunstein, J., Morris, D. R., Nobe, K.,
      Richards, N. E., Eds.) The Electrochemical Society, Princeton,
      N.J. (1976) 78.

[45] YONCO, R. M., MARONI, V. A., STRAIN, J. E., DE VAN, J. H., J. Nucl. Mater. (1979) in press.

[46] HOFFMAN, E. E., ASTM Spec. Tech. Publ. 272, Newer Metals (1960) 195.

[47] KONOVALOV, E. E., SELIVERSTOV, N. I., EMELYANOV, V. P., Russ. Met. 3 (1969) 77.

[48] HOBART, E. M., BJORK, R. G., Anal. Chem. 39 (1967) 202.

[49] YONCO, R. M., HOMA, M. I., Argonne National Laboratory, work in progress.

[50] FEDOROV, P. I., SU, M. T., Acta Chim. Sinica 23 (1957) 30.

[51] DOWN, M. G., HALEY, M. J., HUBBERSTEY, P., PULHAM, R. J., THUNDER, A. E., J. Chem. Soc. Chem. Comm. (1978) 52.

[52] ARONSON, S., SALZANO, F. J., Inorg. Chem. 8 (1969) 1541.

[53] OWEN, J. H., RANDALL, D., Proc. Int. Conf. on Radiation Effects and Tritium Technology for Fusion Reactors (WATSON, J. S., WIFFEN, F. W., Eds.) Rep. CONF-750989 (1976) III-433.

[54] VELECKIS, E., Argonne National Laboratory, work in progress.

[55] MYLES, K. M., MRAZEK, F. C., SMAGA, J. A., SETTLE, J. C., Proc. Symp. and Workshop on Advanced Battery Research and Design, Argonne National Laboratory Rep. ANL-76-8 (1976) 50B.

[56] YAO, N. P., HEREDY, L. A., SAUNDERS, R. C., J. Electrochem. Soc. 118 (1971) 1039.

[57] HULTGREN, R., DESAI, P. D., HAWKINS, D. T., GLEISER, M., KELLEY, K. K., "Selected Values of the Thermodynamic Properties of Binary Alloys," Am. Soc. Metals, Metals Park, OH (1973) 1078.

[58] SABOUNGI-BLANDER, M. L., MARR, J., Argonne National Laboratory, personal communication.

[59] WEN, C. J., BAUKAMP, B. A., HUGGINS, R. A., Extended Abstracts, The Electrochem. Soc. 78-1 (1978) Abstract No. 171.

[60] WEPPNER, W., HUGGINS, R. A., J. Electrochem. Soc. 125 (1978) 7.

[61] SHARMA, R. A., SEEFORTH, R. N., J. Electrochem. Soc. 123 (1976) 1763.

[62] LAI, S., McCOY, L. R., Extended Abstracts, The Electrochem. Soc., 75-2 (1975) Abstract No. 21.

[63] TALBOT, J. B., "A Study of Tritium Removal from Fusion Reactor Blankets of Molten Salt and Lithium-Aluminum," Oak Ridge National Laboratory Rep. ORNL/TM-5104 (1976).

[64] TALBOT, J. B., SMITH, F. J., LAND, J. F., BARTON, P., J. Less-
     Common Metals 50 (1976) 23.

[65] YUNKER, W., "Extraction of Tritium from Lithium Aluminate,"
     Hanford Engineering Development Laboratory Rep. HEDL-TME 76-81
     (1976).

DISCUSSION

J. MAGILL: As regards the use of solid lithium alloy blankets, is there any
mechanism for allowing aluminium or lead to escape from the blanket into the
reaction chamber? If this were possible, the aluminium or lead impurity would
emit high-energy radiation, draining the energy content in the chamber and
effectively shutting off the thermonuclear reaction.

E. VELECKIS: I cannot think of any mechanism by which the components
of the blanket materials could migrate into the plasma chamber across the barrier
provided by the first wall.

D.D. SOOD: You mention that liquid lithium is the leading candidate
material for the blanket in fusion reactors. In view of the problems relating to the
pumping of liquid lithium through strong magnetic fields, why are the solid alloys
not considered more favourably? Could you also comment on the possible use of
lithium-bearing molten salt as the blanket?

E. VELECKIS: Magnetohydrodynamic losses in pumping lithium across the
magnetic field lines can be reduced to $\sim 1\%$ of the total power output. The
advantages offered by liquid lithium (e.g. high lithium atom density, excellent
heat transfer properties and low melting point) make it a preferred choice for a
fusion blanket.

Fused fluorides (e.g. Li_2BeF_4) have been considered as blanket materials.
Recent neutronic calculations, however, show that adequate tritium breeding gains
cannot be maintained with these salts when account is taken of the volume occupied
by the cooling tubes and other structures within the blanket chamber.

H.R. IHLE: We observed large differences in hydrogen solubility between
liquid and solid Li-Pb alloys, the hydrogen solubility in liquid alloys being much
higher than in solid. Have you noticed the same thing?

E. VELECKIS: Yes, we have also observed very low hydrogen solubilities per
unit pressure (Sieverts' law constants) for the solid lithium alloys. One could
speculate that in the two cases you mention the hydrogen absorption process is
governed by different mechanisms: LiH precipitation in the case of solid alloys
and extensive solution formation in the case of liquid ones.

EXPERIMENTAL STUDY OF GASEOUS LITHIUM DEUTERIDES AND LITHIUM OXIDES

Implications for the use of lithium and Li_2O as breeding materials in fusion reactor blankets

H.R. IHLE, H. KUDO*, C.H. WU
Institut für Nuklearchemie,
Kernforschungsanlage Jülich GmbH,
Jülich,
Federal Republic of Germany

Abstract

EXPERIMENTAL STUDY OF GASEOUS LITHIUM DEUTERIDES AND LITHIUM OXIDES: IMPLICATIONS FOR THE USE OF LITHIUM AND Li_2O AS BREEDING MATERIALS IN FUSION REACTOR BLANKETS.

In addition to LiH, which has been studied extensively by optical spectroscopy, the existence of a number of other stable lithium hydrides has been predicted theoretically. By analysis of the saturated vapour over dilute solutions of the hydrogen isotopes in lithium, using Knudsen effusion mass spectrometry, all lithium hydrides predicted to be stable were found. Solutions of deuterium in lithium were used predominantly because of practical advantages for mass spectrometric measurements. The heats of dissociation of LiD, Li_2D, LiD_2 and Li_2D_2, and the binding energies of their singly charged positive ions were determined, and the constants of the gas/liquid equilibria were calculated. The existence of these lithium deuterides in the gas phase over solutions of deuterium in lithium leads to enrichment of deuterium in the gas above 1240 K. The enrichment factor, which increases exponentially with temperature and is independent of concentration for low concentrations of deuterium in the liquid, was determined by Rayleigh distillation experiments. It was found that it is thermodynamically possible to separate deuterium from lithium by distillation. One of the alternatives to the use of lithium in (D,T)-fusion reactors as tritium-breeding blanket material is to employ solid lithium oxide. This has a high melting point, a high lithium density and still favourable tritium-breeding properties. Because of its rather high volatility, an experimental study of the vaporization of Li_2O was undertaken by mass spectrometry. It vaporizes to give lithium and oxygen, and LiO, Li_2O, Li_3O and Li_2O_2. The molecule Li_3O was found as a new species. Heats of dissociation, binding energies of the various ions and the constants of the gas/solid equilibria were determined. The effect of using different materials for the Knudsen cells and the relative thermal stabilities of lithium-aluminium oxides were also studied.

---

* Permanent address: Japan Atomic Energy Research Institute, Tokai-Mura, Japan.

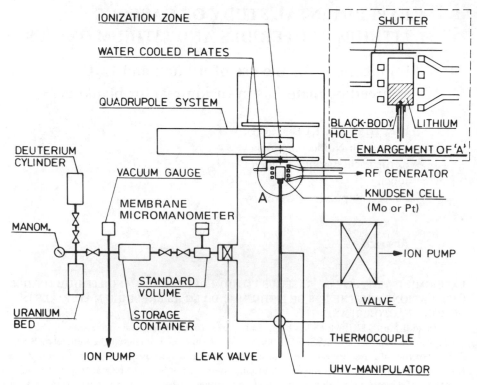

FIG.1.   Schematic diagram of the experimental arrangement (UHV: ultra-high vacuum).

1. INTRODUCTION

Lithium metal is the most frequently cited blanket material for (D,T)-fusion reactors because it is neutronically favourable for the breeding of tritium [1]. Tritium, however, in low steady-state concentration will be difficult to remove from liquid lithium, as can be estimated from the early measurements of the decomposition pressures of LiH, LiD and LiT [2]. A new mass spectrometric investigation of the saturated vapour over dilute solutions of hydrogen isotopes in liquid lithium was therefore undertaken, in particular to study the significance of gaseous lithium hydrides, which had been theoretically predicted to be stable [3,4].

Lithium oxide (Li_2O) has been proposed as an alternative solid breeding material, having less favourable breeding properties than lithium but having a very low tritium solubility [5].   Hence, the vaporization of lithium oxide was also studied by mass spectrometry because of its relatively high volatility and the expectation of a complex vapour composition. Severe attack of metal claddings by lithium oxide was anticipated, particularly because of the high oxygen pressure at elevated temperatures.

2. EXPERIMENTAL

In the studies of the saturated vapour present over solutions of hydrogen isotopes in liquid lithium, deuterium was mainly used because of the practical advantages for the mass spectrometric measurements.

The apparatus used for this work, and also, with minor modifications, for the vaporization studies of lithium oxide, is shown in Fig.1. The effusate from the Knudsen cell is analysed by a quadrupole mass filter. For work with the solutions of hydrogen isotopes in lithium, molybdenum cells were used, while for work with Li_2O, platinum cells were mostly used. The cell temperature was measured with an optical pyrometer and a Pt-Rh/Pt thermocouple. Calibration was undertaken at the triple points of lithium, silver and gold. The accuracy of the temperature measurements is believed to be ± 2 K.

The deuterium used in the measurements was purified by formation and decomposition of uranium deuteride. A gas delivery system permitted the admission of $\sim 2 \times 10^{-9}$ moles D_2 with an accuracy of $\pm 1\%$, which can be obtained by measuring a pressure of $\sim 10^{-3}$ torr in a standard volume to within $\pm 1\%$ by a membrane micromanometer. Thus solutions of deuterium in lithium down to at least an atomic fraction of deuterium of $\sim 5 \times 10^{-7}$ can be prepared in the ultra-high vacuum chamber by reaction of pure lithium contained in a Knudsen cell with a known amount of gas. A shutter between the orifice of the Knudsen cell and the ion source makes it possible to distinguish between residual background and signals due to molecules effusing from the cell. The known vapour-pressure data for lithium were used to calibrate the pressure over the temperature range of the measurements. At the triple point of lithium (T = 453.70 K) for example, the effusion rate, Z, is $2.15 \times 10^8 \, s^{-1}$, corresponding to a pressure, p_{Li}, of 9.3×10^{-14} atm in the Knudsen cell (orifice diameter 0.8 mm).

3. RESULTS AND DISCUSSION

3.1. Dilute solutions of deuterium in lithium

Using the experimental arrangement discussed, the validity of Sievert's Law:

$$p_{D_2}^{\frac{1}{2}} \propto x_D$$

where p is the pressure and x_D is the atomic fraction of deuterium, was confirmed for extremely dilute solutions of deuterium in lithium, and values for Sievert's constant were measured between 773 and 973 K over a concentration range of $0.5 \times 10^{-6} \leqslant x_D \leqslant 10^{-4}$. However, at these low concentrations of

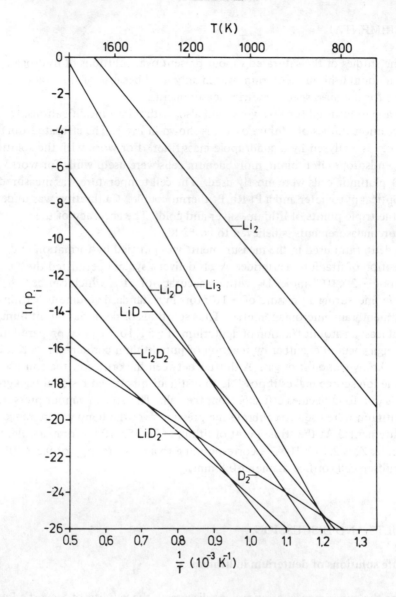

FIG.2.   Partial pressures, p_i, of molecules in the saturated vapour, over a solution of deuterium in liquid lithium at an atomic fraction of deuterium, x_D, of 10^{-5}, as functions of temperature. (Pressure is measured in atmospheres.)

TABLE I. CONSTANTS A AND B FOR THE EQUATION:

$$\ln K_i = -\frac{A}{T} + B$$

K_i		A	B
p_{Li}		1.869×10^4	11.6
p_{Li_2}		2.463×10^4	14.0
p_{Li_3}		3.594×10^4	17.8
K_{s,D_2}	$= p_{D_2}^{\frac{1}{2}}/x_D$	6.381×10^3	6.5
K_{LiD}^*	$= p_{LiD}/x_D$	2.368×10^4	14.8
$K_{Li_2D}^*$	$= p_{Li_2D}/x_D$	2.958×10^4	20.0
$K_{LiD_2}^*$	$= p_{LiD_2}^{\frac{1}{2}}/x_D$	8.800×10^3	8.2
$K_{Li_2D_2}^*$	$= p_{Li_2D_2}^{\frac{1}{2}}/x_D$	1.225×10^4	11.5

deuterium (or tritium), of interest for fusion reactor blankets, and at high temperature, it is not D_2 which is the predominant deuterium-bearing species in this system, but rather the lithium deuterides Li_2D and LiD (as may be seen from Fig. 2). Li_2D, LiD_2, Li_2D_2 and Li_3, which had been predicted theoretically to be stable, were found for the first time as molecular species during these experiments [6].

The partial pressures of LiD and Li_2D vary proportionally with the atomic fraction of deuterium in the liquid, whereas those of LiD_2, Li_2D_2 and D_2, i.e. species containing two deuterium atoms, vary proportionally with the square of the deuterium atomic fraction. Therefore, they are the richer in the gas phase the higher the concentration of deuterium.

Constants for the gas/liquid equilibria as functions of temperature are given in Table I. The resulting heats of dissociation and the gas equilibria from which they were derived are given in Table II. The ionization potentials and binding energies of their singly charged positive ions are summarized in Table III. From these results it can be deduced that, at high temperatures, the ratio of the deuterium atomic fraction in the gas to that in the liquid exceeds unity and deuterium enriches in the gas upon distillation.

TABLE II. ENTHALPIES OF ISOMOLECULAR EXCHANGE REACTIONS AND DERIVED HEATS OF DISSOCIATION

Reaction	Enthalpy ΔH_0^0 (kcal/mol)	Molecule	Heat of dissociation[a]	
			D_0^0 (kcal/mol)	D_T^{0*} (kcal/mol)
$Li_2(g) = 2\ Li(g)$	25.5 ± 0.2	Li_2	25.5 ± 0.3	24 ± 1.5
$2\ Li(s) = Li_2(g)$	51.5 ± 0.3			
$2\ Li_2(g) = Li_3(g) + Li(g)$	10.7 ± 0.2	Li_3	41.3 ± 0.3	40 ± 3
$3\ Li(s) = Li_3(g)$	72.9 ± 0.4			
$Li_2(g) + D_2(g) = 2\ LiD\ (g)$	15.5 ± 0.7	LiD	57.3 ± 0.9	52 ± 4
$Li_2D\ (g) + Li\ (g) = LiD\ (g) + Li_2(g)$	10.8 ± 0.5	Li_2D	89.7 ± 3	82.6 ± 5
$Li_2D\ (g) + Li_2(g) = LiD\ (g) + Li_3(g)$	21.0 ± 0.3			
$Li_3(g) + LiD_2(g) = 2\ Li_2D\ (g)$	$-(15.7 \pm 1.5)$	LiD_2	128.0 ± 4	130 ± 6
$LiD_2(g) + Li_2(g) = LiD\ (g) + Li_2D\ (g)$	7.3 ± 1			
$Li_2D_2(g) + Li\ (g) = LiD_2(g) + Li_2(g)$	15.4 ± 0.6	Li_2D_2	168.0 ± 8	155 ± 8
$Li_2D_2(g) + Li_2(g) = Li_3(g) + LiD_2(g)$	26.5 ± 1			

[a] D_T^{0*}: 'Second Law' values, derived from temperature dependence of gas/liquid equilibrium constants.

TABLE III. IONIZATION POTENTIALS AND ION BINDING ENERGIES

Molecule	Ionization potential (eV)	Binding energy of ion, D_0^0 (kcal/mol)	Dissociation products
Li_2	4.86 ± 0.1	$37.7 \pm 2,0$	$Li^+ + Li + e^-$
Li_3	4.35 ± 0.2	65.5 ± 4.3	$Li^+ + 2Li + e^-$
LiD	7.7 ± 0.1	3.22 ± 0.23	$Li^+ + D + e^-$
Li_2D	4.5 ± 0.2	109 ± 6	$Li^+ + Li + D + e^-$
LiD_2	$5.8 - 6.4$	$105 - 119$	$Li^+ + 2D + e^-$
Li_2D_2	$6.4 - 6.8$	136 ∓ 145	$Li^+ + Li + 2D + e^-$

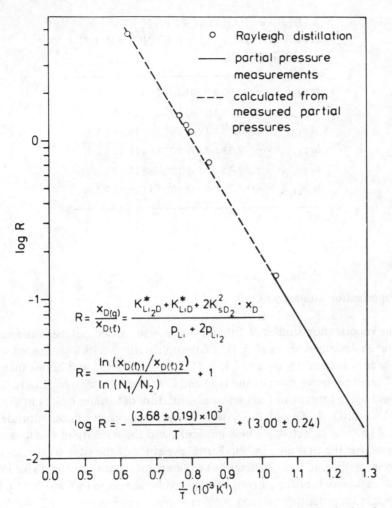

FIG.3. *Distribution factor, R, observed and calculated from vapour pressure data (see text).*

Rayleigh distillation experiments were therefore performed in the same apparatus in order to obtain an independent proof of this effect. The results are shown in Fig.3 and compared with data from partial-pressure measurements. The distribution factor

$$R \equiv x_{D(g)}/x_{D(\ell)}$$

rises exponentially with temperature and is, at low concentrations, independent of concentration [7]. Thermodynamically, therefore, it is possible to separate deuterium at very low concentrations from lithium by distillation.

TABLE IV. PARTIAL PRESSURES
OVER SOLID Li_2O AS A FUNCTION OF
TEMPERATURE

$$\ln p_{Li} = -(41.90 \pm 0.84)\ 10^3/T + (16.32 \pm 0.55)$$

$$\ln p_{Li_2} = -(61.35 \pm 6.04)\ 10^3/T + (20.06 \pm 3.72)$$

$$\ln p_{LiO} = -(45.74 \pm 2.66)\ 10^3/T + (14.30 \pm 1.75)$$

$$\ln p_{Li_2O} = -(47.53 \pm 0.47)\ 10^3/T + (19.42 \pm 0.31)$$

$$\ln p_{Li_3O} = -(59.48 \pm 1.67)\ 10^3/T + (20.89 \pm 1.05)$$

$$\ln p_{Li_2O_2} = -(69.29 \pm 4.31)\ 10^3/T + (25.88 \pm 2.70)$$

3.2. Vaporization studies on Li_2O

The vaporization studies on lithium oxide were carried out using the same apparatus as described above (Fig.1). Platinum Knudsen cells were loaded with Li_2O, which was carefully degassed at ~ 1000 K under a vacuum better than 2×10^{-8} torr to remove most of the H_2O and CO_2 from the sample. Li_2O vaporizes to give lithium and oxygen, and a mixture of lithium oxides in the gas phase. LiO, Li_2O, Li_3O and Li_2O_2 were found to be present in the saturated vapour. Li_3O was detected as a new molecule, and the stability of Li_2O_2 was determined for the first time [8, 9]. Partial pressures of the gaseous species in equilibrium with solid Li_2O as functions of temperature are given in Table IV. Heats of dissociation of the gaseous lithium oxides are given in Table V [10, 11, 12], while ionization potentials and ion binding energies are presented in Table VI [10, 11, 12].

The partial pressures of $Li_2O(g)$ from our work are compared with results from other authors [11–13] who also used platinum as Knudsen cell material (Fig.4).

If, instead of platinum, molybdenum or tantalum is used to contain the solid lithium oxide, the partial pressure of Li_2O is reduced by almost one order of magnitude. In Fig.5, the lithium pressures over solid Li_2O are shown as a function of temperature, obtained with platinum, molybdenum, tantalum or graphite as cell materials. The very pronounced increase in the lithium pressure in the presence of molybdenum or graphite is probably caused by formation of the volatile oxides Li_2MoO_4 and CO, with an associated reduction of oxygen pressure in the system. The pressure of Li_3O is less strongly affected by the presence of molybdenum or tantalum, as may be seen from Fig.6.

TABLE V. HEATS OF DISSOCIATION OF LiO(g), Li_2O(g), Li_3O(g) AND Li_2O_2 (g)

Species	D_0 (kcal/mol)	Reference	
LiO	77.2 ± 2.0	this work	
	80.5 ± 1.5	Hildenbrand	[10]
	80.9	White et al.	[11]
	81.5	Berkowitz et al.	[12]
Li_2O	178.3 ± 1.1	this work	
	178.9	White et al.	[11]
	171.5	Berkowitz et al.	[12]
Li_3O	229.0 ± 10	this work	
Li_2O_2	293.0 ± 12	this work	

TABLE VI. IONIZATION POTENTIALS AND ION BINDING ENERGIES OF GASEOUS LITHIUM OXIDES

Molecule	Ionization potential (eV)	Ion binding energy (kcal/mol)	Dissociation products	Reference	
LiO	8.96 ± 0.2	187.7 ± 6	$Li + O^+$	this work	
	8.45 ± 0.2			Hildenbrand	[10]
	8.60 ± 0.2			White et al.	[11]
	9.0 ± 0.2			Berkowitz et al.	[12]
Li_2O	6.41 ± 0.2	344.0 ± 6	$2 Li + O^+$	this work	
	6.19 ± 0.2			Hildenbrand	[10]
	6.90 ± 0.3			White et al.	[11]
	6.80 ± 0.2			Berkowitz et al.	[12]
Li_3O	4.54 ± 0.2	437.6 ± 7.0	$3 Li + O^+$	this work	
Li_2O_2	7.88 ± 0.2	425.0 ± 8.0	$2 Li + O + O^+$	this work	

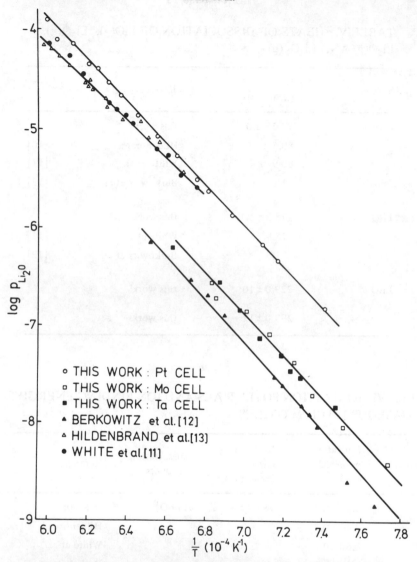

FIG.4. *Partial pressure of* Li_2O *over solid* Li_2O *as a function of temperature, using Knudsen cells of different materials. (Pressure is measured in atmospheres.)*

    If the resulting partial pressures obtained using platinum cells are considered to be reference data (no attack of the platinum was observed), it is possible to make at least an estimate, from the change in the pressures, of the degree to which lithium oxide and the cell materials react chemically at high temperatures. Mass spectrometric analysis provides a sensitive test.

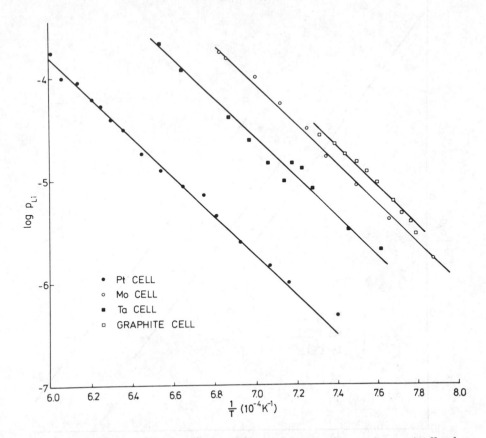

FIG.5.   Partial pressure of lithium over solid Li_2O as a function of temperature, using Knudsen cells of different materials. (Pressure is measured in atmospheres.)

3.3. Related studies

Further studies concerned the relative thermal stabilities of the lithium-aluminium oxides [14]; one of these oxides (γ-LiAlO$_2$) is used as a target for tritium production on a technical scale in fission reactors.

The lithium pressure over lithium-aluminium oxides in the presence of their thermal decomposition products is shown in Fig. 7. From the very small differences in lithium pressures between the co-existing phases Li_2O and β-Li$_5$AlO$_4$, and β-Li$_5$AlO$_4$ and γ-LiAlO$_2$, it can be seen that the lithium-aluminium oxide with the highest lithium content is only slightly more resistant to thermal decomposition than Li_2O itself. The heats of formation of the lithium-aluminium

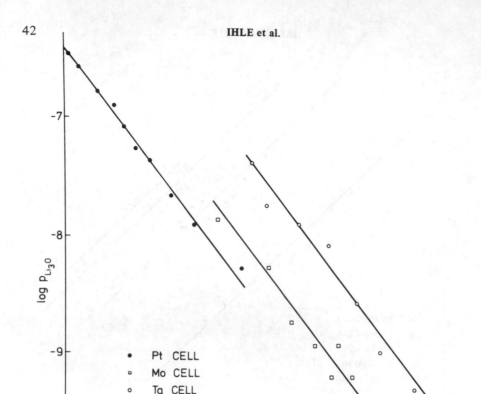

FIG.6. *Partial pressure of* Li_3O *over solid* Li_2O *as a function of temperature, using Knudsen cells of different materials. (Pressure is measured in atmospheres.)*

oxides from the constituent oxides were deduced to be as follows:

$$Li_2O(s) + \frac{1}{5}\, \alpha\text{-}Al_2O_3(s) \rightarrow \frac{2}{5}\, \beta\text{-}Li_5AlO_4(s) \qquad \Delta H^0_{r,298.15\,K} = -4\ \text{kcal/mol}$$

$$Li_2O(s) + \quad \alpha\text{-}Al_2O_3(s) \rightarrow 2\ \gamma\text{-}LiAlO_2(s) \qquad \Delta H^0_{r,298.15\,K} = -25\ \text{kcal/mol}$$

$$Li_2O(s) + 5\alpha\text{-}Al_2O_3(s) \rightarrow \quad 2\ LiAl_5O_8(s) \qquad \Delta H^0_{r,298.15\,K} = -40\ \text{kcal/mol}$$

It may be seen that the thermal stabilities of the lithium-aluminium oxides are in inverse order to their lithium contents and, hence, their tritium-breeding properties.

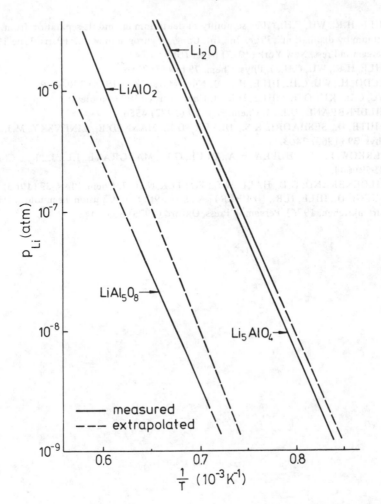

FIG.7. Partial pressure of lithum over lithium-aluminium oxides. (Pressure is measured in atmospheres.)

REFERENCES

[1] LEE, J.D., "Tritium breeding and direct energy conversion", The Chemistry of Fusion
     Technology (GRUEN, D.M., Ed.), Plenum Press, New York (1972).
[2] HEUMANN, F.K., SALMON, O.N., The Lithium Hydride, Deuteride and Tritide Systems,
     Knolls Atomic Power Laboratory, Schenectady, NY, Rep. KAPL-1667 (1956).
[3] COMPANION, A.L., J. Chem. Phys. 48 (1968) 1186.
[4] TYNDALL, J.R., COMPANION, A.L., J. Chem. Phys. 52 (1970) 2036.
[5] SAKO, K., DHTA, M., SEKI, T., YAMATO, H., HIRAOKA, T., TANAKA, T., ASAMI, N.,
     MORI, S., Japan Atomic Energy Research Institute Rep. JAERI-M-5502 (1973).

[6] IHLE, H.R., WU, C.H., "The solubility of deuterium in, and its separation from, liquid lithium by distillation", Proc. 2nd Int. Congr. Hydrogen in Metals (Paris, June 1977), Pergamon Press, New York (1977) paper ID8.

[7] IHLE, H.R., WU, C.H., J. Phys. Chem. **79** (1975) 2386.

[8] KUDO, H., WU, C.H., IHLE, H.R., J. Nucl. Mater **78** 2 (1978) 380–389.

[9] WU, C.H., KUDO, H., IHLE, H.R., J. Chem. Phys. (1979, in press).

[10] HILDENBRAND, D.L., J. Chem. Phys. **57** (1972) 4556.

[11] WHITE, D., SESHADRI, K.S., DEVER, D.F., MANN, D.E., LINEVSKY, M.J., J. Chem. Phys. **39** (1963) 2463.

[12] BERKOWITZ, J., CHUPKA, W.A., BLUE, G.D., MARGRAVE, J.L., J. Phys. Chem. **63** (1959) 644.

[13] HILDENBRAND, D.L., HALL, W.F., POTTER, N.D., J. Chem. Phys. **39** (1963) 296.

[14] GUGGI, D., IHLE, H.R., NEUBERT, A., Proc. 9th Symp. Fusion Technology (Garmisch-Partenkirchen, 1976), Pergamon Press, Oxford (1976) 635-644.

Section I

BASIC THERMODYNAMIC STUDIES

EXTRAPOLATION PROCEDURES FOR CALCULATING HIGH-TEMPERATURE GIBBS FREE ENERGIES OF AQUEOUS ELECTROLYTES

P.R. TREMAINE
Atomic Energy of Canada Research Company,
Whiteshell Nuclear Research Establishment,
Pinawa, Manitoba,
Canada

Abstract

EXTRAPOLATION PROCEDURES FOR CALCULATING HIGH-TEMPERATURE
GIBBS FREE ENERGIES OF AQUEOUS ELECTROLYTES.

Methods for calculating high-temperature Gibbs free energies of mononuclear cations
and anions from room-temperature data are reviewed. Emphasis is given to species required
for oxide solubility calculations relevant to mass transport situations in the nuclear industry.
Free energies predicted by each method are compared with selected values calculated from
recently reported solubility studies and other literature data. Values for monatomic ions
estimated using the assumption $\overline{C}_p^0(T) = \overline{C}_p^0(298)$ agree best with experiment to 423 K.
From 423 to 523 K, free energies from an electrostatic model for ion hydration are more
accurate. Extrapolations for hydrolysed species are limited by a lack of room-temperature
entropy data and expressions for estimating these entropies are discussed.

INTRODUCTION

Several methods for predicting thermodynamic data for
aqueous electrolytes at elevated temperatures have been reported
over the past 15 years [1-7]. This paper briefly reviews the
various methods and critically compares the standard Gibbs free
energies predicted by each with the limited experimental data now
available. The comparison is restricted to ionic species which
we have encountered in studying the solubility of sparingly
soluble oxides relevant to primary and secondary coolant chemistry control in CANDU-PHW[1] reactors. However, this is a sufficiently large cross-section of the available high-temperature
data that the conclusions are probably valid for most ions.

---

[1] CANada Deuterium Uranium—Pressurized Heavy Water system.

EMPIRICAL METHODS

Standard partial molal Gibbs free energy functions for aqueous ions are defined by equation 1,

$$\bar{G}^\circ(T) - \bar{G}^\circ(298) = - \int_{298}^{T} \bar{S}^\circ(T)dT$$

$$= -\bar{S}^\circ(298)[T-298.15] - T \int_{298}^{T} (\bar{C}_p^\circ(T)/T)dT + \int_{298}^{T} \bar{C}_p^\circ(T)dT \qquad (1)$$

where, by convention [4], $[\bar{G}^\circ(H^+,T) - \bar{G}^\circ(H^+,298)]$, $\bar{S}^\circ(H^+,T)$ and $\bar{C}_p^\circ(H^+,T)$ are defined to be zero. Most extrapolation methods assume $\bar{S}^\circ(298)$ is known and attempt to estimate either $\bar{S}^\circ(T)$ or $\bar{C}_p^\circ(T)$. The latter is preferable since $\bar{G}^\circ(T) - \bar{G}^\circ(298)$ is less sensitive to small errors in $\bar{C}_p^\circ$. This is because $\bar{S}^\circ(T)$ is proportional to the first derivative of free energy, $\bar{S}^\circ(T) = -(\partial\bar{G}^\circ(T)/\partial T)_p$, while $\bar{C}_p^\circ(T)$ is related to the second derivative, $\bar{C}_p^\circ(T) = -T(\partial^2\bar{G}^\circ(T)/\partial T^2)_p$.

The most accurate empirical method is to extrapolate $\bar{C}_p^\circ(T)$, if values are known over a reasonable temperature range. The best expression is thought [5] to be

$$\bar{C}_p^\circ(T) = A + BT + C/T^2 \qquad (2)$$

Many ionic heat capacities pass through a maximum near 373 K [4,8-15] and, ideally, experimental data from higher temperatures should be included in the fit. Because of this maximum, the assumption $\bar{C}_p^\circ(T) = \bar{C}_p^\circ(298)$ yields reasonably accurate free energies to temperatures as high as 423 K [3,4,6]. Since the advent of flow microcalorimetry [16], values of $\bar{C}_p^\circ(298)$ have been obtained for a great many species [17-20]. When necessary, approximate values can be estimated from $\bar{S}^\circ(298)$ using the Criss-Cobble principle [2], described below. It has been suggested [4] that a heat capacity extrapolation of the form $\bar{C}_p^\circ(T) = \bar{C}_p^\circ(298)[T/298.15]$ yields an even better free energy extrapolation.

The Criss-Cobble entropy correspondence principle [1,2] has been widely used to calculate $\bar{S}^\circ(T)$ using various numerical methods [1,21-23]. The correspondence principle relates $\bar{S}^\circ(T)$ to $\bar{S}^\circ(298)$ using the expression

$$\bar{S}^\circ(T) = -z\bar{S}^{abs}(H^+,T) + a(T) + b(T)[\bar{S}^\circ(298) - 20.9z] \qquad (3)$$

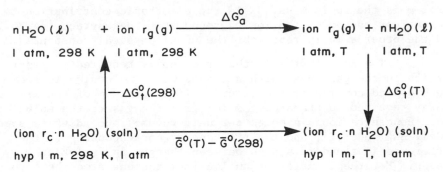

FIG.1. *A generalized charging cycle for calculating free energy functions from electrostatic models. Here r_g and r_c are the gas phase and crystalline ionic radii, respectively.*

Here, z is the ionic charge; a(T) and b(T) are temperature-dependent coefficients obtained for anions, cations, hydroxy-anions and oxyanions from the data available in 1963. The coefficients a(T), b(T) and the single ion entropy for H^+, $\bar{S}^{abs}(H^+,T)$, were derived largely from entropy data below 423 K but have been extrapolated to 573 K.

ELECTROSTATIC MODELS

A more fundamental approach is to use an electrostatic model for the hydrated ion to calculate the variations in $\bar{G}°(T)$. Calculations based on the Born model and a more recent, semi-continuum model have been reported [3,6]. The generalized trans-fer cycle used in such calculations is shown in Figure 1. Here, we assume that the aquo-ion consists of the bare ion, of crystal-lographic radius r_c, hydrated by n water molecules to form a spherically symmetric complex which resides in a cavity in bulk liquid water. From Figure 1, the free energy function is given by the expression

$$\bar{G}°(T) - \bar{G}°(298) = \Delta G_a° + \Delta G_t°(T) - \Delta G_t°(298) \qquad (4)$$

The term $\Delta G_a°$ is the free energy required to transfer one mole of ideal ion gas from 298 K to T. Integration of the Sackur-Tetrode expression for the entropy of an ideal gas [24] yields the ex-pression

$$\Delta G_a° = \left\{ \frac{5}{2} R - R\ln \left[(298.15 \text{ ek})^{5/2} (2\pi m)^{3/2}/h^3 \right] \right\} (T-298.15)$$

$$- \frac{5}{2} RT\ln(T/298.15) - \int_{298.15}^{T} S_e \, dT \qquad (5)$$

Here, m is the ionic mass, S_e is the electronic contribution to
the entropy caused by the ion's ground state degeneracy [24,25],
and the other symbols represent the usual physical constants [24].

The term $\Delta G_t^\circ(T)$ is the free energy required to react
the ion in the gas phase with n liquid water molecules to form
the hydrated complex in the liquid. It is the sum of the free
energy required to transfer the ion from the gas to the solution
(both concentrations in moles per unit volume) according to some
model, $\Delta G_t^\circ(\text{model},T)$, and the free energy corresponding to the
difference in standard states between the gas and solution
phases. For the transfer of an ion from the one atmosphere to
the hypothetical one mol kg^{-1} reference states,

$$\Delta G_t^\circ(T) = \Delta G_t^\circ(\text{model},T) + RT\ell n(\rho RT/1000) \tag{6}$$

where ρ is the density of water at T.

Mathematically, the simplest approach is to assume that
n = 0, so that $\Delta G_t^\circ(\text{model},T)$ is given by the Born expression [26]

$$\Delta G_t^\circ(\text{model},T) = -(1 - 1/\varepsilon)(ze)^2/(2r_c) \tag{7}$$

where ε is the dielectric constant of water and ze is the charge
on one ion. The gas phase ionic radius, r_g in Figure 1, is
assumed equal to r_c. Usually r_c must be used as an adjustable
parameter in order to obtain reasonable hydration energies from
equation 7. Helgeson [3] and Cobble and Murray [7] have used
this approach to estimate the temperature dependence of free
energies of complexation reactions and hydration, respectively.
Comprehensive attempts to apply the Born model to the calculation of free energy functions for ionic species are now under-
way [7,27].

An alternative is to use a more sophisticated semi-
continuum model for ionic hydration, such as that proposed by
Goldman and Bates [28]. For the zero point energy calculation,
this model assumes that the ion is surrounded by a symmetric
cluster of n water molecules, represented as polarizable spheres
containing an off-centered dipole, as shown in Figure 2. Vibra-
tional, rotational and translational partition functions are then
used to calculate $\Delta G_t^\circ(\text{model},T)$. Values of $\bar{G}^\circ(T) - \bar{G}^\circ(298)$ have
been obtained from this model [6] by selecting a value of n which
reproduces the ionic entropy at 298 K and by assuming that n is
independent of temperature. The assumption that n is constant
was required because no experimental measurements of the absolute
entropy or free energy of single ions above 298 K have been
reported. Errors due to this assumption were minimized, for the
cations, by using a scheme based on the model's predictions for
Na$^+$ to split the literature free energies for anion-cation pairs
into single ion components.

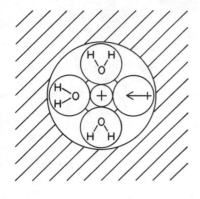

FIG.2.   *The Goldman-Bates semi-continuum model for a monovalent cation having a hydration number of 6, residing in a spherical cavity in bulk water. For the electrostatic calculation, water molecules are represented as polarizable spheres containing two point charges which form an off centered dipole, as shown in the sphere on the right.*

The Goldman-Bates model is applicable only to ions with rare gas electronic structures and the trivalent lanthanides and actinides, where the water-ion interactions can be considered entirely electrostatic.   The method was also found to yield erratic results for anions and large cations such as Cs^+.

EXPERIMENTAL DATA FOR MONATOMIC IONS

Experimental Gibbs free energies for comparison with the predictions were calculated from high-temperature solubility [29-33], EMF [34] and heat capacity [8-15] data.   Details of the calculation are given elsewhere [6,31-33].   Data at 298.15 K are from the U.S. National Bureau of Standards [35] and CODATA [36] compilations.   The heat capacity functions for the solid phases required for the calculations are from Kelley [37].   Room-temperature, standard molal ionic heat capacities were taken from references 17 to 20.   The value for Fe^{2+}, -23 $J \cdot mol^{-1} \cdot K^{-1}$, was interpolated from the heat capacities of Mn^{2+} and Co^{2+} [17,20].

Free energy functions for a number of ions, calculated using various empirical extrapolation methods, are plotted in Figures 3 and 4.   The results for the divalent ions are particularly significant since the experimental data are recent and were not used in the Criss-Cobble [1,2] or earlier $\bar{C}_p^\circ$ studies [3,4].   All the empirical methods agree with the experimental data to within their statistical scatter below about 473 K.   The $\bar{C}_p^\circ$ extrapolations are clearly superior to the Criss-Cobble method.   Also, the assumption $\bar{C}_p^\circ(T) = \bar{C}_p^\circ(298)[T/298.15]$ yields no consistent improvement over the assumption that $\bar{C}_p^\circ(298)$ is constant, even though Khodakovskiy et al. [4] reported better

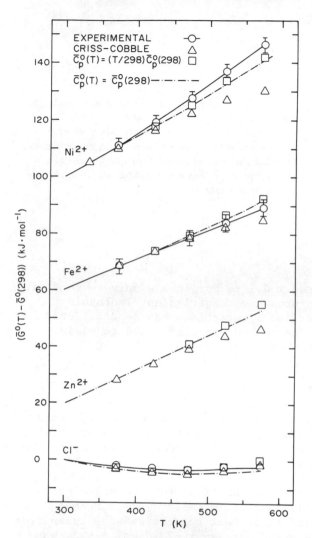

FIG.3. *Gibbs free energy functions for some transition metals and Cl⁻, calculated using various empirical methods and plotted on the conventional scale.*

results using the former $\bar{C}_p^o$ method. The discrepancy between this work and Khodakovskiy's study may be due to the much more limited data then available.

Results from the Goldman–Bates model for ionic hydration [6] are compared with those from the empirical methods in Figure 4. Above about 423 K, free energy functions from the model appear to be more accurate than those from any of the empirical methods for the ions to which it can be applied. Above 523 K, compressibility effects become important [38] and no method appears completely satisfactory.

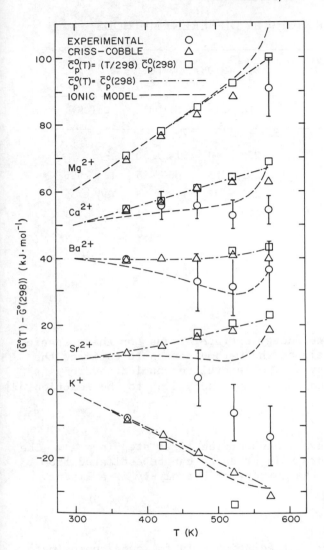

FIG.4. Gibbs free energy functions for alkaline earth cations and K^+, calculated using different methods and plotted on the conventional scale.

HYDROLYZED SPECIES

The empirical methods can be used to calculate free energy functions for hydrolyzed species, $M(OH)_n^{z-n}$, in the rare cases where the entropies are known[2]. Because of the low solubility of the oxides and hydroxides, few entropy data are available for higher hydrolysis products, particularly anions,

---

[2] In principle, an electrostatic model could also be used; however, the calculations would be much more complex than those for monatomic ions, due to the electrostatic asymmetry of the hydrolyzed species [39].

TABLE I. ENTROPIES FOR $M(OH)_3^-$ BY DIFFERENT METHODS

	$\bar{S}_{298}^{\circ}(M(OH)_3^-)$ $(J \cdot mol^{-1} \cdot K^{-1})$		
METHOD	$Fe(OH)_3^-$	$Ni(OH)_3^-$	$Zn(OH)_3^-$
Equation 9	222 ± 50	114 ± 50	-
Equation 10	131 ± 25	88 ± 25	126 ± 25
Cobble [41]	207 ± 35	198 ± 35	190 ± 35
Connick and Powell [45]	112 ± 21	112 ± 21	112 ± 21
Experimental*	123 ± 7	-87 ± 20	46

*   Mean entropies over the range 298 K to 573 K.

and the accuracy of free energy extrapolations for these species depends almost completely on the method chosen to estimate the room-temperature entropy.   The general mononuclear hydrolysis reaction involves ligand replacement according to the reaction[39]

$$M(H_2O)_v^{z+} \rightleftharpoons M(OH)_n(H_2O)_{v-n}^{z-n} + nH^+ \qquad (8)$$

Some entropy data are available for the first step (n = 1).   The entropies of higher hydrolysis products can be estimated from these by assuming that the entropy of hydrolysis is a linear function of n, so that

$$\bar{S}^{\circ}(M(OH)_n^{z-n}) = \bar{S}^{\circ}(M^{z+}) + n[\bar{S}^{\circ}(MOH^{z-1}) - \bar{S}^{\circ}(M^{z+})] \qquad (9)$$

It is important to note that equation 9 is followed approximately by many, but by no means all, cations [39,40]. Baes and Mesmer [39] have reported an empirical expression for estimating $\bar{S}^{\circ}(MOH^{z-1})$ to an accuracy of ± 25 J mol^{-1} K^{-1}. Combining their expression with equation 9 yields the relationship

$$\bar{S}^{\circ}(M(OH)_n^{z-n}) = \bar{S}^{\circ}(M^{z+}) + n[16.9\ A + 34.9\ z/d + \bar{S}^{\circ}(H_2O)] \qquad (10)$$

where z is the ionic charge, d is the M-O distance (nm) in the MOH^{z-1} complex, and A is a constant for each of four groups of cations defined in references [39] and [40]. For transition metals and tetravalent actinides, A is equal to -19.8 ± 1. Alternatives are to use the older empirical expression reported by Cobble [41]

or to establish a correlation between the entropies and values of
z-n for a number of different ligands complexed with the specific
cation in question [42]. Entropies for species of the form
$M(OH)_3^-$ and $M(OH)_4^{2-}$ have also been estimated [31,43,44] from
Connick and Powell's expression [45] for oxyanions, using the
formulae MO_2^{2-} and HMO_2^- to describe the hydrolyzed ions.

As an example, entropies for the species $Fe(OH)_3^-$,
$Ni(OH)_3^-$ and $Zn(OH)_3^-$, calculated by some of the different methods,
are tabulated in Table 1, along with experimental values for the
mean entropy between 298 K and 573 K. Room-temperature entropies
for M^{2+} and MOH^+ were taken from references [33, 46 and 47]. The
experimental values were calculated from solubility constants for
the oxides at 573 K [31-33] and the hydroxides at 298 K
[33,46,48]. Clearly, the experimental data are not sufficiently
accurate to identify one method as superior. The difficulty lies
in making sufficiently accurate solubility measurements on well
characterized, crystalline (and hence sparingly soluble) oxides
or hydroxides at high pH and over a wide temperature range.
Either the room-temperature or high-temperature solubilities used
to calculate these entropy data could be in error. Values for
$\bar{S}°(MOH^{z-1})$ calculated from equation 10 agree with experiment as
well as any other method. Since this equation is based on
hydrolysis data, rather than complexation or oxyanion entropies,
we believe it is the method of choice.

CONCLUSIONS

The experimental data used in this study include most
of the transition metal and noble gas structure cation data
reported for temperatures above 373 K. Thus, although the study
is by no means exhaustive, the results are undoubtedly valid for
these classes of ions and probably for most other systems as
well. Of the methods considered here, the assumption that
$\bar{C}_p°(T) = \bar{C}_p°(298)$ appears to yield the most accurate free energy
functions for monatomic ions up to about 423 K. The Criss-Cobble
method is less accurate but could, no doubt, be improved by
incorporating experimental data published after 1963. The ionic
hydration model gives superior results in the range 423 K < T <
523 K for ions in which covalent bonding is absent. More accurate, experimental partial molal heat capacities, free energies
and single ion entropies, at elevated temperatures, are needed to
improve these methods.

The major uncertainty in extrapolating data for hydrolyzed species is due to the poorly defined room-temperature
entropies and free energies. These uncertainties undoubtedly
outweigh any errors due to a specific extrapolation procedure.

As a result, potential-pH diagrams or solubility data extrapolated from a lower temperature must be considered very qualitative
unless an analysis of the effects of uncertainties in the low-
temperature data are included [46].

REFERENCES

[1] CRISS, C.M., COBBLE, J.W., J. Amer. Chem. Soc. 86 (1964) 5385.

[2] CRISS, C.M., COBBLE, J.W., J. Amer. Chem. Soc. 86 (1964) 5390.

[3] HELGESON, H.C., J. Phys. Chem. 71 (1967) 3121.

[4] KHODAKOVSKIY, I.L., RYZHENKO, B.N., NAUMOV, G.B., Geokhimiya 1968 1486.

[5] MATSUI, T., KO, H.C., HEPLER, L.G., Can. J. Chem. 52 (1974) 2912.

[6] TREMAINE, P.R., GOLDMAN, S., J. Phys. Chem. 82 (1978) 2317.

[7] COBBLE, J.W., MURRAY, R.C., J.C.S. Faraday Disc., in press.

[8] CRISS, C.M., COBBLE, J.W., J. Amer. Chem. Soc. 83 (1961) 3223.

[9] JEKEL, E.C., CRISS, C.M., COBBLE, J.W., J. Amer. Chem. Soc. 86 (1964) 5404.

[10] MITCHELL, R.E., COBBLE, J.W., J. Amer. Chem. Soc. 86 (1964) 5401.

[11] GARDNER, W.L., MITCHELL, R.E., COBBLE, J.W., J. Phys. Chem. 73 (1969) 2025.

[12] AHLUWALIA, J.C., COBBLE, J.W., J. Amer. Chem. Soc. 86 (1964) 5381.

[13] GARDNER, W.L., JEKEL, E.C., COBBLE, J.W., J. Phys. Chem. 73 (1969) 2017.

[14] LIU, C.T., LINDSAY, W.T., J. Solution Chem. 1 (1972) 45.

[15] RÜTERJANS, H., SCHREINER, F., SAGE, U., ACKERMANN, Th., J. Phys. Chem. 73 (1969) 986.

[16] PICKER, P., LEDUC, P.-A., PHILIP, P.R., DESNOYERS, J.E., J. Chem. Thermodynamics 3 (1971) 631.

[17] SPITZER, J.J., SINGH, P.P., OLOFSSON, I.V., HEPLER, L.G., J. Solution Chem. 7 (1978) 623.

[18] DESNOYERS, J.E., de VISSER, C., PERRON, G., PICKER, P., J. Solution Chem. 5 (1976) 605.

[19] SPITZER, J.J., SINGH, P.P., McCURDY, K.G., HEPLER, L.G., J. Solution Chem. 7 (1978) 81.

[20] ROUX, A., MUSBALLY, G.M., PERRON, G., DESNOYERS, J.E., SINGH, P.P., WOOLLEY, E.M., HEPLER, L.G., Can. J. Chem. 56 (1978) 24.

[21] HELGESON, H.C., Amer. J. Sci. 267 (1969) 729.

[22] MACDONALD, D.D., BUTLER, P., Corrosion Sci. 13 (1973) 259.

[23] TAYLOR, D.F., J. Electrochem. Soc. 125 (1978) 808.

[24] McQUARRIE, D.A., "Statistical Mechanics", Harper and Row, New York (1976).

[25] SPEDDING, F.H., RARD, J.A., HABENSCHUSS, A., J. Phys. Chem. 81 (1977) 1069.

[26] FRIEDMAN, H.L., KRISHNAN, C.V., in "Water, A Comprehensive Treatise", Vol. 3, F. Franks, ed., Plenum, New York (1973).

[27] HELGESON, H.C., KIRKHAM, D.H., Amer. J. Sci. 276 (1976) 97.

[28] GOLDMAN, S., BATES, R.G., J. Amer. Chem. Soc. 94 (1972) 1976.

[29] MARSHALL, W.L., J. Inorg. Nucl. Chem. 37 (1975) 2155.

[30] BOOTH, H.S., BIDWELL, R.M., J. Amer. Chem. Soc. 72 (1950) 2567.

[31] KHODAKOVSKIY, I.L., YELKIN, A.Ye., Geokhimiya 1975 1490.

[32] SWEETON, F.H., BAES, C.F., J. Chem. Thermodynamics 2 (1970) 479.

[33] TREMAINE, P.R., LEBLANC, J.C., J. Chem. Thermodynamics (in press).

[34] GREELY, R.S., SMITH, W.T., STOUGHTON, R.W., LIETZKE, M.H., J. Phys. Chem. 64 (1960) 652.

[35] WAGMAN, D.D., EVANS, W.H., PARKER, V.B., HALOW, I., BAILEY, S.M., SCHUMM, R.H., U.S. Nat. Bur. Standards Tech. Note 270-3 (1968) and 270-4 (1969).

[36] CODATA TASK GROUP ON KEY VALUES FOR THERMODYNAMICS, J. Chem. Thermodynamics 9 (1977) 705; ibid 8 (1976) 603.

[37] KELLEY, K.K., U.S. Bureau Mines Bull. 584 (1960).

[38] COBBLE, J.W., J. Amer. Chem. Soc. 86 (1964) 5394.

[39] BAES, C.F., MESMER, R.E., "The Hydrolysis of Cations", Wiley-Interscience, New York (1976).

[40] AHRLAND, S., in "Structure and Bonding", Springer-Verlag, New York 15 (1973) 167.

[41] COBBLE, J.W., J. Chem. Phys. 21 (1953) 1446.

[42] LANGMUIR, D., Geochim. Cosmochim. Acta 42 (1978) 547.

[43] COWAN, R.L., STAEHLE, R.W., J. Electrochem. Soc. 118 (1972) 557.

[44] BROOK, P.A., Corrosion Sci. 12 (1972) 297.

[45] CONNICK, R.E., POWELL, R.E., J. Chem. Phys. 21 (1953) 2206.

[46] TREMAINE, P.R., VON MASSOW, R., SHIERMAN, G.R., Thermochim. Acta 19 (1977) 287.

[47] LARSON, J.W., CERUTTI, P., GARBER, H.K., HEPLER, L.G., J. Phys. Chem. 72 (1968) 2903.

[48] REICHLE, R.A., McCURDY, K.G., HEPLER, L.G., Can. J. Chem. 53 (1975) 3841.

DISCUSSION

L.R. MORSS: In your Fig.3 you include experimental partial molal free
energy data for transition-metal ions. Were the experimental measurements
made only at 25°C, or have experimental measurements been utilized to
calculate the free energy at high temperatures?

P.R. TREMAINE: Data for Fe^{2+} and Ni^{2+} were obtained from Fe_3O_4 and
NiO solubility data at elevated temperatures (my Refs [32] and [33], respectively).

AQUEOUS ACTINIDE COMPLEXES:
A THERMOCHEMICAL ASSESSMENT

J. FUGER
Institute of Radiochemistry,
University of Liège,
Liège,
Belgium

I.L. KHODAKOVSKIJ
Vernadsky Institute of Geochemistry and Analytical Chemistry,
Academy of Sciences of the USSR,
Moscow,
Union of Soviet Socialist Republics

V.A. MEDVEDEV
Institute for High Temperatures,
Academy of Sciences of the USSR,
Moscow,
Union of Soviet Socialist Republics

J.D. NAVRATIL
Rockwell International,
Rocky Flats Plant,
Golden, Colorado,
United States of America

Abstract

AQUEOUS ACTINIDE COMPLEXES: A THERMOCHEMICAL ASSESSMENT.
The scope and purpose of an assessment of the thermodynamic properties of the
aqueous actinide complexes are presented. This work, which at present is limited to
inorganic ligands and three selected organic ligands (formate, acetate and oxalate), is part of
an effort established by the International Atomic Energy Agency to assess the thermodynamic
properties of the actinides and their compounds. The problems involved in this work are
illustrated by discussing the present status of the assessment as related to a few complex species.

1. INTRODUCTION

For some time there has been a need to compile and critically assess the
thermodynamic properties of the actinide elements and compounds. In an

effort to answer this need, the International Atomic Energy Agency has established a collaborative effort involving various thermodynamicists throughout the world. As a result of this effort three publications have already become available and a further eleven are in various stages of preparation [1]. The most recent additions to the series are those dealing with the actinide gaseous ions, the actinide organic complexes and the actinide complex aqueous ions. The present paper deals with the scope and status of the latter effort.

In this compilation, complexes between the various actinide ions and the following ligands are being considered:

OH^-, F^-, Cl^-, Br^-, I^-, ClO_2^-, ClO_3^-, ClO_4^-, BrO_3^-, IO_3^-, $S_2O_3^-$, SO_3^{2-}, SO_4^{2-}, $H_2NSO_3^-$,

N_3^-, H_2NOH, NO_2^-, NO_3^-, PO_4^{3-}, $(P_nO_{3n+1})^{(n+2)-}$, CN^-, SCN^-, CO_3^{2-}, $H_2SiO_4^{2-}$,

$Mo(CN)_8^{4-}$, CrO_4^{2-}, HCO_2^-, $H_3C_2O_2^-$ and $C_2O_4^{2-}$.

At the present stage, the assessment is limited to inorganic ligands and three selected organic ligands. The properties being considered include the stability constant as a function of ionic strength and temperature, and ΔH and ΔS at zero ionic strength whenever possible. In tabulating literature data, only experimental results will be given and estimates as well as recalculated data will be ignored. Our recommended values will be tabulated. More than five hundred references have been consulted during the present work, which was initiated in late 1977 and is expected to be completed by the end of 1980. As in previous parts of this series, the selection of a best value will be discussed and justified, and reasons given for rejecting data. In addition, our estimates of the thermo-dynamic properties, based on interrelationships between analogous systems, will be given when it is felt that this can be done reliably. Another essential aim of this assessment is to indicate those areas in which additional research is required.

In order to illustrate the problems peculiar to this work, we have chosen to discuss some specific aspects of this assessment as they relate to a number of complexes involving OH^-, F^- and CO_3^{2-}, these ions being selected here because of their importance in the nuclear fuel cycle.

2. DISCUSSION

2.1. Hydroxyl complexes

A critical assessment [2] covering the entire Periodic Table has recently been devoted to the hydrolysis of cations. As the present work is only devoted to the actinides, we intend to discuss in more detail the individual data before

selecting values when we feel it appropriate. Although in the final text we intend
to report the data with the OH⁻ ligand as complexation constants, we prefer at
the present stage to refer to hydrolysis constants, i.e. involving H_2O, as we have
not yet selected our final data related to the thermodynamics of water dissociation
at various ionic strengths and temperatures.

The literature data on the first hydrolysis of the trivalent actinide ions,
according to the reaction:

$$M^{3+} + H_2O \rightarrow MOH^{2+} + H^+; \qquad \beta_1^* = \frac{[MOH^{2+}] \cdot [H^+]}{[M^{3+}]} \qquad (I)$$

appear quite limited and contradictory, as shown in Table I.[1]

It is not too surprising that no data are available on the hydrolysis of U^{3+}
in view of the marked instability of this ion in aqueous solution. For Ac^{3+} the
only information is based on the pH of precipitation of $Ac(OH)_3$ [14], indicating
that Ac^{3+} is a distinctly less acidic cation than La^{3+}. In the case of Pu^{3+}, the
reported data, all at relatively low ionic strength, extend over more than three
orders of magnitude. In the case of americium and curium (and probably the
heavier actinides if more data were available!) the scatter is of two orders of
magnitude. It appears worth noting that the most negative values are reported
from acidimetric titration techniques (gl) which are quite straightforward but
involve relatively high actinide ion concentrations ($> 10^{-3}$ M) and, therefore,
the possibility that polynuclear species are formed cannot be ruled out [15].
Conversely, the determinations by electromigration techniques (tp) and
solvent extraction (dis) were carried out at actinide concentrations $\leqslant 10^{-5}$ M
and $\sim 10^{-7}$ M, respectively, and therefore are not open to the same criticism.
However, it is not always obvious how these experimental data should be
interpreted, especially in the case of solvent extraction, where assumptions have
to be made about the extraction mechanism. A similar scatter of data related
to the technique used has also been observed for the first hydrolysis constant
of the trivalent lanthanides [2, 15–17]. It is therefore obvious that this area
requires further careful study. Of course the status of more hydrolysed aqueous
species of the trivalent actinides is still more uncertain.

Much attention has been given to the hydrolysis of the actinides in their
other valency states, especially M^{4+} and MO_2^{2+}. We shall briefly evoke the case
of UO_2^{2+} and of the other MO_2^{2+} ions. The EMF measurements of Baes and
Meyer [18], supported by several other studies, indicate that the hydrolytic

---

[1]   In this table and throughout the assessment, the symbols used to describe methods
employed in the original investigations as well as the medium are those of Sillen and
Martell [3]. Unless otherwise indicated, concentrations are given in moles per litre (mol·ltr⁻¹
or M). Throughout the text the temperature 298 K refers to 298.15 K and other temperatures
are reported accordingly.

TABLE I. LITERATURE DATA FOR THE FIRST HYDROLYSIS OF THE TRIVALENT ACTINIDE IONS

$$M^{3+} + H_2O \rightarrow MOH^{2+} + H^+; \qquad \beta_1^* = \frac{[MOH^{2+}] \cdot [H^+]}{[M^{3+}]}$$

M	T (K)	Medium[a]	(mol·ltr⁻¹)	$\log_{10} \beta_1^*$	Method[b]	Ref.	Year
Np³⁺	296	(Na, H) Cl	0.3	−7.43 ± 0.12	gl	[4]	1974
Pu³⁺	296	(H, Li) ClO₄ (HCl)	0.2 / 0.1	−3.8 ± 0.2	dis	[5]	1975
	R.T.	(Na, H) ClO₄	0.069	−7.22	gl	[6]	1949
	R.T.	(Na, H) Cl	0.024	−7.37	gl	[6]	1949
Am³⁺	288	(K)Cl	0.005	−3.05	tp	[7]	1969
	296	(Li)ClO₄	0.1	−5.92	dis	[8]	1969
	298[c]	(Li)ClO₄	0.1	−5.30 ± 0.07	dis	[9, 10]	1975, 73
	298	(NH₄)ClO₄	0.005	−3.3 ± 0.1	tp	[11]	1972
Cm³⁺	296	(Li)ClO₄	0.1	−6.05 ± 0.1	dis	[12]	1969
	296	(Li)ClO₄	0.1	−5.92	dis	[8]	1969
	298[c]	(Li)ClO₄	0.1	−5.40 ± 0.07	dis	[9, 10]	1975, 73
	298	(NH₄)ClO₄	0.005	−3.4 ± 0.1	tp	[11]	1972
Bk³⁺	296	(Li)ClO₄	0.1	−5.66	dis	[8]	1969

M	T (K)	Medium[a] (mol·ltr⁻¹)		$\log_{10} \beta_1^*$	Method[b]	Ref.	Year
Cf^{3+}	296	0.1	(Li)ClO$_4$	-5.62	dis	[8]	1969
	298^c	0.1	(Li)ClO$_4$	-5.05 ± 0.07	dis	[9, 10]	1975, 73
Es^{3+}	298^c	0.1	(Li)ClO$_4$	-5.15 ± 0.07	dis	[9, 10]	1975, 73
Fm^{3+}	298	0.1	(Li)ClO$_4$	-3.80 ± 0.2	dis	[9, 13]	1975, 72

[a] 0.3 (Na, H)Cl Concentration of the anion held constant at 0.3 mol·ltr⁻¹;
0.1 (HCl) Concentration of substance held constant at 0.1 mol·ltr⁻¹.

[b] Methods — gl: acidimetric titration with glass-electrode; dis: solvent extraction; tp: electromigration.

[c] Data also reported at six other temperatures ranging from 283 to 323 K, from which ΔH and ΔS were reported. They will not be discussed here.

TABLE II. THERMODYNAMIC DATA ASSOCIATED WITH THE FORMATION OF THE PRINCIPAL HYDROLYTIC SPECIES OF THE URANYL ION

$$p\, UO_2^{2+} + q\, H_2O \rightarrow (UO_2)_p(OH)_q^{(2p-q)+} + q\, H^+; \qquad \beta^*_{pq} = \frac{[(UO_2)_p(OH)_q]^{(2p-q)} \cdot [H^+]^q}{[UO_2^{2+}]^p}$$

Species	T (K)	Medium[a] (mol·kg⁻¹)	$\log_{10} \beta^*_{pq}$	ΔH (kcal·mol⁻¹)	ΔS_1 (cal·mol⁻¹·K⁻¹)	Method[b]	Ref.	Year
$UO_2(OH)^+$	298	0.5 NaNO₃	-5.70 ± 0.3	11 ± 2	11 ± 5	EMF	[18]	1962
	392.4	0.5 NaNO₃	-4.19 ± 0.07					
$(UO_2)_2(OH)_2^{2+}$	298	0.5 NaNO₃	-5.92 ± 0.04	10.2 ± 0.5	7.1 ± 2.0	EMF	[18]	1962
	392.4	0.5 NaNO₃	-4.51 ± 0.03					
	298	3ᶜ NaClO₄	-6.02 ± 0.02	9.5 ± 0.1	4.3 ± 0.3	cal	[23]	1968
$(UO_2)_3(OH)_5^+$	298	0.5 NaNO₃	-16.22 ± 0.08	25.1 ± 1.5	10 ± 5	EMF	[18]	1962
	392.4	0.5 NaNO₃	-12.74 ± 0.14					
	298	3ᶜ NaClO₄	-16.54 ± 0.03	24.4 ± 0.5	6 ± 5	cal	[23]	1968

[a] 0.5 NaNO₃: constant concentration of substance stated (0.5 mol·kg⁻¹ NaNO₃).

[b] Methods – EMF: potentiometric; cal: calorimetric.

[c] mol·ltr⁻¹.

behaviour of UO_2^{2+} is best accounted for in nitrate and perchlorate media by the formation of three major hydrolytic ions, $UO_2(OH)^+$, $(UO_2)_2(OH)_2^{2+}$ and $(UO_2)_3(OH)_5^+$, while other polymeric species may appear in chloride media [19–21]. Solubility studies of β–$UO_2(OH)_2$ in water [22] indicate that above 323 K the dissolution is accompanied by the formation of undissociated $UO_2(OH)_2^\circ$ molecules. It should be noted, as a general remark, that the determination with certainty of all the species present in such complicated systems is a very difficult task and that the accuracy of the experimental data does not warrant the existence of all the species claimed by some authors.

Table II summarizes the thermodynamic data for the formation of the principal hydrolytic products of the uranyl ion. The agreement between the results of the temperature-dependence measurements (298 and 367 K) of the stability constants by Baes and Meyer [18] and the more accurate enthalpy titration data of Arnek and Schlyter [23] is remarkable, especially in view of the fact that, in the temperature-dependence measurements, ΔH was considered as constant over a temperature range of 70 K. As we shall see later, such a neglect of the ΔC_p of the reactions may, in some instances, lead to completely erroneous results.

The most recent data indicate that the hydrolysis of NpO_2^{2+} [24] and PuO_2^{2+} [25–28] follow schemes similar to that of UO_2^{2+}. We shall however not attempt here the detailed discussion of these data.

2.2. Fluoride complexes

The data related to these complexes are very numerous, especially for the actinides in their 4+ valency states. Among these, the results on thorium are the most interesting to comment upon, since a number of measurements of stability constants in identical media were reported by several groups of authors. We shall, however, restrict this brief discussion to the monofluoro complexes. Thus, for the reaction:

$$M^{4+} + F^- \rightarrow MF^{3+}; \qquad \beta_1 = \frac{[MF^{3+}]}{[M^{4+}] \cdot [F^-]} \qquad \text{(II)}$$

Table III summarizes the literature data related to the stability of thorium monofluoro complexes. Where needed, the original data reported for the reaction

$$M^{4+} + HF \rightarrow MF^{3+} + H^+ \qquad \text{(III)}$$

have been recalculated to obtain β_1 for reaction (II) using for the pK of hydrofluoric acid the data selected by Smith and Martell [29] or interpolated from them using the van't Hoff equation for values at temperatures other than 298 K.

TABLE III. LITERATURE DATA FOR THE STABILITY CONSTANTS
OF THE THORIUM MONOFLUORO COMPLEXES

$$M^{4+} + F^- \rightarrow MF^{3+}; \qquad \beta_1 = \frac{[MF^{3+}]}{[M^{4+}] \cdot [F^-]}$$

T (K)	Medium[a] $(mol \cdot ltr^{-1})$		$log_{10} \beta_1$	Method[a]	Ref	Year
293	4	(HClO₄)	8.10 ± 0.05	F electr.	[30]	1969
293	4	(HClO₄)	8.04 ± 0.06	dis		
298	3	(Na,H)ClO₄	7.82 ± 0.05	F electr.	[31]	1971
298	2	(HClO₄)	7.56 ± 0.03	dis	[32]	1975
276	1	(HClO₄)	7.52 ± 0.03			
298	1	(HClO₄)	7.47 ± 0.03	F electr.	[33]	1976
320	1	(HClO₄)	7.56 ± 0.06			
283	1	(HClO₄)	7.50 ± 0.03			
298	1	(HClO₄)	7.45 ± 0.04	dis	[33]	1976
328	1	(HClO₄)	7.48 ± 0.04			
298	0.5	(NaClO₄)	7.58 ± 0.05	redox	[34]	1949
298	0.5	(NaClO₄)	7.63 ± 0.05[c]	dis	[35]	1951
298	0.5	(HClO₄)	7.56 ± 0.05[c]	dis	[36]	1950
278	0.01	(NH₄NO₃)	8.11 ± 0.02[c]			
298	0.01	(NH₄NO₃)	8.08 ± 0.02[c]	F electr.	[37]	1970
318	0.01	(NH₄NO₃)	7.95 ± 0.02[c]			

[a] 4 (HClO₄) Ionic strength held constant at the value stated (4 mol·ltr⁻¹) by addition of inert salt shown in parentheses.

[b] Methods − F electr.: potentiometric method using fluorine elctrode; dis: solvent extraction; redox: potentiometric method with redox electrode.

[c] Our estimated uncertainty limits.

Obviously the various data in Table III are in general agreement. We have therefore attempted to obtain a best value for the stability constant of ThF³⁺ at zero ionic strength (β_1^0) from all the literature values at 298 K using a modification of a Debye-Hückel relation proposed by Vasil'ev [38]:

$$\log \beta^0 = \log \beta - \frac{\Delta Z^2 A \mu^{\frac{1}{2}}}{1 + 1.6 \mu^{\frac{1}{2}}} - b\mu \qquad (1)$$

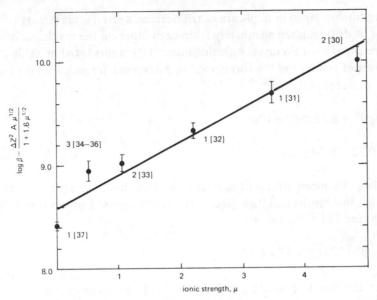

$FIG.1.$ *Variation of* $\left(\log \beta - \dfrac{\Delta Z^2 \cdot A \cdot \mu^{\frac{1}{2}}}{1 + 1.6\,\mu^{\frac{1}{2}}} \right)$ *with ionic strength,* μ, *for* ThF^{3+} *at 298 K from all available experimental data (number of results and references are indicated for each point).*

in which A is the Debye-Hückel constant (0.5115 at 298 K), ΔZ^2 is the variation of the sum of the squares of the ionic charges in the reaction under investigation, -8 in the present case, b is an empirical parameter, and 1.6 is a constant valid for all systems, taken instead of the more classical 0.33a (a being the distance of closest approach of the ions).

It is obvious that the selection of 1.6 for this constant factor for all systems is too simple for a rigorous treatment of the problem, but it may well be used for practical calculations of stability constants, since the uncertainties in the experimental data are usually far greater than those arising from activity coefficients of the electrolytes. In any case, Vasil'ev's equation has been shown to account for the variation of the stability constants of multicharged complexes over a wide range of ionic strengths [38]. Thus a plot of:

$$\log \beta - \frac{\Delta Z^2 \; A \; \mu^{\frac{1}{2}}}{1 + 1.6\,\mu^{\frac{1}{2}}} \quad \text{versus} \quad \mu$$

should yield a straight line with slope b. Figure 1 shows this treatment for ThF^{3+}. We have also included the data at 293 K [30] as this would not introduce a significant discrepancy.

As results obtained in duplicate or triplicate at a given ionic strength could not be distinguished adequately from each other on the graph, we have only represented their averages, with the uncertainty limits listed in Table III. The number of results and the corresponding references are indicated for each point. The best intercept and slope:

$$\log_{10} \beta^0 = 8.58 \pm 0.06 \ (2\sigma)$$

$$b = 0.32 \pm 0.04 \ (2\sigma)$$

$$\tag{2}$$

are calculated by means of a least-squares treatment that takes into account all the individual results and their given uncertainty limits. Conservatively, we shall adopt for ThF^{3+} the value

$$\log_{10} \beta_1^0 \ (298) = 8.58 \pm 0.15 \tag{3}$$

When it comes to extracting enthalpy and entropy changes from temperature-dependence measurements, the situation is far more delicate, as the reported uncertainties of the constants are by no means small with respect to the trend to be measured. Table IV summarizes the available thermodynamic data for the actinide-IV monofluoro complexes, as they are reported in the literature.

In the case of the thorium complexes, the temperature data of Choppin and Unrein [33] are obviously quite consistent with the more precise calorimetric result of these authors, which is unfortunately the only calorimetric value reported for the tetravalent actinide monofluoro complexes. On the other hand, these data do not seem to agree with the negative enthalpy change reported by Baumann [37], even if one takes into account the difference in ionic strength. The relative consistency of the entropy change throughout the actinide series only reflects the fact that the enthalpy term is of minor importance and that all these complexes have stability constants of the same order of magnitude. It is obvious that in this area, as in many others, more calorimetric results are needed.

2.3. Carbonate complexes

We shall focus our attention in this section on the uranyl carbonate complexes because of their interest not only in some aspects of the nuclear fuel cycle but also in the study of hydrothermal uranium deposits.

The stability at room temperature of the two species $UO_2(CO_3)_2^{2-}$ and $UO_2(CO_3)_3^{4-}$ has been determined using a wide range of techniques, including spectrophotometric [41–43], potentiometric [41, 44–46], ion exchange [47],

TABLE IV. THERMODYNAMIC DATA ASSOCIATED WITH THE
FORMATION OF ACTINIDE-IV MONOFLUORO COMPLEXES AT
298 K ACCORDING TO THE REACTION:
$M^{4+} + F^- \rightarrow MF^{3+}$

Species	Medium [a] $(mol \cdot ltr^{-1})$	ΔH $(kcal \cdot mol^{-1})$	ΔS $(cal \cdot mol^{-1} \cdot K^{-1})$	Method [b]	Ref.	Year
ThF^{3+}	1 (HClO$_4$)	0.26 ± 0.32	34.8 ± 1.2	dis		
		0.56 ± 0.44	36.0 ± 1.5	F electr.	[33]	1976
		0.72 ± 0.14		cal		
ThF^{3+}	0	−1.2	39	F electr.	[37]	1970
UF^{3+}	1 (HClO$_4$)	0.18 ± 0.94	36.1 ± 3.2	F electr.	[33]	1976
NpF^{3+}	1 (HClO$_4$)	0.76 ± 0.44	37.0 ± 1.5	dis	[33]	1976
PuF^{3+}	2 (HClO$_4$)	1.31	39.2	dis	[39,40]	1977,76

[a] 1 (HClO$_4$) Ionic strength held constant at the value stated (1 mol·ltr^{-1}) by addition of
the inert salt shown in parentheses.

[b] Methods — dis: solvent extraction; F electr.: potentiometric method using fluorine
electrode; cal: calorimetric.

solvent extraction [48, 49] and solubility [42, 50, 51] studies. We cannot
discuss these here in detail. The various data, although somewhat scattered,
are generally in agreement; for the reactions:

$$UO_2^{2+} + 2CO_3^{2-} \rightarrow UO_2(CO_3)_2^{2-}; \quad \beta_2 \qquad\qquad\qquad (IV)$$

$$UO_2^{2+} + 3CO_3^{2-} \rightarrow UO_2(CO_3)_3^{4-}; \quad \beta_3 \qquad\qquad\qquad (V)$$

extrapolation to zero ionic strength yields as best values:

$$\log_{10} \beta_2^0 (298) = 17.0 \pm 0.3 \qquad\qquad\qquad\qquad (4)$$

$$\log_{10} \beta_3^0 (298) = 21.6 \pm 0.3 \qquad\qquad\qquad\qquad (5)$$

In 1972, Sergeyeva et al. [52] carried out an extensive study of the
solubility of UO_2CO_3 (cryst.) in aqueous solutions of carbon dioxide at various

pH-values over the temperature range 298–473 K. In this investigation, care was taken to ascertain the identity of the solid phases and the effective attainment of the equilibrium between the solid and the solution, essential requirements which, unfortunately, are not always met in solubility studies. The results revealed for the first time the existence of the undissociated complex $UO_2CO_3^o$, which showed a rather broad field of existence in neutral and weakly acid solutions. For the reaction:

$$UO_2^{2+} + CO_3^{2-} \rightarrow UO_2CO_3^o \tag{VI}$$

at zero ionic strength, Sergeyeva et al. [52] report the relation:

$$\log_{10} \beta_1^0 = \frac{2809.1}{T} - 9.7893 + 0.034337\,T \tag{6}$$

The above relation at 298 K yields $\log_{10}\beta_1^0$ (298) = 9.87 ± 0.07, ΔH_1^0 (298) = 1.1 kcal·mol^{-1}, ΔS_1^0 (298) = 48.8 cal·mol^{-1}·K^{-1} and ΔC_p^0 (298) = 93.7 cal·mol^{-1}·K^{-1}. Incidentally, this large ΔC_p^0 value furnishes a good example in which a ΔH value obtained from an average slope of $\log_{10}\beta$ versus 1/T, even over a relatively narrow range of temperatures, would yield completely erroneous results.

The study of Sergeyeva et al. [52] also yielded for the formation of $UO_2(CO_3)_2^{2-}$ at zero ionic strength the values of $\log_{10} \beta_2^0$ (298) = 16.7 and $\log_{10} \beta_2^0$ (323) = 16.95, in agreement with the other studies on this species. A more recent solubility study [53] also supports the conclusions of Sergeyeva et al. about the stability of $UO_2CO_3^o$ in solution.

In 1977, Devina et al. [54] reported the calorimetric measurements at several temperatures of the enthalpy of solution of uranyl nitrate hexahydrate in dilute perchlorate media (0.1M to 0.18M) containing sodium carbonate. At 298 K they obtained for the reaction:

$$UO_2^{2+} + 3CO_3^{2-} \rightarrow UO_2(CO_3)_3^{4-}; \quad \beta_3$$

a value of ΔH_3 (298) = −9.64 kcal·mol^{-1}. With log β_3^0 (298) from Eq.(5), this value being assumed to be valid, within ±0.2 kcal·mol^{-1}, at infinite dilution, yields the entropy change ΔS_3^0.

The presently recommended thermodynamic data for the formation of the uranyl carbonate complexes are listed in Table V. Rather conservative uncertainty limits have been retained.

The uranyl ion has also been reported to form hydroxylcarbonate complexes, but the consideration of these species is beyond the scope of this paper.

TABLE V. RECOMMENDED THERMODYNAMIC DATA FOR THE
FORMATION OF URANYL CARBONATE COMPLEXES AT 298 K
ACCORDING TO REACTION:
$$UO_2^{2+} + q\,CO_3^{2-} \rightarrow UO_2(CO_3)_q^{2(q-1)-}$$

Species	ΔG^0 (kcal·mol^{-1})	ΔH^0 (kcal·mol^{-1})	ΔS^0 (cal·mol^{-1}·K^{-1})
$UO_2CO_3^{\circ}$	-13.4 ± 0.4	1.1 ± 0.4	48.8 ± 2.0
$UO_2(CO_3)_2^{2-}$	-22.8 ± 0.5		
$UO_2(CO_3)_3^{4-}$	-29.5 ± 0.5	-9.64 ± 0.20	66.6 ± 1.8

3. CONCLUSIONS

We hope that the examples selected have helped to highlight the difficulties
of our present endeavour. Too often literature data have to be disregarded
owing to the lack of proper definition of all experimental parameters. Also, but
this is probably commonplace, authors tend to under-evaluate the uncertainty
limits of their data or simply disregard them. Finally, if the selection of best
values for the stability constants (or ΔG values) at 298 K appears feasible for
many systems, the situation is dramatically different for the other thermodynamic
parameters. We are, therefore, taking this opportunity to call for the additional
calorimetric measurements which are so crucially needed.

REFERENCES

[1] INTERNATIONAL ATOMIC ENERGY AGENCY, The Chemical Thermodynamics of
Actinide Elements and Compounds (MEDVEDEV, V., RAND, M.H., WESTRUM, Jr., E.F.,
Eds; OETTING, F.L., Exec. Ed.), IAEA, Vienna (1976 →), published as Parts 1–14.
Part 1: The Actinide Elements: OETTING, F.L., RAND, M.H., ACKERMANN, R.J.,
(1976).
Part 2: The Actinide Binary Alloys: FUGER, J., OETTING, F.L., (1976).
Part 3: Miscellaneous Actinide Compounds: CORDFUNKE, E.H.P., O'HARE, P.A.G.,
(1978).
Part 4: The Actinide Chalcogenides: GRØNVOLD, F., DROWART, J., WESTRUM Jr., E.F.,
(in preparation).
Part 5: The Actinide Binary Alloys: CHIOTTI, P., AKHACHINSKIJ, V.V., ANSARA, I.,
RAND, M.H., (1980).
Part 6: The Actinide Carbides: HOLLEY, C.E., RAND, M.H., STORMS, E.K.,
(in preparation).

Part 7: The Actinide Pnictides: POTTER, P.E., MILLS, K.C., TAKAHASHI, Y.,
   (in preparation).
Part 8: The Actinide Halides: HUBBARD, W.N., FUGER, J., OETTING, F.L.,
   PARKER, V.B., *(in preparation)*.
Part 9: The Actinide Hydrides: FLOTOW, H.E., YAMAUCH, S., *(in preparation)*.
Part 10: The Actinide Oxides: RAND, M.H., ACKERMANN, R.J., GRØNVOLD, F.,
   OETTING, F.L., PATTORET, A., *(in preparation)*.
Part 11: Selected Ternary Systems: POTTER, P.E., HOLLECK, H., SPEAR, K.E.,
   (in preparation).
Part 12: The Actinide Complex Aqueous Ions: FUGER, J., KHODAKOVSKIJ, I.L.,
   MEDVEDEV, V.A., NAVRATIL, J.D., *(in preparation)*.
Part 13: The Actinide Gaseous Ions: HILDENBRAND, D.L., GURVICH, L.V.,
   YUNGMAN, V.S., FRED, M., *(in preparation)*.
Part 14: The Actinide Organic Complexes *(in preparation)*.

[2] BAES, Jr., C.F., MESMER, R.E., The Hydrolysis of Cations, Wiley and Sons, New
     York (1976).
[3] SILLEN, L.G., MARTELL, A.E., Stability Constants of Metal—Ion Complexes, The
     Chemical Society, London, Special Publications No.17 (1964) and No.25 (1971).
[4] MEFOD'EVA, M.P., KROT, N.N., AFANAS'EVA, T.V., GEL'MAN, A.D., Izv. Akad.
     Nauk SSSR, Ser. Khim. **23** (1974) 2370; *English trans.* Bull. Acad. Sci. USSR, Div.
     Chem. Sci. **23** (1974) 2285.
[5] HUBERT, S., HUSSONNOIS, M., GUILLAUMONT, R., J. Inorg. Nucl. Chem. **37**
     (1975) 1255.
[6] KRAUS, K.A., DAM, J.R., The Transuranium Elements (SEABORG, G., et al., Eds.),
     McGraw Hill, New York **14B** (1949) 466.
[7] MARIN, B., KIKINDAI, T., C.R. Hebd. Séances Acad. Sci., Ser. C **268** (1969) 1.
[8] DESIRÉ, B., HUSSONNOIS, M., GUILLAUMONT, R., C.R. Hebd. Séances Acad. Sci.,
     Ser. C **268** (1969) 448.
[9] HUBERT, S., HUSSONNOIS, M., BRILLARD, L., GUILLAUMONT, R., Transplutonium
     Elements (Proc. 4th Int. Symp. Baden-Baden, 1975), North Holland Publ. Co.,
     Amsterdam (1976) 109.
[10] HUSSONNOIS, M., HUBERT, S., BRILLARD, L., GUILLAUMONT, R., Radiochem.
      Radioanal. Lett. **15** (1973) 47.
[11] SHALINETS, A.B., STEPANOV, A.V., Radiokhimya **14** (1972) 280; *English trans.*
      Sov. Radiochem. **14** (1972) 290.
[12] GUILLAUMONT, R., FERREIRA de MIRANDA, C., GALIN, M., C.R. Hebd. Séances
      Acad. Sci., Ser. C, **268** (1969) 140.
[13] HUSSONNOIS, M., HUBERT, S., RUBIN, L., GUILLAUMONT, R., BOUISSIERES, G.,
      Radiochem. Radioanal. Lett. **10** (1972) 231.
[14] ZIV, D.M., SHESTAKOVA, I.A., Radiokhimiya **7** (1965) 175; *English trans.*
      Sov. Radiochem. **7** (1965) 176.
[15] STEPANOV, A.V., SHVEDOV, V.P., Zh. Neorg. Khim. **10** (1965) 1000; *English trans.*
      Russ. J. Inorg. Chem. **10** (1965) 541.
[16] GUILLAUMONT, R., DESIRÉ, B., GALIN, M., Radiochem. Radioanal. Lett. **8** (1971) 189.
[17] BOUHLASSA, S., HUBERT, S., GUILLAUMONT, R., Radiochem. Radioanal. Lett.
      32 (1978) 247.
[18] BAES, Jr., C.F., MEYER, N.J., Inorg. Chem. **1** (1962) 780.
[19] RUSH, R.M., JOHNSON, J.S., KRAUS, K.A., Inorg. Chem. **1** (1962) 378.
[20] DUNSMORE, H.S., HIETANEN, S., SILLEN, L.G., Acta Chem. Scand. **17** (1963) 2644.

[21] ABERG, M., Acta Chem. Scand. **24** (1970) 2901.

[22] NIKITIN, A.A., SERGEYEVA, E.I., KHODAKOVSKIJ, I.L., NAUMOV, C.B., Geokhimiya **3** (1972) 297.

[23] ARNEK, R., SCHLYTER, K., Acta Chem. Scand. **22** (1968) 1331.

[24] CASSOL, A., MAGON, L., TOMAT, G., PORTANOVA, R., Inorg. Chem. **11** (1972) 515.

[25] SCHEDIN, U., Acta Chem. Scand. **25** (1971) 747.

[26] CASSOL, A., MAGON, L., PORTANOVA, R., TONDELLO, E., Radiochim. Acta **17** (1972) 28.

[27] MUSANTE, Y., PORTHAULT, M., Radiochem. Radioanal. Lett. **15** (1973) 299.

[28] SCHEDIN, U., Acta Chem. Scand., Ser.A **29** (1975) 333.

[29] SMITH, R.M., MARTELL, A.E., Critical Stability Constants. Vol.4: Inorganic Complexes, Plenum Press, New York (1976).

[30] NOREN, B., Acta Chem. Scand. **23** (1969) 331.

[31] KLOTZ, P., MUKHERJI, A., FELDBERG, S., NEWMAN, L., Inorg. Chem. **10** (1971) 740.

[32] PATIL, S.K., RAMAKRISHNA, V.V., Inorg. Nucl. Chem. Lett. **41** (1975) 421.

[33] CHOPPIN, G.R., UNREIN, P.J., Transplutonium Elements (Proc. 4th Int. Symp. Baden-Baden, 1975), North Holland Publ. Co., Amsterdam (1976) 97.

[34] DOGDEN, H.W., ROLLEFSON, G.K., J. Am. Chem. Soc. **71** (1949) 2600.

[35] ZEBROSKI, A.H., ALTER, H.W., HEUMANN, F.K., J. Am. Chem. Soc. **73** (1951) 5646.

[36] DAY, Jr., R.B., STOUGHTON, R.W., J. Am. Chem. Soc. **72** (1950) 5662.

[37] BAUMANN, E.W., J. Inorg. Nucl. Chem. **32** (1970) 3823.

[38] VASIL'EV, V.P., Zh. Neorg. Khim. **7** (1962) 1788; *English trans.* Russ. J. Inorg. Chem. **7** (1962) 924.

[39] BAGAWDE, S.V., RAMAKRISHNA, V.V., PATIL, S.K., J. Inorg. Nucl. Chem. **39** (1976) 1340, 2085.

[40] BAGAWDE, S.V., RAMAKRISHNA, V.V., PATIL, S.K., Radiochem. Radioanal. Lett. **31** (1977) 56.

[41] BULLWINKEL, E.P., USAEC Rep. RMO-2614 (1954).

[42] BLAKE, C.A., COLEMAN, C.F., BROWN, K.B., HILL, D.G., LOWRIE, R.S., SCHMITT, J.M., J. Am. Chem. Soc. **78** (1956) 5978.

[43] SCANLAN, J.P., J. Inorg. Nucl. Chem. **39** (1977) 635.

[44] PARAMONOVA, V.I., NIKOL'SKIJ, B.P., NIKOLAEVA, V.M., Zh. Neorg. Khim. **7** (1962) 1028; *English trans.* Russ. J. Inorg. Chem. **7** (1962) 528.

[45] BILON, A., Commissariat à l'énergie atomique Rep. CEA-R-3611 (1968).

[46] TSYMBAL, C., Commissariat à l'énergie atomique Rep. CEA-R-3476 (1969).

[47] PARAMONOVA, V.I., NIKOLAEVA, N.M., Radiokhimiya **4** (1962) 84.

[48] O'CINNEIDE, S., SCANLAN, J.P., HYNES, M.J., Chem. Ind. (London) (1972) 340.

[49] O'CINNEIDE, S., SCANLAN, J.P., HYNES, M.J., J. Inorg. Nucl. Chem. **37** (1975) 1013.

[50] KLYGIN, A.E., SMIRNOVA, Y.D., Zh. Neorg. Khim. **4** (1959) ~30; *English trans.* Russ. J. Inorg. Chem. **4** (1959) 16.

[51] BABKO, A.K., KODENSKAYA, V.S., Zh. Neorg. Khim. **5** (1960) 2568; *English trans.* Russ. J. Inorg. Chem. **5** (1960) 1241.

[52] SERGEYEVA, E.I., NIKITIN, A.A., KHODAKOVSKIJ, I.L., NAUMOV, G.B., Geokhimiya **11** (1972) 1340; *English trans.* Geochem. Int. **11** (1972) 900.

[53] PIROZHKOV, A.V., NIKOLAEVA, N.M., Izv. Sib. Otd. Akad. Nauk SSSR, Ser. Khim. Nauk **12** (1976) 55.

[54] DEVINA, O.A., KHODAKOVSKIJ, I.L., EFIMOV, M.E., MEDVEDEV, V.A., 7th All Union Conference on Calorimetry (Moscow, 1977), Atomizdat, Moscow (1977) 231.

DISCUSSION

A.S. STREZOV: I should like to ask a question concerning thorium monofluoro complexes. At lower ionic strengths only one complex may be formed, but at high ionic strengths higher complexes can appear. Is there any evidence for the higher complexes and, if so, how do they affect your data?

J. FUGER: At very low fluoride concentrations, using a supporting electrolyte to keep the ionic strength constant, we can indeed define conditions in which only the monofluoro complex is formed. With increasing fluoride concentration polyfluoro species rapidly appear. The data presented here are taken from the literature; however, they take into account, where applicable, the formation of polyfluoro species.

L.R. MORSS: You have pointed out that ΔH values for hydroxyl complex formation are, in general, poorly characterized. Usually, equilibrium is reached slowly for complexes, especially polynuclear complexes, but calorimetric measurements must be carried out over rather short time periods. Do you have any suggestions for ways to resolve this dilemma, or can you suggest concentration limits to ensure that equilibrium is reached promptly?

J. FUGER: If you look at the existing data on hydroxyl complexes throughout the Periodic Table, you will find that calorimetric measurements (mostly enthalpy titrations) can, in a good number of instances, yield fairly reliable results on polynuclear species, provided no solid phase appears in the process.

P.R. TREMAINE: In Ref.[2] of your paper, Baes and Mesmer report correlations which predict polynuclear and first mononuclear stability constants. Did you find that the actinide values you report agree with these correlations?

J. FUGER: The correlations given by Baes and Mesmer are indeed based on a wide range of experimental data. Our analysis of the data on the actinide hydroxyl complexes has not however reached its final stage. So it would be premature for us to say anything definite on that point.

REGULARITY IN THE CHANGES OF THE THERMODYNAMIC FUNCTIONS ASSOCIATED WITH THE FORMATION OF MONONUCLEAR COMPLEXES*

A.S. STREZOV, M.Kh. MIKHAJLOV,
V.Ts. MIKHAJLOVA, M.I. TASKAEVA
Bulgarian Academy of Sciences,
Institute of Nuclear Research and Nuclear Energy,
Sofia,
Bulgaria

Abstract

REGULARITY IN THE CHANGES OF THE THERMODYNAMIC FUNCTIONS ASSOCIATED WITH THE FORMATION OF MONONUCLEAR COMPLEXES.

Regularity in the changes of the free energy, ΔG, enthalpy, ΔH, and entropy, ΔS, have been determined, a regularity that is associated with the complex-formation processes in metal–ligand systems whose stability constants of the consecutive mononuclear complexes ML, ML_2, ML_3, ML_4 ... ML_n satisfy the relation: $\beta_n = A(a^n/n!)$ for $n = 1, 2, 3 ... N$, where β_n is the overall stability constant of the ML_n complex, n is the number of ligands ($1 \leqslant n \leqslant N$), and A and a are constants. The validity of the above equation for a given metal–ligand system leads to linear dependences of the quantities A and a upon $1/T$. This fact facilitates the calculation of the β_n constants and their variation with temperature, and hence enables the determination of the changes of all thermodynamic functions with higher precision. The feasibility and utility of the above method are demonstrated by applying it to several metal–ligand systems.

1. INTRODUCTION

The creation of more effective and precise methods of analysis of nuclear materials depends, in many cases, upon a knowledge of the complex-formation processes, which are the main factors governing the chemical behaviour of a great number of systems containing metal ions and ligands in aqueous and non-aqueous solutions. This type of knowledge is required for the analytical control of nuclear fuel, for the chemistry of fission products and transuranium elements, and for studies of the chemical behaviour of radionuclides in the marine environment.

From this point of view the determination of the variations of the free energy, ΔG, enthalpy, ΔH, and entropy, ΔS, associated with the formation of metal

---

* This work is part of a programme financially supported by the IAEA under Research Contract No.2044/RB.

complexes in solution is needed not only to provide a better understanding of the chemistry of the solution itself, but also in many practical applications.

The enthalpy changes can be obtained either calorimetrically or from the variations with temperature of the stability constants of the corresponding complexes, while the free energy changes can be calculated from the stability constants alone. The calorimetric method is generally preferred as it is more precise and accurate [1–3], but sometimes the concentrations necessary for a calorimetric measurement are much higher than those demanded by many methods used for the determination of the stability constants [3, 4]. The evaluation of ΔG, ΔH and ΔS on the basis of the variations with temperature of the stability constants also requires high precision which, in many cases, is hardly attainable.

The purpose of our work is to show that the disadvantages of a determination of ΔG, ΔH and ΔS based upon the variation of the β_n constants with temperature can be reduced considerably where a correlation exists between the stability constants of the consecutive mononuclear complexes in a given metal–ligand system. In this case the thermodynamic functions change in a regular manner and this makes it possible to obtain their values more precisely, while fewer measurements are needed to achieve a given precision.

2.   CALCULATION OF THE THERMODYNAMIC FUNCTIONS

2.1. General equations

We shall consider a system containing a complex-forming metal, M, and a ligand, L, in solution. The stepwise complex-formation process can be represented by the following conventional equations:

$$M + L \rightleftharpoons ML; \qquad k_1 = \frac{[ML]}{[M] \cdot [L]} \tag{1}$$

$$ML_{n-1} + L \rightleftharpoons ML_n; \qquad k_n = \frac{[ML_n]}{[ML_{n-1}] [L]} \tag{2}$$

$$\beta_n = k_1 k_2 k_3 \ldots \ldots k_n = \frac{[ML_n]}{[M] \cdot [L]^n} \tag{3}$$

where n is the number of ligands in the ML_n complex ($1 \leqslant n \leqslant N$), N is the co-ordination number of the complex-forming metal M, k_n and β_n are its stepwise and overall stability constants, respectively, and square brackets denote concentrations.

The following relations are used for the determination of the free energy, ΔG, enthalpy, ΔH, and entropy, ΔS, connnected with the complex-formation process [5]:

$$\Delta G_n = - RT \ln \beta_n \tag{4}$$

$$\Delta H_n = - \frac{R \, d \ln \beta_n}{d(1/T)} \tag{5}$$

$$\Delta S_n = \frac{\Delta H_n - \Delta G_n}{T} \tag{6}$$

and

$$\Delta g_n = - RT \ln k_n \tag{7}$$

$$\Delta h_n = - \frac{R \, d \ln k_n}{d(1/T)} \tag{8}$$

$$\Delta s_n = \frac{\Delta h_n - \Delta g_n}{T} \tag{9}$$

Equations (5) and (8) are valid when $\Delta C_p = 0$. In many cases it is found, when only a narrow temperature interval is considered, that this assumption is justified.

2.2. Thermodynamic functions for metal–ligand systems whose stability constants obey a correlation formula

Let us now consider the above system assuming that it follows the recently developed mathematical model of the complex-formation process [6]. Then the overall stability constants of the consecutive complexes of this system are correlated according to the equation:

$$\beta_n = A \frac{a^n}{n!} \qquad (n = 1, 2, 3 \ldots \ldots N) \tag{10}$$

where A and a are characteristic parameters of the system ($A \geqslant 0$, $a \geqslant 0$). The validity of the above correlation formula makes it possible to describe the system behaviour with only two instead of n parameters. The existence of a correlation between the stability constants allows the constants in Eqs (1) and (2) to be transformed into:

$$k_1 = Aa \tag{11}$$

$$k_n = a/n \quad (n = 2, 3 \ldots \ldots N) \tag{12}$$

If we introduce k_1 of Eq. (11) into Eqs (7, 8, 9), the following relations are obtained for the changes of free energy, enthalpy and entropy associated with the formation of the ML complex:

$$\Delta g_1 = - RT \ln (Aa) \tag{13}$$

$$\Delta h_1 = \frac{RT^2 \, d \ln (Aa)}{dT} \tag{14}$$

$$\Delta s_1 = \frac{\Delta h_1 - \Delta g_1}{T} \tag{15}$$

Substitution of k_n of Eq. (12) into Eqs (7, 8, 9) will give the following equations for Δg_n, Δh_n and Δs_n associated with the formation of the complex ML_n (for $n > 1$):[1]

$$\Delta g_n = - RT \ln (a/n) \tag{16}$$

$$\Delta h_n = \frac{RT^2 \, d \ln (a/n)}{dT} \tag{17}$$

$$\Delta s_n = \frac{\Delta h_n - \Delta g_n}{T} \tag{18}$$

Using Eqs (16, 17, 18) the following relations (valid for $n > 1$) are obtained:

$$\Delta g_{n+1} - \Delta g_n = RT \ln \left(\frac{n}{n+1}\right) \tag{19}$$

$$\Delta h_{n+1} - \Delta h_n = 0 \tag{20}$$

$$\Delta s_{n+1} - \Delta s_n = R \ln \left(\frac{n}{n+1}\right) \tag{21}$$

---

[1] It will be noted that Eq. (9) remains unchanged (cf. Eq. (15) and Eq. (18)), bearing in mind that n = 1 in Eq. (15).

From Eq. (20) it follows that:

$$\Delta h_2 = \Delta h_3 = \Delta h_4 = \ldots\ldots = \Delta h_n = \Delta h = \text{const.} \tag{22}$$

which means that the enthalpy change for each step of the complex-formation
process is constant ($= \Delta h$). (This fact is a consequence of assumption (iii) in
Ref. [6] concerning the equivalence of the co-ordination sites of the complex-
forming metal, M.)

Then Δh_1 can be represented as:

$$\Delta h_1 = \Delta h' + \Delta h \tag{23}$$

where the term $\Delta h'$ could be regarded as the part of Δh_1 relevant to the structural
changes of the hydration shell of the metal ion, M, occurring when the temperature
in the system changes.

Combining Eqs (19) and (21), we get:

$$\Delta g_{n+1} - \Delta g_n = T (\Delta s_{n+1} - \Delta s_n) \tag{24}$$

Equation (24) shows that the difference between the free energy changes of two
successive steps ($n > 1$) of the complex-formation process depends only on the
difference between the entropy changes of these steps.

If we consider Eq. (17), the solution of this differential equation is given by:

$$\ln a = -\frac{\Delta h_n}{RT} + C = -\frac{\Delta h}{RT} + C \tag{25}$$

C being the constant of integration. Writing:

$$a = (\exp C) \left(\exp\left[-\frac{\Delta h}{RT} \right] \right) = a_0 \left(1 - \frac{\Delta h}{RT} + \frac{(\Delta h)^2}{2!(RT)^2} - \ldots \right) \tag{26}$$

and retaining only two terms of the expansion (26) (for $\dfrac{\Delta h}{RT} \ll 1$) we get:

$$a \approx a_0 \left(1 - \frac{\Delta h}{RT} \right) = a_0 - \frac{b}{T} \tag{27}$$

where $a_0 = e^C = \text{constant}$, and $b = a_0 \Delta h/R = \text{constant}$.

Equation (27) shows that the quantity a could be expressed with sufficient
accuracy as a linear function of $1/T$.

TABLE I. COMPARISON BETWEEN THE EXPERIMENTAL AND CALCULATED
$\bar{n}$ FUNCTIONS FOR THE ZINC–IMIDAZOLE SYSTEM STUDIED AT 4.5°C

| L | $\bar{n}_{exp}$ | Ref. [14][a] | | | This work[a] | |
(10⁻² M)		$\bar{n}_{calc}$ (footnote b)	$\Delta\bar{n}_{calc}$ (%) (footnote b)		$\bar{n}_{calc}$ (footnote c)	$\Delta\bar{n}_{calc}$ (%) (footnote c)
0.0207	0.14	0.12	−14.29		0.12	−14.29
0.0471	0.27	0.28	3.70		0.26	−3.70
0.095	0.54	0.57	5.56		0.57	5.56
0.191	1.07	1.16	8.41		1.15	7.48
0.347	1.86	1.94	4.84		1.93	3.76
0.568	2.68	2.64	−1.49		2.64	−1.49
0.95	3.18	3.20	0.63		3.19	0.31
1.83	3.63	3.61	−0.55		3.61	−0.55
2.71	3.76	3.74	−0.53		3.74	−0.53
4.35	3.86	3.85	−0.29		3.85	−0.26

[a] $S_{[14]} = 6.92 \times 10^{-3}$; $S_{this\ work} = 4.60 \times 10^{-3}$.
[b] Obtained using the β_n values reported in Ref. [14]: $\beta_1 = 578$; $\beta_2 = 1.81 \times 10^5$; $\beta_3 = 4.09 \times 10^7$;
   $\beta_4 = 6.45 \times 10^9$.
[c] Obtained using the following β_n values: $\beta_1 = 575$; $\beta_2 = 1.84 \times 10^5$; $\beta_3 = 3.93 \times 10^7$;
   $\beta_4 = 6.28 \times 10^9$. These are correlated according to Eq. (10), where A = 0.90 and a = 640.

Now if we combine Eq. (14) with Eq. (17), bearing in mind the result of
Eq. (22), we get:

$$\Delta h_1 - \Delta h = \frac{RT^2\, d \ln A}{dT} \tag{28}$$

If the procedure used for Eq. (17) is applied to Eq. (28), the following relation
will be obtained:

$$A \approx e^{C_1}\left(1 - \frac{\Delta h_1 - \Delta h}{RT}\right) = A_0\left(1 - \frac{\Delta h'}{RT}\right) \tag{29}$$

where $A_0 = e^{C_1}$ = constant.

Equation (29) shows that the parameter A could also be approximated as a
linear function of $1/T$.

TABLE II.   COMPARISON BETWEEN THE EXPERIMENTAL AND CALCULATED
$\bar{n}$ FUNCTIONS FOR THE ZINC–IMIDAZOLE SYSTEM STUDIED AT 24°C

L $(10^{-2}\,M)$	$\bar{n}_{exp}$	Ref. [14][a]		This work[a]	
		$\bar{n}_{calc}$ (footnote b)	$\Delta\bar{n}_{calc}$ (%) (footnote b)	$\bar{n}_{calc}$ (footnote c)	$\Delta\bar{n}_{calc}$ (%) (footnote c)
0.0263	0.12	0.10	− 16.67	0.11	− 8.33
0.062	0.26	0.25	− 3.85	0.25	− 3.85
0.132	0.51	0.55	7.84	0.55	7.84
0.275	0.95	1.18	24.21	1.17	23.16
0.435	1.78	1.79	0.56	1.76	− 1.12
0.66	2.51	2.40	− 4.38	2.37	− 5.58
1.05	3.07	2.97	− 3.26	2.96	− 3.58
2.09	3.43	3.50	2.04	3.51	2.33

[a] $S_{[14]} = 3.25 \times 10^{-2}$; $S_{this\ work} = 1.47 \times 10^{-2}$.
[b] Obtained using the β_n values reported in Ref. [14]: $\beta_1 = 382$; $\beta_2 = 8.87 \times 10^4$; $\beta_3 = 1.52 \times 10^7$; $\beta_4 = 1.51 \times 10^9$.
[c] Obtained using the β_n values: $\beta_1 = 392$; $\beta_2 = 8.98 \times 10^4$; $\beta_3 = 1.37 \times 10^7$; $\beta_4 = 1.57 \times 10^9$. These are correlated according to Eq.(10), where A = 0.86 and a = 458.

In real metal–ligand systems, the linear dependences for A and a (Eqs (27) and (29)) derived above will be valid as long as there is no change in the mechanism governing the complex-formation processes. It should be pointed out that Eqs (27) and (29) were worked out assuming that $\Delta C_p \to 0$. Now we shall show that this requirement is obligatorily fulfilled when the stability constants of the complexes in the system obey the correlation formula (10) at a given temperature.

Let us assume that the correlation of the stability constants is found to be valid for a given temperature, T, but the condition $\Delta C_p \to 0$ is not valid, so that a second approximation, ΔC_p = const., is accepted [5]. As was shown before, the existence of the correlation at the temperature, T, requires a to be a linear function of 1/T. Therefore we can write [5]:

$$\ln\left(a_0 - \frac{b}{T}\right) = -\frac{\Delta H_J}{RT} + \Delta C_p \ln\left(\frac{T}{R}\right) - \frac{J}{R} \qquad (30)$$

If we rearrange Eq. (30) into the form:

$$a_0 - \frac{b}{T} = \left(1 - \frac{\Delta H_J}{RT}\right) T^{(\Delta C_p/R)} e^{-(J/R)} \qquad (31)$$

TABLE III. COMPARISON BETWEEN THE EXPERIMENTAL
AND THEORETICAL COMPLEXITY FUNCTIONS FOR
THE THALLIUM−ETHYLENE-BIS(3-MERCAPTOPROPIONATE)
SYSTEM STUDIED AT 30 AND 40°C

IIIa. Studies at 30°C

L (M)	F_{exp}	Ref. [15][a]		This work	
		F_{calc} (footnote b)	ΔF_{calc} (%) (footnote b)	F_{calc} (footnote c)	ΔF_{calc} (%) (footnote c)
0.02	1.1277	1.1314	0.33	1.1105	−1.52
0.04	1.314	1.315	0.08	1.303	−0.84
0.06	1.599	1.596	−0.19	1.611	0.75
0.10	2.630	2.625	−0.19	2.696	2.51
0.12	3.614	3.462	−4.20	3.538	−2.10

[a] $S_{[15]} = 1.78 \times 10^{-3}$; $S_{this\ work} = 7.15 \times 10^{-4}$.
[b] Obtained using the β_n values given in Ref. [15]: $\beta_1 = 6$; $\beta_2 = 10$; $\beta_3 = 925$.
[c] Obtained using the following β_n values: $\beta_1 = 4$; $\beta_2 = 63$; $\beta_3 = 666$. These are correlated according to Eq. (10), where A = 0.1266 and a = 31.6

IIIb. Studies at 40°C

L (M)	F_{exp}	Ref. [15][a]		This work	
		F_{calc} (footnote b)	ΔF_{calc} (%) (footnote b)	F_{calc} (footnote c)	ΔF_{calc} (%) (footnote c)
0.02	1.1261	1.1188	−0.65	1.0984	−2.46
0.04	1.276	1.278	0.16	1.268	−0.63
0.06	1.506	1.510	0.26	1.534	1.86
0.10	2.350	2.330	−0.98	2.467	4.84
0.12	3.368	2.985	−11.37	3.186	−5.40

[a] $S_{[15]} = 1.3 \times 10^{-2}$; $S_{this\ work} = 3.1 \times 10^{-3}$.
[b] Obtained using the β_n values given in Ref. [15]: $\beta_1 = 5.5$; $\beta_2 = 8$; $\beta_3 = 700$.
[c] Obtained using the following β_n values: $\beta_1 = 3.6$; $\beta_2 = 55$; $\beta_3 = 557$. These are correlated according to Eq. (10), where A = 0.119 and a = 30.4.

TABLE IV. COMPARISON BETWEEN THE EXPERIMENTAL
AND THEORETICAL COMPLEXITY FUNCTIONS FOR
THE INDIUM–ETHYLENE-BIS(3-MERCAPTOPROPIONATE)
SYSTEM STUDIED AT 30 AND 40°C

IVa. Studies at 30°C

L (M)	F_{exp}	Ref. [16][a]		This work	
		F_{calc} (footnote b)	ΔF_{calc} (%) (footnote b)	F_{calc} (footnote c)	ΔF_{calc} (%) (footnote c)
0.04	1.80	1.78	− 1.11	1.82	1.11
0.06	3.29	3.32	0.91	3.38	2.73
0.08	6.16	6.15	− 0.16	6.23	1.14
0.10	11.00	10.70	− 0.29	10.78	− 2.00
0.12	17.42	17.37	− 0.29	17.40	− 0.11

[a] $S_{[16]} = 9.6 \times 10^{-4}$; $S_{this\ work} = 7.0 \times 10^{-4}$.
[b] Obtained using the β_n values given in Ref. [16]: $\beta_1 = 2$; $\beta_2 = 100$; $\beta_3 = 8500$.
[c] Obtained using the following β_n values: $\beta_1 = 1.6$; $\beta_2 = 142$; $\beta_3 = 8196$. These are correlated
according to Eq.(10), where A = 0.0094 and a = 173.6.

IVb. Studies at 40°C

L (M)	F_{exp}	Ref. [16][a]		This work	
		F_{calc} (footnote b)	ΔF_{calc} (%) (footnote b)	F_{calc} (footnote c)	ΔF_{calc} (%) (footnote c)
0.04	1.92	1.92	0.00	1.92	0.00
0.06	3.55	3.56	0.28	3.59	1.13
0.08	6.53	6.55	0.31	6.61	1.22
0.10	11.46	11.30	− 1.40	11.37	− 0.78
0.12	18.33	18.21	− 0.65	18.28	− 0.27

[a] $S_{[16]} = 2.6 \times 10^{-4}$; $S_{this\ work} = 1.7 \times 10^{-4}$.
[b] Obtained using the β_n values given in Ref.[16]: $\beta_1 = 4$; $\beta_2 = 130$; $\beta_3 = 8600$.
[c] Obtained using the following β_n values: $\beta_1 = 2.5$; $\beta_2 = 177$; $\beta_3 = 8354$. These are correlated
according to Eq. (10), where A = 0.01765 and a = 141.6.

TABLE V.   THERMODYNAMIC FUNCTIONS CALCULATED FOR THE ZINC–IMIDAZOLE SYSTEM AT 24°C

Complex	Ref. [14]			This work		
	ΔG (kcal·mol^{-1})	ΔH (kcal·mol^{-1})	ΔS (cal·mol^{-1}·K^{-1})	ΔG (kcal·mol^{-1})	ΔH (kcal·mol^{-1})	ΔS (cal·mol^{-1}·K^{-1})
ML	− 3.51	− 3.48	0.10	− 3.52	− 3.22	1.01
ML$_2$	− 6.72	− 5.99	2.46	− 6.73	− 6.02	2.39
ML$_3$	− 9.76	− 8.31	4.88	− 9.70	− 8.83	2.39
ML$_4$	− 12.50	− 11.82	2.29	− 12.50	− 11.63	2.93

and differentiate with respect to T, we get:

$$b = \frac{T^{(\Delta C_p/R)} e^{-(J/R)}}{R} \left(\Delta H_J - \left(T - \frac{\Delta H_J}{R} \right) \Delta C_p \right) \tag{32}$$

Equation (32) clearly shows that only when $\Delta C_p \to 0$ can b be constant, otherwise b would be a function of the temperature and, naturally, a could not then be a linear function of 1/T. On the contrary, if $\Delta C_p \neq 0$ for the temperature, T, then the correlation formula cannot be proved for this temperature. It can easily be shown, that the same reasoning applies for the parameter A.

Having in mind that all investigations in aqueous solutions are carried out over temperature ranges of less than 50°C, it is obvious that the probability of the mechanism of complex formation changing drastically between two experimental temperatures T_1 and T_2 (i.e. within the temperature interval $T_1 \to T_2$) is very small. Thus, the validity of the correlation formula for the two temperatures provides a strong indication that the parameters A and a are linear functions of 1/T over the whole interval $T_1 \to T_2$. These linear dependences of A and a simplify the evaluation of all stability constants over the interval $T_1 \to T_2$ (e.g. only two experiments are needed), and also facilitate and make more precise the determination of all thermodynamic-function changes over the same interval.

3.   RESULTS AND DISCUSSION

In previous papers [6–11] the application of the formula of Eq. (10) was shown, connecting the values of the overall stability constants of metal complexes in aqueous solutions for 34 metal–ligand systems, studied at one temperature. In Ref. [12], empirical linear dependences between the parameters A and a and 1/T were established for the system cadmium–thiodipropionate, studied experimentally at 30, 40 and 50°C [13]. If we consider Eqs (27) and (29) and the conditions of their application, the linear dependences evaluated for A and a in the cadmium–thiodipropionate system are well founded.

Now we shall show the advantages of the proposed method for determining the changes in the thermodynamic functions when the stability constants of the complexes are obtained at two temperatures. The subjects of the present analyses are the following metal–ligand systems:

zinc–imidazole [14];

thallium–ethylene-bis(3-mercaptopropionate);

indium–ethylene-bis(3-mercaptopropionate) [15, 16].

The system Zn^{2+}–imidazole has been studied potentiometrically at 4.5 and 24°C, with $\mu = 0.16M$, and the evaluation of the stability constants was carried out by the method of Bjerrum [17]. The systems Tl^- and In–ethylene-bis(3-mercaptopropionate) were studied polarographically at 30 and 40°C and $\mu = 0.5M$ $NaClO_4$. The concentration of the ligand varies from 0.02 to 0.12M. The method of DeFord and Hume [18] was applied for the determination of the complexity-function values, while the stability constants were obtained by the method of Leden [19].

The evaluation of the stability constants obtained in the present study was performed by the method given elsewhere [6–12]. The application of the correlation formula is estimated by the S-factor, defined as [20]:

$$S = const. \cdot S_{min} = \frac{1}{k} \sum_i \frac{[(X_{calc})_i - (X_{exp})_i]^2}{(X_{exp})_i^2} \tag{33}$$

where k $(= j - p - 1)$ is the number of degrees of freedom in the system, j is the number of data points, and p is the number of parameters. X_{exp} and X_{calc} are the values of the experimental and theoretical functions used for the determination of the stability constants (complexity function, F, formation function, $\bar{n}$, etc.). It is obvious that low values of the S-factor demonstrate a good fit between the experimental and theoretical functions.

The results for the functions $\bar{n}$ and F determined using the sets of stability constants of various authors and those obtained by us are compared in Tables I–IV. The percentage deviation of each point from the experimentally obtained functions, as well as the corresponding S-factors obtained by the present analysis are also given.

These data show that the stability constants obtained by us fit the experimental data better than the original authors' stability constants. Hence it can be seen that

TABLE VI. CALCULATED THERMODYNAMIC FUNCTIONS FOR THE THALLIUM— AND INDIUM—ETHYLENE-BIS(3-MERCAPTOPROPIONATE) SYSTEMS AT 40°C

| Complex | Tl—ethylene-bis(3-mercaptopropionate) | | | | | | In—ethylene-bis(3-mercaptopropionate) | | | | | |
| | Ref. [15] | | | This work | | | Ref. [16] | | | This work | | |
	ΔG (kcal·mol^{-1})	ΔH (kcal·mol^{-1})	ΔS (cal·mol^{-1}·K^{-1})	ΔG (kcal·mol^{-1})	ΔH (kcal·mol^{-1})	ΔS (cal·mol^{-1}·K^{-1})	ΔG (kcal·mol^{-1})	ΔH (kcal·mol^{-1})	ΔS (cal·mol^{-1}·K^{-1})	ΔG (kcal·mol^{-1})	ΔH (kcal·mol^{-1})	ΔS (cal·mol^{-1}·K^{-1})
ML	−1.0	−1.6	−1.8	−0.8	−2.0	−3.8	−0.86	13.0	44.0	−0.57	8.4	28.6
ML$_2$	−1.3	−4.3	−9.7	−2.5	−2.6	−0.3	−3.04	4.7	24.0	−3.22	4.25	23.5
ML$_3$	−4.1	−5.2	−3.5	−3.9	−4.4	1.6	−5.66	0.4	19.0	−5.62	0.4	19.2

Eq. (10) has the advantage that it makes possible the description of the behaviour of a system with fewer parameters, giving at the same time a better correspondence with experiment.

Using Eqs (16–21), the changes of the thermodynamic functions ΔG, ΔH and ΔS for the three systems have been obtained, and these are presented in Tables V and VI. These data show that all thermodynamic functions change in a regular manner with n; this is a consequence of the correlation between the stability constants, which is found to be valid for these systems.

Using the linear dependences of A and a on $1/T$, all stability constants over the interval 4.5–24°C for the Zn^{2+}–imidazole system, and over the interval 30–40°C for the Tl− and In−ethylene-bis(3-mercaptopropionate) systems can be obtained and, hence, all the thermodynamic-function changes for the above temperature intervals also.

Our data for the Zn^{2+}–imidazole system are similar to those of Ref. [14], which is due to the fact that the stability constants obtained by the original authors are very close to those of the correlation formula (10).

4.   SUMMARY

The above analysis has shown that, for systems obeying Eq. (10), one has to determine only the values of parameters A and a at two temperatures. Then, from the linear plots of A and a versus $1/T$, the values of all k_n and β_n constants can be calculated for the temperature interval considered, together with all thermodynamic-function changes.

This procedure is much simpler and more precise than the conventional one of determining the variation of n stability constants with T. The above method also makes the experiment more precise and less time consuming, which is of prime importance in radiochemical studies in which isotopes of very high radio-toxicity are handled.

REFERENCES

[1] McAULEY, A., NANCOLLAS, G., J. Chem. Soc. (1963) 989.
[2] CHOPPIN, G.R., SCHNEIDER, J.K., J. Inorg. Nucl. Chem. 32 (1970) 3283.
[3] de CARVALHO, Rg.G., CHOPPIN, G.R., J. Inorg. Nucl. Chem. 29 (1967) 725, 737.
[4] AHRLAND, S., The Chemistry of Nonaqueous Solvents, Vol. 5, Academic Press, New York (1977).
[5] EREMIN, E., Osnovy Khimicheskoj Termodinamiki, Vysshaya Shkola, Moscow (1974).
[6] MIHAILOV, M.H., J. Inorg. Nucl. Chem. 36 (1974) 107.
[7] MIHAILOV, M.H., MIHAILOVA, V.Ts., KHALKIN, V.A., J. Inorg. Nucl. Chem. 36 (1974) 115.

[8] MIHAILOV, M.H., MIHAILOVA, V.Ts., KHALKIN, V.A., J. Inorg. Nucl. Chem. **36**
     (1974) 121.

[9] MIHAILOV, M.H., MIHAILOVA, V.Ts., KHALKIN, V.A., J. Inorg. Nucl. Chem. **36**
     (1974) 127.

[10] MIHAILOV, M.H., MIHAILOVA, V.Ts., KHALKIN, V.A., J. Inorg. Nucl. Chem. **36**
     (1974) 133.

[11] MIHAILOV, M.H., MIHAILOVA, V.Ts., KHALKIN, V.A., J. Inorg. Nucl. Chem. **36**
     (1974) 141.

[12] MIHAILOV, M.H., MIHAILOVA, V.Ts., KHALKIN, V.A., J. Inorg. Nucl. Chem. **36**
     (1974) 145.

[13] RAWAT, P.C., GUPTA, C.M., J. Inorg. Nucl. Chem. **34** (1972) 951.

[14] ESDALL, J.T., FELSENFELD, G.A., GOODMAN, D.S., GURD, F.R.N., J. Am. Chem.
     Soc. **76** (1954) 3272.

[15] SAXENA, R.S., CHATURVEDI, U.S., J. Inorg. Nucl. Chem. **34** (1972) 2964.

[16] SAXENA, R.S., CHATURVEDI, U.S., J. Inorg. Nucl. Chem. **34** (1972) 3272.

[17] BJERRUM, J., Metal Amine Formation in Aqueous Solution, Haase and Son,
     Copenhagen (1941).

[18] DeFORD, D.H., HUME, D.N., J. Am. Chem. Soc. **73** (1951) 5321.

[19] LEDEN, I., Thesis, Lund University (1943).

[20] SANDELL, A., Acta Chem. Scand. **23** (1969) 478.

IAEA-SM-236/12

THERMODYNAMIQUE DE SOLUBILISATION DE LA MAGNETITE EN MILIEU BASIQUE

I. LAMBERT, J. MONTEL
CEA, Centre d'études nucléaires de Saclay,
Gif-sur-Yvette

P. BESLU, A. LALET
CEA, Centre d'études nucléaires de Cadarache,
Saint-Paul-lez-Durance,
France

Abstract–Résumé

THERMODYNAMICS OF MAGNETITE SOLUBILIZATION IN BASIC MEDIUM.

Magnetite is solubilized in the presence of hydrogen in accordance with the reactions:

$$Fe_3O_4 + 6H^+ + H_2 \rightleftharpoons 3Fe^{2+} + 4H_2O;$$

$$Fe_3O_4 + 3H^+ + H_2 \rightleftharpoons 3FeOH^+ + H_2O;$$

$$Fe_3O_4 + H_2 + 2H_2O \rightleftharpoons 3HFeO_2^- + 3H^+.$$

The variations in its solubility as a function of temperature and pH-value can be calculated from thermodynamic data relating to each of the species involved. To determine exactly the value of the free energy of formation of the species $HFeO_2^-$, solubility measurements were made in an autoclave at temperatures between 25°C and 300°C, in a 10^{-2}N aqueous solution of potash. On the basis of these measurements the following value was obtained: $\Delta G^0_{HFeO_2^-} \simeq -88$ kcal·mol^{-1}. These measurements also demonstrated the variation in the average size of magnetite crystals in relation to the experimental conditions and also the effect of size on solubility. For example, in an autoclave treatment at high temperature, solubility decreases with time and the crystals increase in size, whereas the opposite phenomenon occurs at ordinary temperatures. The rather poor effectiveness of efforts made to decontaminate a primary circuit during programmed coolings can be linked to these observations.

THERMODYNAMIQUE DE SOLUBILISATION DE LA MAGNETITE EN MILIEU BASIQUE.

La magnétite se solubilise en présence d'hydrogène selon les réactions:

$$Fe_3O_4 + 6H^+ + H_2 \rightleftharpoons 3Fe^{2+} + 4H_2O;$$

$$Fe_3O_4 + 3H^+ + H_2 \rightleftharpoons 3FeOH^+ + H_2O;$$

$$Fe_3O_4 + H_2 + 2H_2O \rightleftharpoons 3HFeO_2^- + 3H^+.$$

Les variations de sa solubilité en fonction de la température et du pH sont calculables à partir des données thermodynamiques relatives à chacune des espèces mises en jeu. Afin de préciser la valeur de l'enthalpie libre de formation de l'espèce $HFeO_2^-$, des mesures de solubilité ont été effectuées en autoclave, entre 25°C et 300°C, en solution aqueuse de potasse 10^{-2} N.

Ces mesures ont permis de déterminer la valeur: $\Delta G^0_{HFeO_2^-} \simeq -88 \text{ kcal} \cdot \text{mol}^{-1}$. Elles ont en outre mis en évidence la variation de la taille moyenne des cristaux de magnétite suivant les conditions expérimentales, et son influence sur la solubilité. Ainsi, lors d'un traitement en autoclave, à température élevée, la solubilité décroît au cours du temps et les cristaux grossissent, alors que le phénomène inverse se produit à la température ordinaire. La médiocre efficacité des tentatives de décontamination d'un circuit primaire lors de refroidissements programmés peut être rattachée à ces observations.

1. INTRODUCTION

Les variations de solubilité de la magnétite avec la température et le pH sont généralement considérées comme l'un des facteurs essentiels qui conditionnent les mouvements des produits de corrosion dans le circuit primaire des réacteurs à eau pressurisée.

Les équilibres chimiques qui régissent la solubilité dans l'eau contenant de l'hydrogène dissous, comme c'est le cas dans un réacteur, sont les suivants :

$$Fe_3O_4 + 6H^+ + H_2 \rightleftharpoons 3Fe^{2+} + 4H_2O \qquad (1) \quad K_1$$

$$Fe_3O_4 + 3H^+ + H_2 \rightleftharpoons 3FeOH^+ + H_2O \qquad (2) \quad K_2$$

$$Fe_3O_4 + H_2 + 2H_2O \rightleftharpoons 3HFeO_2^- + 3H^+ \qquad (3) \quad K_3$$

Chacune des constantes d'équilibre ci-dessus est reliée aux enthalpies libres de formation des espèces mises en jeu par la relation :

$$- RT \operatorname{Ln} K = \Delta G = \sum \Delta G_f \text{ (produits)} - \sum \Delta G_f \text{ (réactifs)}$$

où ΔG_f représente l'enthalpie libre de formation d'une espèce à la température considérée.

Certains auteurs ont calculé la solubilité de la magnétite en fonction de la température et du pH à partir des valeurs (issues de la littérature) des enthalpies libres standard de formation des espèces mises en jeu (McDONALD [1], LEWIS [2], BESLU et coll. [3]). Parallèlement, des mesures expérimentales de solubilité ont été effectuées [4, 5, 6]. SWEETON et BAES [4] ont déduit de leurs résultats les constantes d'équilibre.

Cependant les différentes méthodes d'approche conduisent à des valeurs assez éloignées, notamment dans le cas de l'espèce $HFeO_2^-$ prédominante en milieu basique et à haute température. Compte tenu des lacunes de la littérature, nous avons repris les mesures de solubilité de la magnétite, par une méthode statique, entre 25°C et 300°C, en commençant par l'étude des milieux basiques, de façon à préciser en priorité les données relatives à cette espèce $HFeO_2^-$.

2. REACTIFS ET MATERIAUX

2.1. Solutions

Les solutions de potasse sont préparées sous azote, à partir de KOH solide et d'eau désionisée et désoxygénée.

2.2. Hydrogène

Nous avons utilisé de l'hydrogène à 99,9995 %, fourni par " l'Air Liquide ", sans purification préalable.

2.3. Magnétite

La magnétite a été préparée par voie sèche, selon la réaction :

$$3Fe + 4H_2O \rightleftarrows Fe_3O_4 + 4H_2$$

La pureté de la magnétite est vérifiée par diffraction de rayons X.

3. METHODE EXPERIMENTALE

3.1. Appareillage

L'appareillage utilisé est schématisé sur la figure 1. Il est essentiellement constitué par un autoclave d'un litre, en titane non allié, muni d'un système d'agitation magnétique. Un tube plongeur permet le prélèvement d'échantillons à travers un fritté de titane de porosité 1 micron. Deux manomètres M_1 et M_2, et une série de vannes en titane permettent le raccordement à une pompe à vide, l'introduction d'hydrogène gazeux et l'introduction de liquide.

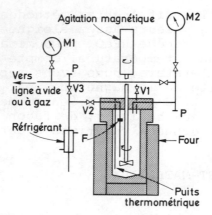

FIG.1.   Appareillage de mesure des solubilités.

Le tube de prélèvement est prolongé à l'extérieur de l'autoclave par un réfrigérant, pour permettre des prélèvements sous forme liquide à haute température.

Le chauffage de l'autoclave est assuré par un four électrique régulé à 0,5°C près, au moyen d'un régulateur commandé par un thermocouple plongeant dans le bain par l'intermédiaire d'un puits thermométrique.

3.2. Réalisation d'une expérience

L'introduction des réactifs dans l'autoclave se fait à l'abri de l'air. De l'hydrogène est introduit sous une pression de 6 atm, ce qui présente l'avantage d'accroître sensiblement les concentrations de fer à mesurer, la solubilité étant proportionnelle à la racine cubique de la pression d'hydrogène.

L'autoclave est ensuite amené à la température désirée et maintenu au moins pendant une semaine entre chaque prélèvement de solution. On arrête l'agitation 5 heures avant le moment du prélèvement, pour laisser décanter la magnétite et limiter ainsi l'entraînement de particules. Puis on effectue le prélèvement à travers la vanne v_2 ; après rinçage du tube de prélèvement par la solution on recueille 10 cm^3 dans une éprouvette contenant 0,2 cm^3 d'acide nitrique 1N, de façon à éviter toute reprécipitation; cet échantillon est analysé aussitôt.

TABLEAU I. INFLUENCE DE LA PRESENCE DE MAGNETITE SOLIDE
SUR L'ANALYSE DU FER DISSOUS

Essai	Temps écoulé depuis la prise d'échantillon	Concentration en fer (µg/kg)
1	0 2 heures 3 heures	18 22 22
2	0 4 jours 6 jours	6 19 27
3	0 1 h 30	41 41
4*	0 1 h 20 3 heures 3 h 50	120 130 155 160
5**	0 1 heure	50 55

* Expérience de solubilisation en bécher; la magnétite n'a pas été filtrée, mais retirée avec un aimant.

** Magnétite introduite dans une solution titrée à 50 µg/kg.

3.3. Analyse

3.3.1. Méthode utilisée

Le fer est analysé en présence d'un réducteur qui le maintient au degré d'oxydation II, par colorimétrie du complexe avec la tripyridyltriazine, complexe dont on mesure l'absorption à 590 nm , à l'aide d'un analyseur automatique.

La sensibilité de l'analyse est de 1 à 2 µg/kg.

Cette méthode permet de doser le fer dissous dans l'échantillon au moment de la mesure, mais des précautions sont nécessaires pour qu'elle soit représentative du fer dissous à l'équilibre dans l'autoclave.

Deux sources d'erreur sont en effet possibles:
la précipitation à partir de la solution, et, en sens
contraire, la dissolution de magnétite de granulo-
métrie inférieure à 1μm qui passerait à travers le
filtre.

Nous avons traité successivement ces deux pro-
blèmes.

3.3.2. Elimination des possibilités d'erreur dues à la précipitation

L'échantillon prélevé est reçu directement dans
une petite quantité d'acide nitrique Merck Suprapur,
calculée de façon que la solution soit finalement
à pH $\sim$ 2 ; cette procédure permet d'éviter la précipitation de magnétite due à la baisse de solubilité
à froid, ainsi que celle d'hydroxyde ferrique, beaucoup moins soluble que la magnétite, due à l'oxydation à l'air de la solution ferreuse.

3.3.3. Influence de la présence de magnétite solide en suspension dans l'eau

Le principe de l'élimination de l'influence de
la présence de magnétite solide repose sur l'hypothèse que celle-ci ne se dissout que lentement,
même en milieu acide, et ne perturbe donc pas une
analyse effectuée aussitôt après le prélèvement.
Cette hypothèse a été vérifiée par les deux séries
d'expériences suivantes :

a) à une solution étalon de 50 μg/kg de fer, on
ajoute une pincée de magnétite, agitée quelques minutes dans la solution (avec un barreau magnétique
sur lequel se rassemble la poudre), puis laissée en
présence de la solution sans agitation;

b) sur le même prélèvement, nous avons effectué
des analyses successives au bout d'intervalles de
temps différents.

Les résultats de ces essais, indiqués dans le
tableau I, ont permis de considérer comme représentative l'analyse effectuée au cours des quelques
minutes faisant suite au prélèvement.

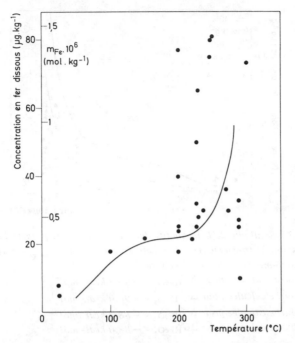

FIG.2. Solubilité de la magnétite dans une solution de potasse $10^{-2} N$.
Première série: $(P_{H_2})_{25°C} = 6\,atm$.

● *points expérimentaux; ——— courbe calculée pour $\Delta G^0_{Fe_3O_4} = -242,7\,kcal\cdot mol^{-1}$,*
$\Delta G^0_{HFeO_2^-} = -92,1\,kcal\cdot mol^{-1}$.

4. MESURES DE SOLUBILITE

Trois séries de mesures ont été effectuées entre 25°C et 300°C, dans une solution de potasse 10^{-2} N, en présence d'hydrogène sous une pression de 6 atmosphères mesurée à froid.

4.1. Dans la première série, en raison d'une imperfection de l'appareillage, la pression d'hydrogène diminuait notablement lors de chaque prélèvement. Il a donc été nécessaire, après chaque mesure, de refroidir l'autoclave afin d'introduire de l'hydrogène à la pression de 6 atmosphères ; la durée de la perturbation de température correspondante était comprise entre 5 et 15 heures, ce qui pouvait sembler relativement court en regard de l'intervalle séparant deux prélèvements (une semaine au moins).

Les résultats relatifs à cette série de mesures sont reportés sur la figure 2.

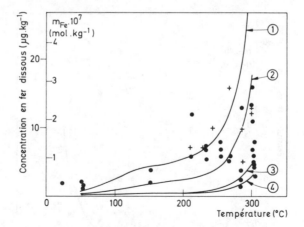

FIG.3.   Solubilité de la magnétite dans une solution de potasse $10^{-2}\,N$.
Deuxième et troisième série de mesures: $5\,atm < (P_{H_2})_{25°C} < 6\,atm$.

● *deuxième série;   + troisième série.*

① *courbe calculée pour $\Delta G^0_{HFeO_2^-} = -90,8\,kcal \cdot mol^{-1}$*
② *courbe calculée pour $\Delta G^0_{HFeO_2^-} = -90,0\,kcal \cdot mol^{-1}$*
③ *courbe calculée pour $\Delta G^0_{HFeO_2^-} = -88,5\,kcal \cdot mol^{-1}$*
④ *courbe calculée pour $\Delta G^0_{HFeO_2^-} = -88,0\,kcal \cdot mol^{-1}$.*

4.2. Lors de la deuxième série de mesures, effectuées
comme précédemment à partir de magnétite préparée
par voie sèche, l'appareil avait été amélioré, et
il était suffisant de réajuster la pression d'hydrogène après 5 prélèvements : celle-ci restait alors
toujours comprise entre 5 et 6 atmosphères (à froid),
ce qui entraîne une incertitude d'environ 6 % sur
les solubilités, incertitude tout à fait acceptable
compte tenu de la précision des mesures.
Dans ce cas, nous avons réussi à limiter à 3 heures la
durée de la perturbation lors des refroidissements
indispensables.
     Les résultats relatifs à ces mesures sont repor-
tés sur la figure 3.

4.3. La troisième série de mesures a été effectuée
dans les mêmes conditions que la deuxième, mais en
remplaçant la magnétite en poudre par une plaque
d'acier au carbone oxydée " in situ " dans l'eau à
300°C, sous une atmosphère d'hydrogène ; les résultats sont également reportés sur la figure 3.

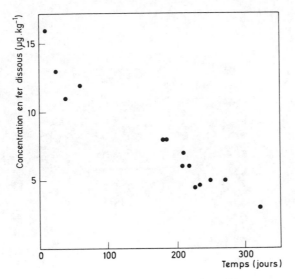

FIG.4. Variation de la solubilité de la magnétite en fonction du temps, à 300°C. Deuxième série.

5. DISCUSSION

5.1. Mesures de solubilité

La comparaison des figures 2 et 3 montre que les solubilités sont moins dispersées et, en moyenne, moins élevées dans la série 2 que dans la série 1. Les résultats de la série 3 sont comparables à ceux de la série 2.

D'autre part la figure 4 donne la variation de la solubilité mesurée à 300°C lors de la série 2, en fonction de la durée de séjour dans l'autoclave : on peut observer une baisse régulière de la solubilité, atteignant un facteur 5 en 300 jours.

Pour interpréter ce phénomène, nous avons émis l'hypothèse que le séjour prolongé dans une solution aqueuse à chaud favorise la formation d'une forme cristalline de magnétite relativement moins soluble, qui serait détruite lors du séjour à basse tempéra - ture. Dans cette hypothèse, les valeurs élevées et dispersées obtenues dans la série 1 s'expliquent par les refroidissements fréquents et relativement longs qui ont suivi chaque expérience et n'ont pas permis

MAGNETITE ①

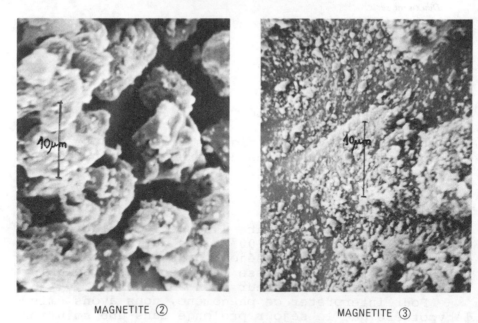

MAGNETITE ②   MAGNETITE ③

FIG.5. Micrographies de magnétite.

la formation d'une forme cristalline stable ; au contraire dans la deuxième série, les refroidissements plus courts et moins fréquents ont sans doute peu perturbé le vieillissement des cristaux.

5.2. Morphologie de la magnétite

En vue d'éprouver la validité de ces hypothèses, nous avons comparé, par diffraction de rayons X et par microscopie électronique: la magnétite préparée par voie sèche avant son introduction dans l'autoclave (magnétite ①) ; la magnétite isolée avec précaution de la solution après la deuxième série d'expériences (magnétite ②); et le produit obtenu après un séjour de 8 semaines à température ordinaire, dans une solution dont le pH est compris entre 7 et 8, des magnétites ① et ② (magnétite ③).

Les micrographies électroniques montrent (fig.5) qu'avant introduction dans l'autoclave, la magnétite ① est formée de grains relativement réguliers, d'une dimension de quelques microns. Quant à la magnétite ② , elle contient des cristaux beaucoup plus gros entourés de petits cristaux, dont la taille est inférieure à 1 micron et quelquefois accolés au gros cristal ; on y trouve également des agglomérats de petits grains, de dimensions supérieures à 1 mm.

La magnétite ③ , qu'elle soit formée à partir des échantillons ① ou ②, est constituée de cristaux très petits, souvent inférieurs à 0,3 micron, à côté de quelques cristaux de taille comparable à celle des précédents.

Un exemple de micrographie représentative de ces différents échantillons est donné par la figure 5.

Les spectres de diffraction de rayons X confirment ces observations : dans tous les cas, le spectre de l'oxyde Fe_3O_4 apparaît nettement, mais la magnétite ③ produit des spectres à raies élargies, ce qui est caractéristique d'une dimension moyenne de grains inférieure à 0,1 micron.

5.3. Interprétation proposée

On sait que, dans une solution, les gros cristaux ont normalement tendance à croître aux dépens des plus petits, de façon à évoluer vers un système d'énergie libre minimale. Ce phénomène explique l'évolution observée à chaud.

Mais dans le cas de la magnétite, il doit se
produire en outre un transport de matière d'origine
chimique : de l'hydrogène participant à la dissolution de la magnétite, il se pourrait que des variations locales de concentration de ce gaz expliquent
l'apparition d'une succession de précipitations et de
redissolutions. A chaud, l'évolution normale des cristaux et l'homogénéisation de la solution (quant à sa
concentration d'hydrogène) doivent être rapides ;
par contre à froid, les précipitations dues à ces
variations locales de concentration d'hydrogène doivent conduire à la formation de petits cristaux dont
le grossissement est lent ; le phénomène observé glo-
balement est alors la formation de petits cristaux
à partir des gros.

Quelques observations, encore incomplètes, semblent montrer que la précipitation de petits cristaux
à froid est plus importante quand le pH diminue.
Des expériences complémentaires restent évidemment
nécessaires pour confirmer ces hypothèses.

6. CALCUL DE L'ENTHALPIE LIBRE DE FORMATION DE L'ION $HFeO_2^-$

A partir des enthalpies libres de formation et
des entropies à 25°C de toutes les espèces mises en
jeu et de leur extrapolation en fonction de la température selon la méthode préconisée par McDONALD
$/\ 1\ /$, le code de calcul POTHY $/\ 3\ /$ permet de calculer les solubilités dans une solution de concentration connue d'acide ou de base, et à une pression
d'hydrogène donnée. La prise en compte des réactions
chimiques (1) (2) et (3) et de l'équation de conser-
vation de charge conduit à écrire une fonction impli-
cite de l'activité (H^+), dont la solution est obtenue
par un processus itératif. La détermination du pH permet alors de calculer la solubilité à l'équilibre des
différents constituants en fonction de la température.

Dans la potasse 10^{-2} M, l'espèce $HFeO_2^-$ est lar-
gement prédominante ; si on connaît l'enthalpie libre
de formation de Fe_3O_4, il est donc possible de calculer des courbes de solubilité correspondant à différentes valeurs de $\Delta G°_{HFeO_2^-}$, et de déterminer ainsi

la valeur de l'enthalpie libre de formation de l'ion
qui correspond le mieux aux valeurs expérimentales.

Le calcul a été effectué en utilisant la valeur
d'entropie standard pour $HFeO_2^-$ préconisée par McDONALD,
soit :

$$S°_{HFeO_2^-} = 15 \text{ cal.mol}^{-1}. \text{ d}°^{-1}$$

La variation de la pression d'hydrogène n'étant
pas très bien définie du fait de la présence de par-
ties froides (tubes de raccordement aux vannes et au
manomètre), le calcul a été effectué dans l'hypothèse
d'une pression constante de 6 atmosphères.

Les courbes calculées avec différentes valeurs
de $\Delta G°_{HFeO_2^-}$ sont portées sur la figure 3.

Comme nous l'avons vu précédemment, les valeurs
expérimentales obtenues diffèrent selon la taille des
cristaux de magnétite ; ces différences sont liées
à l'existence d'un terme d'énergie libre d'interface
entre la magnétite et la solution aqueuse, terme dont
la contribution devient négligeable lorsque les cristaux sont suffisamment gros.

Si les solubilités les plus faibles que nous
ayons mesurées correspondent bien à la limite basse,
ce qui demandera à être vérifié, la valeur d'enthalpie libre de formation de $HFeO_2^-$ doit être proche de
$\Delta G°_{HFeO_2^-} = -88 \text{ kcal.mol}^{-1}$ sans qu'il soit possible

d'indiquer la précision sur cette valeur, car, dans
le domaine des faibles solubilités, la solubilité
est peu sensible à la valeur de $\Delta G°$.

Cependant, pour rendre compte de la solubilité
de la magnétite obtenue par oxydation d'un acier,
avant qu'elle ait eu le temps d'évoluer, il est nécessaire d'adopter une valeur plus négative, de l'ordre
de $-90,5 \text{ kcal.mol}^{-1}$, ce qui est une manière fictive
de tenir compte d'un terme d'énergie interfaciale.

7.   APPLICATION A LA CHIMIE DE L'EAU DES REACTEURS

Pour comprendre les phénomènes régissant les mouvements des produits de corrosion activés dans un circuit primaire de réacteur, en vue de dégager des règles
d'exploitation minimisant l'activité des produits de
corrosion ainsi transportés et le débit de dose des
rayonnements dont ils sont responsables, il est es-
sentiel de bien connaître les valeurs des solubilités.

Le fait que celles-ci varient notablement avec
la taille des cristaux complique les prévisions. Cependant on peut remarquer qu'il existe une différence

de température d'environ 60°C entre les parois froides et les parois chaudes du réacteur, ce qui entraîne
en permanence des dissolutions et des reprécipitations [1]. Or, du fait de l'importance des débits de
liquide (12 à 13 tonnes par seconde) et des vitesses
linéaires élevées (supérieures à 5 mètres par seconde),
cette précipitation ne doit pas pouvoir conduire
à une formation de gros cristaux, contrairement à
ce qui a lieu en autoclave dans les mêmes conditions
de température.

La solubilisation par injection d'acide borique
au cours du refroidissement du réacteur a été envi-
sagée /¯7, 8_7 comme moyen de décontamination du
circuit primaire ; on observe cependant que le relâ-
chement de produits de corrosion décroît fortement
après un certain temps. Une explication possible
serait la solubilisation, dans un premier temps,
d'une couche superficielle de petits cristaux formés
comme indiqué précédemment, la couche sous-jacente
d'oxyde plus ancien, formée de gros cristaux, étant
beaucoup moins soluble.

Selon ce qui a été dit au chapitre 5.3, on
pourrait s'attendre, à froid, à une évolution des gros
cristaux vers une forme plus soluble ; cette évolution
a effectivement été observée /¯8_7, mais elle
est lente vis-à-vis du temps disponible pour la décontamination.

8. CONCLUSION

L'influence importante de la taille des cristaux de magnétite sur leur solubilité dans les solutions aqueuses a été mise en évidence, ainsi que les
variations de cette taille suivant les conditions
expérimentales. A chaud, les cristaux ont tendance à
grossir et la solubilité à diminuer, ce qui représente leur comportement normal lorsqu'ils sont en
équilibre avec une solution saturée.

A froid, c'est cependant le phénomène inverse
qui se produit : la taille moyenne des cristaux
diminue, sans doute à cause de dissolutions et de
reprécipitations liées à des variations locales de
la pression d'hydrogène ; la cinétique de grossissement des petits grains ainsi formés est alors très
lente. L'écart de solubilité à 300°C entre les différentes formes peut atteindre un facteur 20.

---

[1] Le calcul montre que le transfert de matière dû à ces phénomènes est généralement
supérieur à 1 mg de fer par seconde.

La forme stable de la magnétite devant corres-
pondre aux solubilités les plus faibles mesurées nous
proposons pour l'enthalpie libre de $HFeO_2^-$ une valeur
proche de

$$\Delta G^\circ_{HFeO_2^-} = -88 \ kcal.mol^{-1}$$

Une étude de la solubilité d'une forme bien
cristallisée dans tout le domaine de température per-
mettra de préciser cette valeur.

L'influence de la taille des grains sur la solubilité permet d'interpréter les variations du relâ-
chement des produits de corrosion au cours du refroidissement des circuits de réacteurs.

REFERENCES

[1] McDONALD, D.D. AECL 4137 (1972).

[2] LEWIS, D. A.E. 432 (1971).

[3] BESLU, P., FREJAVILLE, G., LALET, A.
International conference on water chemistry
of nuclear reactor systems. Bournemouth 1977.
BNES. London (1978).

[4] SWEETON, F.H., BAES, C.F.
J. of Chem. Thermodyn. 2, (1970) 419.

[5] STYRIKOVICH, M.A. et coll.
Teploehnergetika 19 9 (1972) 127.

[6] KANERT, G.A., GRAY, G.W., BALDWIN, W.G.
A.E.C.L. 5528 (1976).

[7] BESLU, P., FREJAVILLE, G., LALET, A., MICAUX,
B.
Seminar on decontamination of nuclear plants
Columbus (OHIO, USA) 1975 (non publié)

[8] BESLU, P., CAMP, J.J., FREJAVILLE, G.,
MARCHAL, A.
Corrosion 78 Conference HOUSTON (USA) 1978.
(non publiée).

DISCUSSION

P.R. TREMAINE: Did you consider using crystalline Fe_3O_4? Also, did
you observe any change with hydrogen pressure?

I. LAMBERT: No, I did not consider using single crystals, but this would
certainly be interesting. Within the limits of our pressure variations (factor of 1.5)
we did not detect any effect because of liquid volume variations in the autoclave,

but the expected effect is much lower than the variation due to changes in the physical state of magnetite.

P.R. TREMAINE: We have completed a study of Fe_3O_4 solubility at 300°C using a flow system, and obtained very similar results. The stoichiometry of the Fe_3O_4 solubility reaction suggests that the solubility should be proportional to $p_{H_2}^{1/3}$, but we did not see any change between solutions saturated with 1 atm H_2 and 0.1 atm H_2 at 25°C. Sandler and Kunig also failed to observe any variation in $NiFe_2O_4$ solubility at these hydrogen concentrations. Could you please comment?

I. LAMBERT: I cannot find any explanation at the moment, but I think it would be interesting to bring about large pressure variations (by a factor 100 or 1000) to test the dependence of solubility on pressure.

M.C. NOE: May I add a few words to the reply? The effect of hydrogen partial pressure on magnetite solubility has not been investigated. In the experimental apparatus it is difficult to keep this parameter very constant because (a) the system is not completely isothermal in the vapour phase, and (b) the volume ratio between vapour and solution varies in each sampling. However, the variation in magnetite solubility was expected to be proportional to the one-third power of the hydrogen pressure.

W. NAGEL: In a quite different area of research, namely corrosion in conventional power stations, we observed that Cl^- influenced magnetite formation. Did you check for Cl^-, or any other impurities?

I. LAMBERT: We did not check chloride since we worked with pure products (water of resistivity > 10 MΩ, iron and KOH of analytical reagent grade). Under such conditions we did not consider the possibility of Cl^- ion formation.

COMPLEXES AQUEUX D'IONS METALLIQUES PRESENTS DANS LE RETRAITEMENT DES COMBUSTIBLES NUCLEAIRES
Complexes avec les ligands orthophénanthroline et dibutylphosphate

C. MUSIKAS, G. LE MAROIS, J. RACINOUX
CEA, Centre d'études nucléaires de Fontenay-aux-Roses,
Division de chimie,
Fontenay-aux-Roses,
France

Abstract–Résumé

AQUEOUS COMPLEXES OF METAL IONS PRESENT IN THE REPROCESSING OF NUCLEAR FUELS: COMPLEXES WITH THE LIGANDS ORTHOPHENANTHROLINE AND DIBUTYL PHOSPHATE.
The authors have studied the aqueous complexes formed by the ligands orthophenanthroline and dibutyl phosphate with the metal ions present in nuclear fuel reprocessing solutions. Orthophenanthroline forms the complexes ML^{3+} and ML_2^{3+} with the trivalent lanthanide and actinide ions. The stability of americium complexes is greater than that of the lanthanide ions of similar ionic radius. The Np(V) ions from a single complex, NpO_2L^+, whose stability is greater than that of the complex AmL^{3+}, contrary to what is normally observed. The dibutyl phosphate ions tend to form compounds of low solubility and having the formula $M(DBP)_n$ (where n is the charge on the metal cation). However, in the case of low and high ligand concentrations, one observes cation and anion complexes, respectively. Uranium forms the soluble complexes $UO_2(DBP)^+$ and $UO_2(DBP)_4^{2-}$. Neptunium(V) is very insoluble: the existence of two not very soluble species, $NpO_2(DBP).xH_2O$ and $NpO_2(DBP)_2Na.yH_2O$, has been demonstrated. In the case of the Ln^{3+} ions, the smaller their ionic radius, the greater their interaction with the DBP^- ions. Very considerable differences are observed between the different members of the series. The existence of the soluble species $Ln(DBP)_2^+$, $Ln(DBP)_4^-$ and $Ln(DBP)_5^{2-}$ has been shown. Iron(III) and chromium(III) form very different complexes. Chromium(III) gives rise mainly to $Cr(DBP)_4^-$ and the relatively soluble $Cr(DBP)_3$, whereas iron is of very low solubility over the whole range of DBP^- ion concentrations which can be obtained in aqueous solutions.

COMPLEXES AQUEUX D'IONS METALLIQUES PRESENTS DANS LE RETRAITEMENT DES COMBUSTIBLES NUCLEAIRES: COMPLEXES AVEC LES LIGANDS ORTHO-PHENANTHROLINE ET DIBUTYLPHOSPHATE.
Les auteurs ont étudié les complexes aqueux formés par les coordinats orthophénanthroline et dibutylphosphate avec les ions métalliques présents dans les solutions de retraitement des combustibles nucléaires. L'orthophénanthroline forme les complexes ML^{3+} et ML_2^{3+} avec les ions lanthanides et actinides trivalents. La stabilité des complexes de l'américium est supérieure à celle des ions lanthanides de rayons ioniques voisins. Les ions Np(V) forment un seul complexe

NpO_2L^+ dont la stabilité est plus grande que celle du complexe AmL^{3+}, contrairement à ce qui est observé habituellement. Les ions dibutylphosphate ont tendance à former des composés peu solubles de formule $M(DBP)_n$ (où n est la charge du cation métallique). Cependant, pour les concentrations faibles et fortes en ligand, on observe respectivement des complexes cationiques et anioniques. L'uranium forme les complexes solubles $UO_2(DBP)^+$ et $UO_2(DBP)_4^{2-}$. Le neptunium(V) est très insoluble, deux espèces peu solubles $NpO_2(DBP) . xH_2O$ et $NpO_2(DBP)_2Na . yH_2O$ ont été mises en évidence. Les ions Ln^{3+} interagissent d'autant plus avec les ions DBP^- que leur rayon ionique est faible. Des différences très notables sont observées entre les différents membres de la série. On a mis en évidence les espèces solubles $Ln(DBP)_2^+$, $Ln(DBP)_4^-$ et $Ln(DBP)_5^{2-}$. Le fer(III) et le chrome(III) forment des complexes très différents. Le chrome(III) donne essentiellement $Cr(DBP)_3$ relativement soluble et $Cr(DBP)_4^-$ alors que le fer est très peu soluble dans toute la gamme des concentrations en ions DBP^- réalisables en solutions aqueuses.

1. INTRODUCTION

Cet article est consacré à des études de complexes dans des solutions aqueuses, entreprises dans un laboratoire associé à un service où sont étudiés les procédés de séparation chimique du retraitement des combustibles nucléaires. Deux préoccupations sont à l'origine de ces études: les séparations actinides-lanthanides, et les propriétés des complexes métalliques résultant de la dégradation du tributyl-phosphate.

2. COMPLEXES AQUEUX DES IONS LANTHANIDES ET ACTINIDES AVEC L'ORTHOPHENANTHROLINE

Les complexes des ions des éléments 4f et 5f avec des ligands à azote comme atome donneur ont été peu étudiés. Cela tient à la faible affinité de ces donneurs pour les ions 4f et 5f réputés pour se lier préférentiellement à des ligands à atomes donneurs plus électronégatifs.

Des complexes orthophénanthroline ($C_{12}H_8N_2$) des ions lanthanides(III), de l'uranium(VI) et du thorium(IV) peu solubles ont été préparés par précipitation à partir de solutions aqueuses [1]. Cependant, l'existence de complexes solubles et les valeurs de leur constante de formation n'ont pas été examinées systématiquement.

2.1. Lanthanides et actinides trivalents

Nous allons d'abord considérer les ions trivalents des actinides et des lanthanides. A ce degré d'oxydation, les ions des éléments 4f et 5f sont réputés pour donner des complexes dont la stabilité est surtout fonction du rayon ionique;

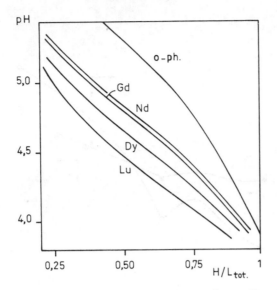

FIG.1.   *Titrage acido-basique de l'orthophénanthroline en présence d'ions lanthanides*
(μ = 1, milieu chlorure, C_L = 0,001M, C_M = 0,05M, θ = 20°C).

c'est le cas par exemple des complexes avec les acides carboxyliques ou les ions
F^- [2]. En étudiant les complexes orthophénanthroline nous avons tenté de
mettre en évidence des effets covalents différents dans les séries 4f et 5f. Cette
différence est connue dans le cas des séries d, et est la conséquence d'une plus
grande extension des orbitales 4d et 5d, comparativement aux orbitales 3d.

Nous avons étudié les complexes des lanthanides par potentiométrie,
spectrophotométrie et partage de l'orthophénanthroline entre deux phases
liquides.

La potentiométrie est basée sur les variations de pH dues à la compétition
des ions Ln^{3+} et H^+ pour le site de coordination de l'orthophénanthroline
($K_a \cong 10^{-5,2}$) suivant l'équation (1), où L représente l'orthophénanthroline et
Ln^{3+} un ion lanthanide.

$$m(LH^+) + Ln^{3+} \rightarrow mH^+ + Ln(L)_m^{3+} \qquad (1)$$

On peut voir sur la figure 1 l'influence de la présence d'ions lanthanides sur
le titrage de l'orthophénanthroline par de l'acide chlorhydrique. A pH constant,
on a la relation (2) entre $\bar{n}$, le nombre moyen de molécules d'orthophénanthroline
coordinées au métal, et les valeurs expérimentales C_L, C_M, H_L, H_M.

$$\bar{n} = \frac{C_L}{C_M} \frac{(H_L - H_M)}{H_L} \qquad (2)$$

MUSIKAS et al.

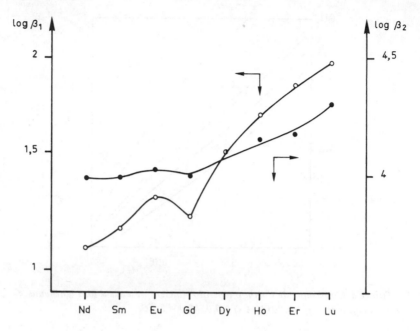

FIG.2. Constantes de formation des complexes orthophénanthroline des ions lanthanides(III).

Dans cette relation

C_L est la concentration en orthophénanthroline sous toutes ses formes.

C_M est la concentration en métal.

H_M, H_L sont les abscisses des courbes de la figure 2 à un pH déterminé en présence et en l'absence de métal.

$\bar{n}$ est le rapport de la concentration de l'orthophénanthroline liée à celle du métal; il peut s'exprimer en fonction des constantes de formation des complexes par la relation (3).

$$\bar{n} = \sum_{i=1}^{i=4} i \cdot \beta_i \cdot L^i \Bigg/ \sum_{i=0}^{i=4} \beta_i \cdot L^i \tag{3}$$

β_i est la constante de formation du complexe orthophénanthroline contenant i molécules de ligand; nous avons limité i à 4 à cause de l'empêchement stérique qui rend improbable l'existence de complexes plus riches en orthophénanthroline.

L est la concentration en orthophénanthroline libre; elle est obtenue en utilisant l'équation (4), où LH^+ est la concentration en ion orthophénanthrolinium, donnée par la valeur de l'abscisse des courbes de la figure 1.

$$C_L = L + LH^+ + \bar{n} \cdot C_M \tag{4}$$

Les essais ont été effectués en milieu chlorhydrique à cause des solubilités limitées soit des ions orthophénanthrolinium, soit des complexes des lanthanides dans d'autres milieux peu complexants (ClO_4^-, NO_3^-, etc.).

Un programme mathématique basé sur la méthode des moindres carrés a été utilisé pour calculer les valeurs des β_i.

Pour l'holmium, on a pu vérifier que l'absorption à 480 nm en présence d'orthophénanthroline correspond à la loi

$$\epsilon_M = \sum_{i=0}^{i=4} \epsilon_i \cdot \beta_i \cdot L^i \Bigg/ \sum_{i=0}^{i=4} \beta_i \cdot L^i \tag{5}$$

ϵ_M est le coefficient d'extinction molaire apparent de l'holium.

ϵ_i correspond aux coefficients d'extinction molaire des différentes espèces en présence, i indiquant le nombre de molécules de ligand du complexe considéré.

Enfin le partage de l'orthophénanthroline entre une phase benzénique en présence et en l'absence de lanthanides nous a permis de mettre en évidence deux complexes, car les coefficients de distribution obéissent à la loi

$$D = \frac{D_0 \cdot L}{L + K_a \cdot (H^+) \cdot L + \bar{n} \cdot C_M} \tag{6}$$

D et D_0 sont les coefficients de partage de l'orthophénanthroline en présence et en l'absence de métal.

K_a est la constante d'acidité de l'orthophénanthroline, L sa concentration en phase aqueuse.

$\bar{n}$ et C_M ont les significations définies plus haut.

Comme on a vérifié que seule l'orthophénanthroline neutre était soluble en phase organique on a

$$L = (C_L)_{org} \cdot D_0 \tag{7}$$

$(C_L)_{org}$ est la concentration en orthophénanthroline en milieu benzénique.

TABLEAU I. CONSTANTES DE FORMATION DES COMPLEXES DES IONS LANTHANIDES(III) AVEC L'ORTHOPHENANTHROLINE

Elément	Potentiométrie		Spectrophotométrie		Extraction par solvant	
	$\log \beta_1$	$\log \beta_2$	$\log \beta_1$	$\log \beta_2$	$\log \beta_1$	$\log \beta_2$
Nd	1,09	3,99			0,79	3,53
Sm	1,17	3,99				
Eu	1,31	4,02				
Gd	1,22	4,00				
Dy	1,50	4,10				
Ho	1,66	4,16	1,67	3,95		
Er	1,78	4,16			1,99	4,03
Lu	1,88	4,3				

Les équations (6) et (7) ont permis d'obtenir $\bar{n}$ en fonction de L ce qui a conduit, après un calcul par la méthode des moindres carrés, aux valeurs des constantes de formation des complexes du néodyme et de l'erbium.

Dans le tableau I on a rassemblé les valeurs des constantes de formation des complexes avec l'orthophénanthroline déduites des différentes méthodes d'étude utilisées. En première approximation les résultats obtenus sont concordants. Sur la figure 2 on peut voir comment varient les constantes de formation en fonction du numéro atomique des lanthanides.

Les complexes orthophénanthroline de l'américium(III) ont été étudiés par spectrophotométrie. Sur la figure 3, on peut voir les variations spectrales de solutions d'américium(III) en fonction de la concentration en orthophénanthroline. La déconvolution des spectres montre que l'on a affaire à deux complexes, $AmL^{3+} \cdot xH_2O$ et $AmL_2^{3+} \cdot yH_2O$, dont les log des constantes de formation ont pour valeur 2,56 et 4,03.

2.2. Complexes du neptunium(V)

On a étudié les complexes orthophénanthroline du neptunium(V) par deux méthodes, une méthode potentiométrique et une méthode spectrophotométrique.

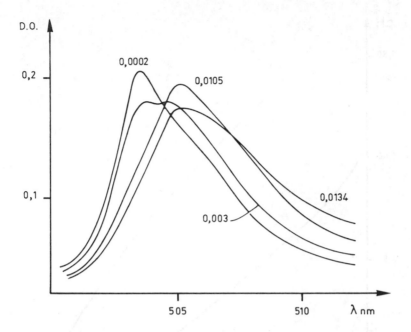

FIG.3. Spectres d'absorption des ions Am(III) en fonction de la concentration en orthophénan-throline (μ = 0,1; milieu chlorure, C_{Am} = 0,00053M, θ = 20°C).

Sur la figure 4 on peut voir les courbes de titrage acido-basique de l'ortho-phénanthroline en présence et en l'absence de neptunium(V). A l'aide de la méthode de calcul décrite au paragraphe 2.1, on a pu déterminer la constante de formation du seul complexe observé, NpO_2L^+; on trouve β_1 = 676.

Sur la figure 5 on a reporté les variations spectroscopiques pour la bande principale des ions NpO_2^+ par addition d'orthophénanthroline. Nous constatons que seules deux bandes d'absorption sont en présence; l'une, connue, est attribuable à l'ion non complexé $NpO_2^+ \cdot xH_2O$, l'autre est attribuable au complexe NpO_2L^+, dont la formation est quantitative en présence d'orthophénanthroline $1,4 \cdot 10^{-2}$ M. Nous en avons déduit la constante de formation du complexe orthophénanthroline du neptunium(V); on trouve β_1 = 602, en accord avec la constante de formation trouvée par potentiométrie.

En conclusion, on peut souligner que l'orthophénanthroline forme avec les ions 4f et 5f des complexes de stabilité bien inférieure à celle des complexes des ions des éléments d. A cause du faible caractère ionique de la liaison orthophénan-throline-cation on peut supposer que la différence d'affinité des ions 4f et 5f trivalents pour le ligand est due à la tendance plus grande des ions 5f à former des

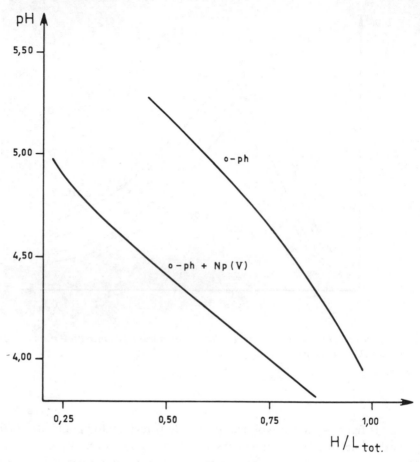

FIG.4.   Titrage acido-basique de l'orthophénanthroline en présence de neptunium(V)
(μ = 0,1; milieu chlorure, C_L = 0,0022M, C_M = 0,0086M, θ = 20°C).

liaisons à électrons délocalisés. Le complexe NpO_2L^+ est plus stable que le
complexe AmL^{3+}, ce qui n'est pas l'ordre habituel observé pour les ligands à
atomes donneurs plus électronégatifs, pour lesquels ce sont les ions trivalents qui
ont le plus d'affinité.

3.   COMPLEXES DIBUTYLPHOSPHORIQUES

Le tributylphosphate se dégrade sous l'action des rayonnements et des
agents chimiques, ce qui conduit à l'obtention de nombreux précipités dans le
retraitement des combustibles nucléaires. Ces précipités n'ont pas été identifiés

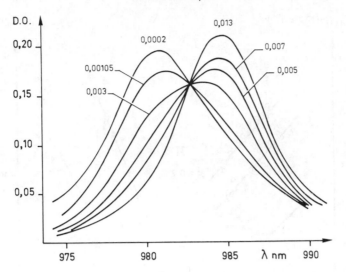

FIG.5. Spectres d'absorption des ions NpO_2^+ en fonction de la concentration en orthophénan-throline ($\mu = 0,1$; milieu chlorure, $\theta = 20°C$).

sans ambiguïté mais on pense qu'ils résultent de l'interaction de l'acide dibutyl-phosphorique ($HPO_4(C_4H_9)_2$), produit principal de dégradation, avec les ions métalliques présents. Les complexes dibutylphosphoriques des ions minéraux n'ont pas été étudiés systématiquement. Cependant, on peut citer les études de Sheka et Sinjavskaja [3—5] et Krutikov et Solovskin [6] concernant les complexes du thorium(IV) et des lanthanides(III). Nous rapportons ici les études effectuées dans notre laboratoire concernant les complexes aqueux de l'uranium(VI), du neptunium(V), des lanthanides(III), du fer(III) et du chrome(III).

3.1. Complexes dibutylphosphoriques de l'uranium(VI)

Le sel $UO_2(DBP)_2$ est peu soluble en solution aqueuse. Nous avons mesuré la solubilité de l'uranium(VI) en fonction de la concentration en ions dibutyl-phosphate. Les résultats sont présentés sur la figure 6.

Nous avons montré par analyse chimique que le sel en équilibre avec la solution était toujours $UO_2(DBP)_2$. Dans ce cas la concentration en uranium $C_{U(VI)}$ peut s'exprimer en fonction des constantes de formation des complexes (équation (8)).

$$C_{U(VI)} = \sum_{i=0}^{i=m} (UO_2^{2+}) \cdot \beta_i \cdot (DBP^-)^i \tag{8}$$

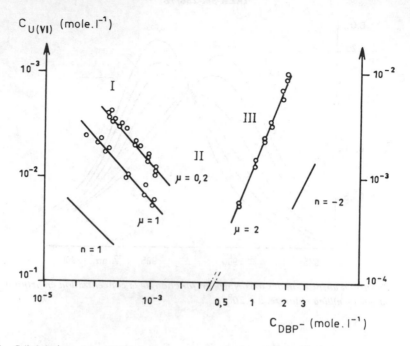

FIG.6. *Solubilité des ions U(VI) dans des solutions aqueuses de dibutylphosphate de sodium.*

Dans cette équation

β_i est la constante de formation du complexe $UO_2(DBP)_i^{(i-2)-}$.

m   est le nombre maximal d'ions DBP^- qui peuvent se coordiner à l'uranium;
on a pris m = 4 pour des raisons d'empêchement stérique.

Comme on a affaire à un équilibre solide-solution on a les relations
supplémentaires

$$K_s = (UO_2^{2+})(DBP^-)^2 = C_{UO_2(DBP)_2}/\beta_2 \tag{9}$$

K_s est le produit de solubilité de $UO_2(DBP)_2$.
$C_{UO_2(DBP)_2}$ est la solubilité de $UO_2(DBP)_2$.

La pente des courbes $\log C_{U(VI)}$ en fonction de $\log C_{DBP^-}$ indique le
nombre moyen d'ions DBP^- échangés au cours de la précipitation comme
l'indique l'équilibre représenté par l'équation (10).

$$UO_2(DBP)_x^{(2-x)+} + n\,DBP^- \rightleftarrows UO_2(DBP)_2 \downarrow \tag{10}$$

avec x + n = 2.

TABLEAU II. CONSTANTES DE FORMATION DES COMPLEXES
DIBUTYLPHOSPHORIQUES DE L'URANIUM(VI)

$UO_2(DBP)^+$	$\log \beta = 4{,}32$
$UO_2(DBP)_2$	$K_s = 7 \cdot 10^{-12} = (UO_2^{2+})\,(DBP^-)^2$
$UO_2(DBP)_4^{2-}$	$\log \beta = 8{,}38$

Dans la zone I de la figure 6 les complexes en solution sont cationiques,
alors que dans la zone III ils sont anioniques, la zone II étant une zone
intermédiaire.

A partir de l'équation (8) nous avons pu, à l'aide des valeurs expérimentales
de solubilité, calculer les constantes de formation des complexes de l'uranium;
les résultats sont consignés dans le tableau II.

Afin de nous assurer que les hypothèses suggérées par la forme des courbes
de solubilité étaient fondées, nous avons utilisé d'autres méthodes physico-
chimiques d'étude des complexes indépendantes, à savoir la polarographie à
l'électrode à goutte de mercure et la spectrophotométrie d'absorption.

Le polarogramme des solutions de perchlorate de sodium en contact avec
le solide $UO_2(DBP)_2$ qui contiennent $4 \cdot 10^{-4}M$ d'ions U(VI) comporte une onde
dont le potentiel à mi-hauteur est 0,09 V, plus négatif que le potentiel normal
apparent du couple U(VI)/U(V) en milieu perchlorique. Ceci indique que
l'uranium redissous l'est sous forme de complexe.

Dans la zone III, la spectrophotométrie montre que l'on a surtout une
espèce, car on n'observe qu'un seul type de spectre. Ceci est en accord avec
l'ajustement obtenu à partir des résultats de mesure de la solubilité, cette espèce
étant $UO_2(DBP)_4^{2-}$.

3.2. Complexes dibutylphosphoriques du neptunium(V)

Les complexes du neptunium(V) ont été étudiés par la méthode de solubilité
utilisée déjà pour l'uranium(VI). La solubilité des ions $N_p(V)$ en fonction de la
concentration en ions DBP^- dans le surnageant est montrée sur la figure 7. Dans
la zone I on a pu vérifier par spectrophotométrie que la solution surnageante ne
contient que des ions NpO_2^+. Par conséquent la solubilité obéit à la simple loi du
produit de solubilité.

$$(NpO_2^+)\,(DBP^-) = K_s \tag{11}$$

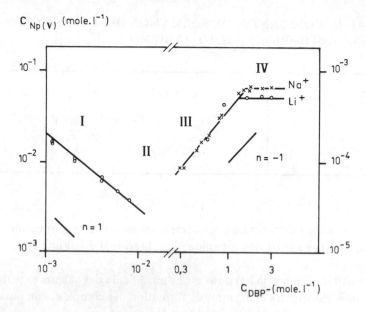

FIG.7.   Solubilité des ions Np(V) dans des solutions aqueuses de dibutylphosphate alcalines.

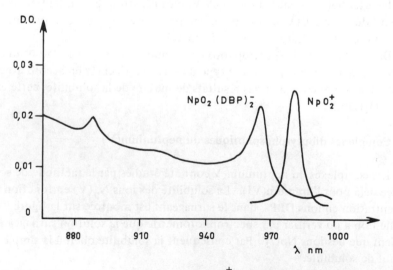

FIG.8.   Spectre d'absorption des solutions d'ions NpO_2^+ en présence et en l'absence de dibutylphosphate de sodium ($NpO_2^+ = 6,3 \cdot 10^{-5}M$, $NpO_2(DBP)_2^- = 6,3 \cdot 10^{-4}M$ dans NaDBP 3M).

TABLEAU III. CONSTANTES DE FORMATION DES COMPLEXES DIBUTYLPHOSPHORIQUES DU NEPTUNIUM(V)

$NpO_2(DBP)$	$K_S = 5,2 \cdot 10^{-7} = (NpO_2^+)(DBP^-)$
$NpO_2(DBP)_2^-$	$\log \beta = 2,76$
$NpO_2(DBP)_2Na$	$K_S = 2,5 \cdot 10^{-10} = (NpO_2^+)(DBP^-)^2(Na^+)$

Dans la zone III, la solution contient des ions Np(V) sous une forme complexée comme on peut le voir sur le spectre d'absorption de la figure 8. La pente de la courbe $\log C_{Np(V)}$ en fonction de $\log C_{DBP^-}$ est une droite de pente 1; on peut donc conclure que l'on a affaire à une espèce soluble de formule $NpO_2(DBP)_2^-$. Dans la zone IV la solubilité est indépendante de la concentration en ions DBP^-, indiquant qu'une nouvelle forme solide est probablement obtenue, car le spectre d'absorption des surnageants ne varie pas lorsqu'on passe de la zone III à la zone IV. Les constantes de formation des complexes du neptunium(V) et les produits de solubilité des deux solides obtenus sont consignés dans le tableau III.

3.3. Complexes dibutylphosphoriques des lanthanides

La solubilité des ions lanthanides(III) dans les solutions aqueuses d'ions dibutylphosphate varie comme celle de l'uranium(VI). Pour les faibles concentrations en ions DBP^- la solubilité décroît avec la concentration car les complexes solubles cationiques sont transformés en complexe neutre $Ln(DBP)_3$, peu soluble. Pour les fortes concentrations on a redissolution par formation de complexes anioniques $M(DBP)_4^-$ et $M(DBP)_5^{2-}$. Sur la figure 9, on peut voir comment varie la solubilité du cérium en fonction de la concentration en dibutylphosphate. Sur la figure 10 on a reporté en coordonnées logarithmiques la solubilité de sept lanthanides en fonction de la concentration en ions dibutylphosphate (dans le domaine des faibles concentrations). La pente des droites obtenues est 1, ce qui indique que l'équilibre solide-solution peut s'écrire

$$Ln(DBP)_2^+ + DBP^- \rightarrow Ln(DBP)_3 \downarrow \qquad (12)$$

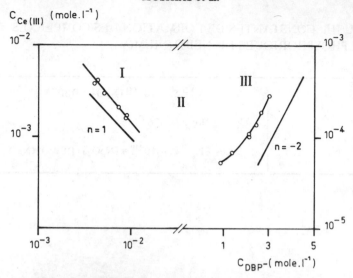

FIG.9. Solubilité du cérium(III) dans des solutions de dibutylphosphate de sodium.

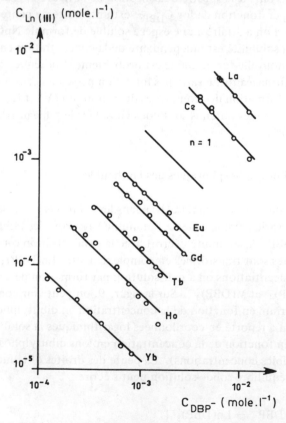

FIG.10. Solubilité des ions Ln(III) dans des solutions de dibutylphosphate de sodium.

TABLEAU IV. PRODUITS DES CONSTANTES β_2 ET K_s POUR LES
COMPLEXES DIBUTYLPHOSPHORIQUES DES IONS LANTHANIDES
TRIVALENTS

Elément	$\log(\beta_2 \cdot K_s)$	Elément	$\log(\beta_2 \cdot K_s)$
La	4,05	Gd	6,45
Ce	4,95	Tb	6,75
Nd	5,8	Ho	7,3
Eu	6,35	Yb	7,8

On voit que l'interaction DBP^--ion lanthanide est fonction du rayon ionique.
Les constantes de formation des complexes peuvent être déduites de l'ajustement
des solubilités expérimentales à la fonction (13), qui exprime la solubilité en
fonction des constantes de formation des complexes où on s'est limité à $Ln(DBP)_5^{2-}$.

$$C_{Ln(III)} = \sum_{i=0}^{i=5} (Ln^{3+}) \cdot \beta_i \cdot (DBP^-)^i \tag{13}$$

Les relations supplémentaires (14), qui expriment qu'une phase solide
$Ln(DBP)_3$ est en équilibre avec la solution, ont également été utilisées

$$K_s = (Ln^{3+})(DBP^-)^3 = C_{Ln(DBP)_3}/\beta_3 \tag{14}$$

Les résultats sont consignés dans le tableau IV. Dans le domaine étudié on
a seulement eu accès aux valeurs des produits $\beta_2 \cdot K_s$.

3.4. Complexes dibutylphosphoriques du chrome(III) et du fer(III)

Les solubilités du chrome(III) et du fer(III) en fonction de la concentration
en ions DBP^- varient comme l'indique la figure 11.

TABLEAU V. CONSTANTES DE FORMATION DES COMPLEXES DIBUTYLPHOSPHORIQUES DU CHROME

$Cr(DBP)_2^+$	$\beta_2 \ll \beta_3$
$Cr(DBP)_3$	$\log \beta = 4{,}48$
$Cr(DBP)_3$	$K_S = 1{,}7 \cdot 10^{-8} = (Cr^{3+})(DBP^{-3})$
$Cr(DBP)_4^-$	$\log \beta = 5{,}94$

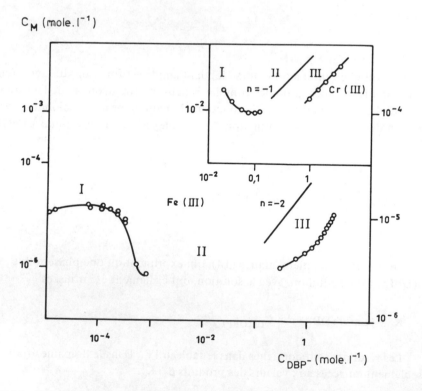

FIG.11. *Solubilité du fer et du chrome dans des solutions de dibutylphosphate de sodium.*

On peut y voir une différence importante pour les deux métaux. Pour le fer, la forme peu soluble $Fe(DBP)_3$ prédomine dans tout le domaine de concentration en dibutylphosphate, alors que pour le chrome on peut observer en plus du complexe neutre $Cr(DBP)_3$, assez soluble, une espèce anionique $Cr(DBP)_4^-$ obtenue dans les milieux les plus concentrés, et un complexe cationique dans les milieux dilués.

La valeur de la constante de formation du complexe $Cr(DBP)_3$ et sa prédominance ont pu être vérifiées par polarographie. En effet, le potentiel de demi-onde de la vague réversible $Cr(III) + e \rightarrow Cr(II)$ est déplacé vers les valeurs négatives à raison de 0,187 V pour une variation de concentration en ions DBP^- d'un facteur 10. Cela indique, comme le montrent les équations (15) et (16), que trois ions DBP^- sont échangés au cours de la réduction du chrome(III). Dans la zone de concentration considérée, les complexes des ions divalents n'étant pas stables [7], on peut simplifier l'équation (15) et utiliser l'équation (16) pour calculer la constante de formation de $Cr(DBP)_3$.

$$E_{\frac{1}{2}C} = E_{\frac{1}{2}S} - 0,06 \log \frac{\sum_{i=0}^{i=n} \beta_i \cdot (DBP)^i}{\sum_{j=0}^{j=n} \beta_j \cdot (DBP)^j} \tag{15}$$

$$E_{\frac{1}{2}C} = E_{\frac{1}{2}S} - 0,06 \log \sum_{i=0}^{i=n} \beta_i \cdot (DBP)^i \tag{16}$$

Les résultats concernant le chrome figurent dans le tableau V.

Le produit de solubilité de $Cr(DBP)_3$ a été déduit de la constante de formation du complexe et de la valeur de la solubilité dans la zone II de la figure 11. La valeur de la constante de formation de $Cr(DBP)_4^-$ a été déduite de la solubilité du chrome dans la zone III et du produit de solubilité de $Cr(DBP)_3$.

La forme de la courbe de solubilité du fer(III) dans la zone des complexes cationiques pourrait être due à l'existence d'espèces solubles polymérisées, dont la formation et la précipitation obéissent aux équations (17) et (18).

$$nFe(DBP)^{2+} + m(DBP^-) \rightarrow (Fe)_n (DBP)_{n+m}^{(2n-m)+} \tag{17}$$

$$(Fe)_n (DBP)_{n+m}^{(2n-m)+} + 2n - m(DBP^-) \rightarrow nFe(DBP)_3 \downarrow \tag{18}$$

L'existence de ces équilibres ne permet pas de calculer de constantes de
formation à partir des courbes de solubilité. Cependant, les pentes de la courbe
de solubilité dans la zone III suggèrent l'existence de complexes anioniques
contenant quatre ou cinq ions dibutylphosphate. La faible solubilité observée
doit être imputée probablement à la solidité des ponts phosphate entre atomes
de fer dus au partage d'ions DBP⁻; cette structure est courante dans les complexes
dibutylphosphoriques.

3.5. Conclusion

On a étudié les complexes formés dans des solutions aqueuses entre les
ions dibutylphosphate et les ions métalliques contenus dans les solutions de
retraitement des combustibles nucléaires. On peut mettre l'accent sur leur
insolubilité, cause de la formation de crasses d'interphase au cours des
extractions. Cependant, des zones de solubilité plus ou moins étendues sont
observées, qui sont dues à la formation de complexes cationiques aux faibles
concentrations en dibutylphosphate et de complexes anioniques aux fortes
concentrations. En solution aqueuse des espèces polynucléaires ne semblent
exister que pour le fer(III), alors qu'on les a souvent mises en cause dans des
solvants moins polaires.

REFERENCES

[1] MacWHINIE, W.R., MILLER, J.D., Adv. Inorg. Chem. Radiochem. 12 (1969) 135.
[2] SILLEN, L.G., MARTELL, A.E., in Stability Constants of Metal-Ion Complexes,
    Chemical Society, London, Special Publ. No.17 (1964) and No.25 (1971).
[3] SHEKA, Z.A., SINJAVASKAJA, E.I., Sov. Radiochem. 7 5 (1965) 595.
[4] SHEKA, Z.A., SINJAVSKAJA, E.I., Russ. J. Inorg. Chem. 9 9 (1964) 1212.
[5] SHEKA, Z.A., SINJAVSKAJA, E.I., Russ. J. Inorg. Chem. 9 8 (1964) 1065.
[6] KRUTIKOV, P.G., SOLOVSKIN, A.S., Russ. J. Inorg. Chem. 15 6 (1970) 825.
[7] RACINOUX, J., MUSIKAS, C., Résultats à publier.

DISCUSSION

A.S. STREZOV: In Table I of your paper you have given two formation
constants for complexes of lanthanide(III) ions with orthophenanthroline. In
most cases reported in the literature the lanthanides form three complexes. Do
you think there is a structural hindrance preventing appearance of the third complex,
or is there some other reason? Perhaps the region of ligand concentration is low?

C. MUSIKAS: In aqueous solutions the solubility of free orthophenanthroline at room temperature does not exceed 0.015 mol/ltr and this is probably the reason why the tris-orthophenanthrolinium lanthanide complexes are not observed — because of competition with water. Steric hindrance occurs only in the case of complexes containing more than four orthophenanthroline molecules, as has been shown by the successful preparation of tetraorthophenanthrolinium lanthanide(III) perchlorates; here it has been suggested that the lanthanides are completely surrounded by nitrogen giving the chromophore LnN_8.

J. FUGER: Could you suggest an explanation for the higher stability of the NpO_2^+-o-phenanthroline complex in comparison with Am^{3+}-o-phenanthroline?

C. MUSIKAS: The unusual order of stability of orthophenanthroline complexes is probably due to a higher covalent contribution in the case of nitrogen-actinide bonds as compared with the oxygen-actinide ion bonds or fluoride-actinide bonds. When the participation of covalency increases, ions of type MO_2^+ form more stable complexes than M^{3+} ions. This inversion was already observed in sulphocyanide complexes, which have probably more covalent character than complexes with donor oxygen ligands.

J. FUGER: To what extent, in your opinion, can the complexes with monobutylphosphoric acid also be responsible for the formation of precipitates during reprocessing?

C. MUSIKAS: The exact nature of precipitates obtained during reprocessing is variable and is a function of the irradiation rate. The diluent probably also plays a non-negligible part, through its products of radiolysis. However, dibutylphosphate is the most common degradation product of mixtures containing tributylphosphate, and knowledge of the interactions of metal ions present in reprocessing with this ligand should improve our understanding of the phenomena of formation of interphase slag.

INVESTIGATIONS ON CAESIUM URANATES
VII. THERMOCHEMICAL PROPERTIES OF $Cs_2U_4O_{12}$

E.H.P. CORDFUNKE
Netherlands Energy Research Foundation,
Petten,
The Netherlands

E.F. WESTRUM Jr.
Department of Chemistry,
University of Michigan,
Ann Arbor, Michigan,
United States of America

Abstract

INVESTIGATIONS ON CAESIUM URANATES: VII. THERMOCHEMICAL
PROPERTIES OF $Cs_2U_4O_{12}$.

The thermochemical properties of $Cs_2U_4O_{12}$ have been evaluated from new experimental data, including the low-temperature heat capacities, the enthalpy of formation at room temperature, and the high-temperature enthalpy increments by drop calorimetry. From the results a section of the Cs−U−O phase diagram at 1000 K has been constructed showing the stability of the compound as a function of caesium and oxygen pressure.

1. Introduction

The cesium−uranium−oxygen system has been the subject of a variety of investigations because of its importance in fast-reactor technology. The fission−product cesium has been observed to migrate axially, and high localized concentrations have been observed at the fuel−blanket interface. Moreover, adjacent to this interface, cladding deformation and breach have been found and attributed to the formation of a low−density Cs−U−O compound (1).

Notwithstanding extensive phase studies of the Cs−U−O system (2,3) no common opinion is to be found in literature concerning the composition of the cesium uranate that might be

TABLE 1. ANALYTICAL RESULTS FOR α-$Cs_2U_4O_{12}$

Sample	ω (Cs) found	calc.	ω [U(IV)] found	calc.	ω (total U) found	calc.
A	18.98 ± 0.14	18.85	16.82 ± 0.16	16.88	67.48 ± 0.07	67.53
B	18.73 ± 0.15	18.85		16.88	67.56 ± 0.05	67.53
C	19.08 ± 0.10	18.85	17.15 ± 0.03	16.88	67.73 ± 0.02	67.53

molar mass $Cs_2U_4O_{12}$ = 1409.92

ω = mass fraction given in wt %

formed in the fuel (4). The conflicting opinions center
around the stability of the possible compounds, mainly
Cs_2UO_4 and $Cs_2U_4O_{12}$, and the conditions required for their
formation. As a contribution to clarify the situation, we
here present the thermodynamic properties of $Cs_2U_4O_{12}$. The
results will serve, in combination with the previously
published thermodynamic properties of Cs_2UO_4 (5), as the
basis for the construction of the phase diagram for the
Cs-U-O system.

2. Experimental

Preparation of $Cs_2U_4O_{12}$

$Cs_2U_4O_{12}$ was prepared by heating of $Cs_2U_4O_{13}$ in an inert atmosphere (e.g. argon) at about 800 $^\circ$C. The latter compound was
prepared by ignition in air of the stoichiometric amounts of
UO_3 and Cs_2CO_3 at about 600 $^\circ$C, as described previously (2).
Three different samples of $Cs_2U_4O_{12}$ were prepared for the
calorimetric measurements and indicated here as the samples
A, B and C. The analytical data of these samples are collected
in Table 1. The cesium content was determined by the atomic
absorption technique, and the uranium was analysed titrimetric-
ally, as described previously (6).

Low-temperature heat-capacity measurements

Heat capacity data over the 5 to 350 K range were made in the
University Mark II adiabatic cryostat by the usual intermittent-heating technique. All determinations of mass,
potential, current, time and temperature were ultimately
referenced to calibrations made by the National Bureau of
Standards.

TABLE 2.   EXPERIMENTS IN LIQUID CESIUM

Starting material	heating temperature (^{o}C)	final product*	U^{4+}/U-total
$Cs_2U_4O_{12}$	650	X + UO_2	
Cs_2UO_4	550	X	0.405
Cs_2UO_4	650	X	
Cs_2UO_4	725	X	
$Cs_2U_2O_7$	650	X + UO_2	
Cs_2UO_4	800	X (+ little $Cs_2U_4O_{12}$)	0.52

*) X = new phase, see X-ray pattern in Table 3

High-temperature enthalpy measurements

The drop calorimetric determinations were made in the isothermal diphenyl ether calorimeter that has been described
before (7).

Enthalpy of solution measurements

For the enthalpy-of-solution measurements the calorimeter
has been used which was described in detail previously (8).
The $Cs_2U_4O_{12}$ samples were dissolved in H_2SO_4 + $Ce(SO_4)_2$
solution which was prepared from reagent grade acid and
analysed by titration with standard NaOH solution. Cerium(IV)
sulphate (J.T. Baker, A.R.) was dissolved in 1.505 $mol \cdot dm^{-3}$
H_2SO_4 solution, and analysed by titration with standard
iron(II) ammonium sulphate.

TABLE 3. X-RAY PATTERN OF Cs_2UO_{4-x}

$Q^*_{obs.}$	$I_{rel.}$	$Q^*_{obs.}$	$I_{rel.}$
155.02	40	1087.09	5
160.69	10	1101.60	5
175.61	< 5	1116.93	5
271.01	30	1124.27	5
317.24	5	1222.42	80
462.21	10	1261.66	100
466.07	10	1378.73	10
591.62	10	1424.26	15
656.08	20	1462.99	5
760.59	5	1698.93	5
783.37	5	1726.51	5
841.38	5	1755.19	5
878.96	< 5	1835.13	10
896.72	< 5	1901.69	5
926.02	100	1988.24	20
945.58	100	2017.79	10
983.88	30	2039.83	5
1077.70	10	2060.04	5

* $Q = 10^4/d^2$

Experiments in liquid cesium

Cesium uranate samples of different Cs/U ratios were heated
in excess of liquid cesium in nickel capsules. These were
filled in the glove box and sealed with a nickel flanged
closure; temperatures up to 800 $^\circ$C were applied.
After heating for some hours, the capsules were opened in the
box and X-ray films (using the Guinier technique) were made.

TABLE 4.   THERMODYNAMIC FUNCTIONS OF $Cs_2U_4O_{12}$

$(cal_{th} = 4.184$ J)

$\frac{T}{(K)}$	$\frac{C_p}{(cal_{th} \cdot K^{-1} \cdot mol^{-1})}$	$\frac{\{S^o(T) - S^o(0)\}}{(cal_{th} \cdot K^{-1} \cdot mol^{-1})}$	$\frac{\{H^o(T) - H^o(0)\}}{(cal_{th} \cdot K^{-1} \cdot mol^{-1})}$	$\frac{-\{G^o(T) - H^o(0)\}/T}{(cal_{th} \cdot K^{-1} \cdot mol^{-1})}$
5	0.35	0.12	0.45	0.029
25	13.96	8.02	135.89	2.582
50	27.31	22.04	660.7	8.825
100	48.84	48.40	2618.8	22.21
150	65.41	71.49	5495	34.86
200	77.04	91.99	9071	46.64
250	85.74	110.17	13153	57.56
300	91.97	126.39	17605	67.70
350	96.74	140.94	22328	77.15
298.15	91.78	125.82	17435	67.34

In some cases the U^{4+}/U-total ratio of the resulting brown-
coloured uranate was determined. A summary of the results
obtained is given in Table 2 (and the data on the X-ray pattern
of Cs_2UO_{4-x} in Table 3).

3. Results

Low-temperature data

The low-temperature heat capacity data for sample A is based
on a gram formula mass of 1409.92 on the basis of the 1961
international atomic weights (9), and presented in chrono-
logical sequence to permit deduction of the approximate
temperature increments employed in the measurements from
the differences in the adjacent mean temperature.
The experimental heat capacities were curve-fitted to polynomials in reduced temperature by the method of least-squares,

and then integrated by digital computer to yield values of
the thermal functions at regular temperature intervals presented in Table 4. Values of the entropy and enthalpy increments below 5 K were obtained from a plot of Cp/T versus
T^2. The magnitude of these extrapolations are only minute
fractions of the totals at 298.15 K. It was assumed that the
zero-point entropy was zero and that no magnetic contribution
was appropriate.

The uncertainty in the thermodynamic functions in Table 4
is considered to be less than 0.1 % from 100 to 350 K. Hence,
the standard entropy at 298.15 K would have a standard deviation of approximately ± 0.01.

High-temperature data

The high-temperature enthalpy data for the samples A and B
agree well (Fig. 1), and we have calculated a combined
fit of the data to obtain the polynomial for the α-form of
$Cs_2U_4O_{12}$ (298 – 898 K):

$$\{H^o(T) - H^o(298.15\ K)\}\ /\ cal \cdot mol^{-1} = 101.273(T/K) +$$
$$+ 8.5971.10^{-3}(T/K)^2 + 12.996.10^5(T/K)^{-1} - 35318$$

The standard deviation is 0.23 %.

In the temperature region (898 – 968 K), in which the β-form
is stable, we obtain:

$$\{H^o(T) - H^o(298.15\ K)\}\ /\ cal \cdot mol^{-1} = 75.9129\ (T/K) +$$
$$+ 26.845.10^{-3}(T/K)^2 - 25020$$

Here, the standard deviation is 0.10 %.

At the transition temperature, 898 K, the enthalpy difference of the two forms, gives the enthalpy of transition:

$$H^o_\beta - H^o_\alpha = 793\ cal \cdot mol^{-1}.$$

Here also the data are based on the 1961 international
atomic weights.

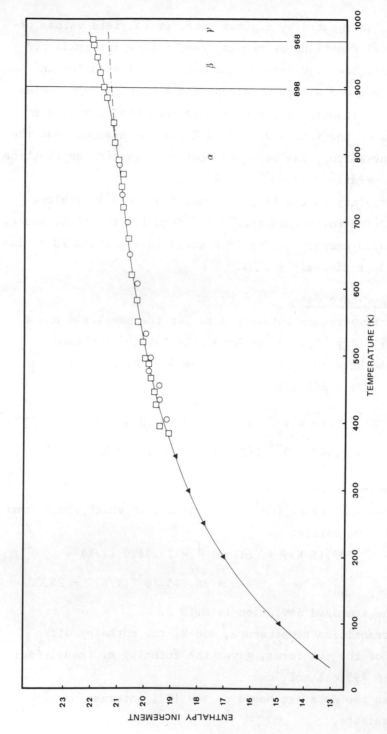

FIG. 1. The enthalpy increment of $Cs_2U_4O_{12}$. ▲ Low-temperature data (smoothed values); □, ○ high-temperature data of samples A and B, respectively.

TABLE 5.   ENTHALPY OF SOLUTION OF $Cs_2U_4O_{12}$ AND Cs_2SO_4 IN
200 cm³ 1.505M H_2SO_4 + 0.0350M $Ce(SO_4)_2$ AT 298.15 K

m(solute)	ε*	$\Delta\Theta$*	ΔH^o(soln)
(g)			(kcal·mol⁻¹)
a. $Cs_2U_4O_{12}$ (sample A)			
0.6277	22.3237	2.2805	- 114.350
0.5966	22.1478	2.1905	- 114.653
0.6269	22.0805	2.3112	- 114.773
0.6321	21.9695	2.3492	- 115.119
		average:	- 114.724 ± 0.55
b. $Cs_2U_4O_{12}$ (sample C)			
0.6329	22.2124	2.3223	- 114.914
0.6454	22.0678	2.3732	- 114.409
0.6397	22.3086	2.3343	- 114.775
		average:	- 114.700 ± 0.43
c. Cs_2SO_4			
0.1720	21.5957	- 0.1368	+ 6.216
0.1641	21.6680	- 0.1309	+ 6.254
0.1650	21.9620	- 0.1285	+ 6.189
		average:	+ 6.220 ± 0.053

* ε = energy equivalent of calorimeter, and $\Delta\Theta$ = temperature change
    of calorimeter (in same arbitrary units).

Enthalpy of formation at 298.15 K

The enthalpy of solution of α-$Cs_2U_4O_{12}$ in 1.505 M H_2SO_4 +
+ 0.0350 M $Ce(SO_4)_2$ solution, as given in Table 5, is for
sample A -(114.72 ± 0.55) kcal·mol⁻¹. For sample C, which
has a slightly different composition, it is found for the
enthalpy of solution -(114.70 ± 0.43) kcal·mol⁻¹.
When the enthalpy of solution of α-$Cs_2U_4O_{12}$ is combined
with auxiliary thermodynamic data for the enthalpies of

TABLE 6.   REACTION SCHEME FOR THE ENTHALPY OF FORMATION OF α-$Cs_2U_4O_{12}$

(soln) refers to 1.505 M H_2SO_4 + 0.0350 M $Ce(SO_4)_2$

$\Delta H_{11} = - \Delta H_1 + \Delta H_2 + \Delta H_3 + \Delta H_4 + \Delta H_5 + \Delta H_6 + \Delta H_7 - \Delta H_8 + \Delta H_9 + \Delta H_{10}$

		$(\text{kcal}\cdot\text{mol}^{-1})$
(1) α-$Cs_2U_4O_{12}$ + $\lvert 4\ H_2SO_4 + 2\ Ce(SO_4)_2 \rvert$ (soln)	$\rightarrow$ $\lvert Cs_2SO_4 + 4\ UO_2SO_4 + 4\ H_2O + Ce_2(SO_4)_3 \rvert$ (soln)	$-\ 114.70 \pm 0.43$
(2) UO_2(s) + 2 $Ce(SO_4)_2$(soln)	$\rightarrow$ $\lvert UO_2SO_4 + Ce_2(SO_4)_3 \rvert$ (soln)	$-\ 53.92 \pm 0.06$
(3) 3 γ-UO_3(s) + 3 H_2SO_4(soln)	$\rightarrow$ $\lvert 3\ UO_2SO_4 + 3\ H_2O \rvert$ (soln)	$-\ 60.69 \pm 0.27$
(4) 3 U(s) + 4½ O_2(g)	$\rightarrow$ 3 γ-UO_3(s)	$-\ 877.5\ \pm 0.60$
(5) U(s) + O_2(g)	$\rightarrow$ UO_2(s)	$-\ 259.3\ \pm 0.2$
(6) Cs_2SO_4(s)	$\rightarrow$ Cs_2SO_4(soln)	$+\ 6.22 \pm 0.05$
(7) 2 Cs(s) + S(s) + 2 O_2(g)	$\rightarrow$ Cs_2SO_4(s)	$-\ 344.89 \pm 0.13$
(8) H_2(g) + S(g) + 2 O_2(g)	$\rightarrow$ H_2SO_4(soln)	$-\ 211.50 \pm 0.010$
(9) H_2(g) + ½ O_2(g)	$\rightarrow$ H_2O(ℓ)	$-\ 68.315\pm 0.01$
(10) H_2O(ℓ)	$\rightarrow$ H_2O(soln)	$-\ 0.012$
(11) 2 Cs(s) + 4 U(s) + 6 O_2(g)	$\rightarrow$ α-$Cs_2U_4O_{12}$(s)	$-\ 1332.21 \pm 0.83$
with $\Delta H_1 = -\ 114.72 \pm 0.55$:		$-\ 1332.19 \pm 0.89$
	average	$-\ 1332.20 \pm 0.85$

TABLE 7.   MOLAR THERMODYNAMIC PROPERTIES OF $Cs_2U_4O_{12}$

$\dfrac{T}{(K)}$	$\dfrac{H^{o}(T) - H^{o}(298.15\ K)}{(kcal \cdot mol^{-1})}$	$\dfrac{\Delta H_f^{o}(T)}{(kcal \cdot mol^{-1})}$	$\dfrac{S^{o}(T)}{(cal \cdot K^{-1} \cdot mol^{-1})}$	$\dfrac{\Delta G_f^{o}(T)}{(kcal \cdot mol^{-1})}$
(a)	$\alpha-Cs_2U_4O_{12}$			
298.15	0.00	− 1332.2	125.82	− 1255.6
300	0.17	− 1332.2	126.39	− 1255.1
400	9.82	− 1332.1	154.08	− 1229.3
500	20.07	− 1330.6	176.94	− 1203.7
600	30.71	− 1329.2	196.33	− 1178.5
700	41.64	− 1327.9	213.18	− 1153.5
800	52.83	− 1326.8	228.11	− 1128.6
898	64.01	− 1326.0	241.29	− 1104.4
(b)	$\beta-Cs_2U_4O_{12}$			
898	64.80	− 1325.2	242.17	− 1104.4
900	65.05	− 1325.1	242.45	− 1103.9
968	73.62	− 1358.9	251.63	− 1086.6

formation of $\gamma-UO_3$ (8), UO_2 (10), anhydrous Cs_2SO_4 (11), $H_2O(\ell)$ (12), H_2SO_4 (12), and with the enthalpy of solution of UO_2 (13) and anhydrous Cs_2SO_4 (Table 5) in the same solvent, the enthalpy of formation of $\alpha-Cs_2U_4O_{12}$ can be calculated, according to the reaction scheme given in Table 6.

For sample A it is found that ΔH_f^{o} (298.15 K) = − (1332.19 ± ± 0.89) kcal·mol^{-1}, and for sample C it is found ΔH_f^{o}(298.15 K)= = − (1332.21 ± 0.83) kcal·mol^{-1}. The agreement is excellent, and we therefore take the average to obtain for the enthalpy of formation of $\alpha-Cs_2U_4O_{12}$ the value:
ΔH_f^{o}(298.15 K) = − (1332.20 ± 0.85) kcal·mol^{-1}.

Thermodynamic functions of $Cs_2U_4O_{12}$

A combination of all data of the present research permits
the evaluation of the chemical thermodynamics of formation
for $\alpha-Cs_2U_4O_{12}$ and $\beta-Cs_2U_4O_{12}$. The results are given in
Table 7.

Experiments in liquid cesium

It has been suggested (14) that in the presence of liquid
cesium a cesium uranate is formed in which uranium has a
valency lower than 6, and probably being different from
$Cs_2U_4O_{12}$. However, this phase has not been characterized
by X-ray diffraction.

Indeed, our experiments clearly indicate that under these
conditions a uranate phase is formed of which the X-ray
pattern (Table 3) does not belong to any of the known
cesium uranates (15). Analysis of the results in Table 2
shows that only for Cs_2UO_4 precipitation of UO_2 did not occur
during equilibration in liquid cesium. From this fact, in
combination with the chemical analysis (Table 3), we conclude
that a non-stoichiometric phase, Cs_2UO_{4-x} is formed, in which
$x = \sim 0.5$. This brown-coloured phase is extremely oxygen-
sensitive; it readily oxidises in the glove box to yellow
Cs_2UO_4.

4. Discussion

The thermodynamic data of $Cs_2U_4O_{12}$, now available, permit one
- in combination with the data of the other cesium uranates -
to predict the in-reactor behaviour. This means that the
composition of the cesium uranate formed at the fuel-
blanket interface in irradiated uranium-plutonium oxide
fuel pins can be derived from a knowledge of the oxygen
potential and temperature.

To this purpose we have calculated sections of the Cs-U-O
system at constant temperatures. In these sections three-
phase equilibria, like:

$$Cs_2U_4O_{12} \rightleftarrows 4 \ UO_2 + 2 \ Cs(g) + 2 \ O_2(g)$$

are represented by lines which limit the two-phase regions
in which the solid uranate exists in equilibrium with cesium
vapour (or liquid).

In calculating the ΔG^o-values for the three-phase equilibria
we have used the high-temperature thermodynamic properties
of UO_2 (16), U_4O_9 (17), U_3O_8 (18), and Cs_2UO_4 (5). For the
other cesium uranates the data estimated by Fee and Johnson (4)
have been taken. We have assumed that phase transitions and
deviations of stoichiometry in the oxides play only a minor
role in the calculations. From the values of ΔG^o (= $-RT \ \ell n \ K_p$)
thus obtained, the sections have been constructed by plotting
$\log p_{Cs}$ against the oxygen potential. The section for 1000 K is
shown in Fig. 2. As is shown in Fig. 2 only Cs_2UO_4 and $Cs_2U_4O_{12}$
can exist in equilibrium with UO_{2+x} and gaseous cesium. It is
evident, however, that at the low oxygen potentials existing at
the fuel-blanket interface ($\Delta \bar{G}^o_{O_2} < -90 \ kcal \cdot mol^{-1}$), only Cs_2UO_4
can be formed. The latter conclusion agrees with that of Fee
and Johnson (4), and disagrees with our previous assumption that
only $Cs_2U_4O_{12}$ is formed in the fuel (2). This assumption was
based on phase studies which, apparently, were performed at the
somewhat higher oxygen pressures at which the $Cs_2U_4O_{12}$ - UO_{2+x}
region is stable.

A second feature in Fig. 2 involves the cesium uranate which
exists in equilibrium with liquid cesium and urania at very low
oxygen potentials. Our phase studies indicate that a non-stoichiometric phase, Cs_2UO_{4-x}, exists under these conditions. This
conclusion differs from Fee and Johnson's who stated only
Cs_2UO_4 to be in equilibrium with liquid cesium. Although their

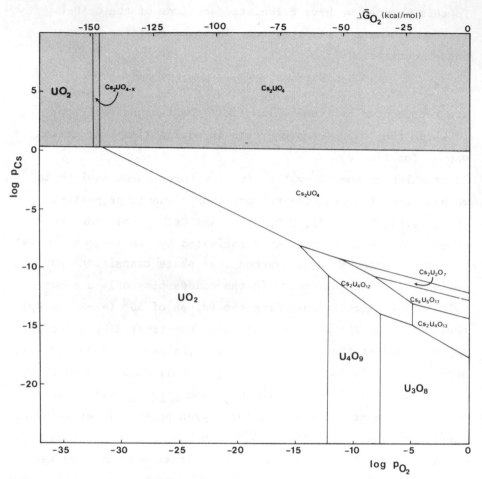

FIG.2. A section of the Cs−U−O phase diagram at 1000 K. In the shaded area the uranium compounds are in equilibrium with liquid cesium. (Pressures, p, are in atmospheres.)

conclusion was indeed based on experiments in liquid cesium, ours differ in that the X-ray patterns were taken with the uranate in contact with the cesium instead of after having distilled off the cesium. In our experience oxidation of the uranate to Cs_2UO_4 takes place during the vacuum distillation.

For the same reason we were not able to analyse the Cs/U ratio in the compound, and we have assumed the composition to be Cs_2UO_{4-x}.

This assumption is based on the experiments in liquid cesium (Table 3), and on the oxidation behaviour of the compound after removing the excess of cesium as a liquid from the sample. If the composition is Cs_3UO_4, as has been postulated by Lorenzelli et al. during this symposium [19], then the formation of Cs_2O should have been observed on the X-ray patterns. We have not found this, but the X-ray patterns are still of a poor quality.

Since the Cs_2UO_{4-x}-phase has not been isolated in pure form we have not yet been able to determine its thermodynamic properties and we have, therefore, taken the EMF-values by Adamson et al. [20], extrapolated to 1000 K, for the construction of the $Cs_2UO_4-Cs_2UO_{4-x}$ boundary in Fig. 2.

Acknowledgements

The authors are much indebted to Dr. G. Prins for fruitful discussions, and to Mrs. F. Kleverlaan, R.P. Muis, W. Ouweltjes and P. van Vlaanderen for technical assistance.

References

[1] FEE, D.C., et al., ANL-report no. 76-126 (1977).

[2] CORDFUNKE, E.H.P., EGMOND, A.B. VAN, and VOORST, G. VAN, J. Inorg. Nucl. Chem. 37 (1975) 1433.

[3] CORDFUNKE, E.H.P., Thermodynamics of Nuclear Materials (Proc. IAEA Symposium, Vienna, 1974) Vol. II, Vienna (1975) 185.

[4] FEE, D.C., and JOHNSON, C.E., J. Inorg. Nucl. Chem. 40 (1978) 1375.

[5] CORDFUNKE, E.H.P., and O'HARE, P.A.G., The Chemical Thermodynamics of Actinide Elements and Compounds: Part III, Miscellaneous Compounds, IAEA, Vienna (1978).

[6] SLANINA, J., et al., Mikrochim. Acta 1 (1978) 519.

|7| CORDFUNKE, E.H.P., MUIS, R.P., and PRINS, G., J. Chem. Thermodynamics, in the press.

|8| CORDFUNKE, E.H.P., OUWELTJES, W., and PRINS, G., J. Chem. Thermodynamics 7 (1975) 1137.

|9| CAMERON, A.E., and WICKERS, E., J. Am. Chem. Soc. 84 (1962) 4175.

|10| HUBER, E.J., Jr. and HOLLEY, C.E., J. Chem. Thermodynamics 1 (1969) 267.

|11| PARKER, V.B., (N.B.S., Washington), private communication.

|12| WAGMAN, D.D., et al., Nat. Bur. Stand. U.S., Techn. Note 270-3 (1968).

|13| CORDFUNKE, E.H.P., OUWELTJES, W., and PRINS, G., J. Chem. Thermodynamics 8 (1976) 241.

|14| AITKEN, E.A., et al., Thermodynamics of Nuclear Materials (Proc. IAEA Symposium, Vienna, 1974) Vol. I, IAEA, Vienna (1975) 187.

|15| EGMOND, A.B. VAN, "Investigations on cesium uranates and related compounds",Thesis, Amsterdam (1976).

|16| RAND, M.H., and MARKIN, T.L., Thermodynamics of Nuclear Materials (Proc. IAEA Symposium, Vienna, 1967), IAEA, Vienna (1968) 637.

|17| MACLEOD, A.C., J. Chem. Thermodynamics 4 (1972) 699.

|18| ACKERMANN, R.J., and CHANG, A.T., J. Chem. Thermodynamics 5 (1973) 873.

|19| LORENZELLI, R., DUDAL, R. LE, and ATABECK, R., this proceedings, p.   (= paper IAEA-SM-236/87)

|20| ADAMSON, M.G., AITKEN, E.A., and JETER, D.W., (Proc. Int. Conf. Liquid Metal Technology in Energy Production, Champion, Pennsylvania, 1976). CONF-760503-P2,866. USERDA, Washington DC (1976).

DISCUSSION

L.R. MORSS: Table 1 of your paper shows that sample C had a composition high in Cs and U(IV), so that its composition more closely approximates $Cs_2U_4O_{11.5}$ than $Cs_2U_4O_{12}$. Yet the heat of solution of sample C is almost identical with that of sample A (Table 5). Should a correction have been applied to the heat of solution to yield the $\Delta H_f^0(Cs_2U_4O_{12})$, and if so, would that not have produced a discrepancy between the enthalpy of formation based upon samples A and C?

E.H.P. CORDFUNKE: The values for the enthalpy of solution of $Cs_2U_4O_{12}$ given in the paper have been corrected for the U^{6+}/U^{4+} ratio.

M.G. ADAMSON: In view of the fact that Na—U—O compounds with $V_U < 6$ (i.e. Na_3UO_4 and $NaUO_3$) can exist in phase fields in which the sodium activity is less than unity, do you not think it possible that the 'lower-valent' compound Cs_2UO_{4-x} also exists at lower caesium activities ($a_{Cs} < 1$), possibly in co-existence with phases such as $Cs_2U_4O_{12}$ or Cs_2UO_4?

E.H.P. CORDFUNKE: At caesium activities lower than unity the Cs_2UO_{4-x} phase does in all probability exist, but not in co-existence with $Cs_2U_4O_{12}$ because of the low oxygen potentials at which Cs_2UO_{4-x} is formed.

ALKALINE EARTH URANATES
An exploration of their thermophysics and stabilities*

E.F. WESTRUM Jr., H.A. ZAINEL**
Department of Chemistry,
University of Michigan,
Ann Arbor, Michigan,
United States of America

D. JAKEŠ
Nuclear Research Centre,
Řež, nr. Prague,
Czechoslovakia

Abstract

ALKALINE EARTH URANATES: AN EXPLORATION OF THEIR THERMOPHYSICS
AND STABILITIES.

An investigation of the thermodynamics of the alkaline-earth monouranates and of
magnesium triuranate has been undertaken. At the present time the low-temperature heat
capacities of the alkaline earth monouranates $MgUO_4$, $CaUO_4$, α-$SrUO_4$, and $BaUO_4$ as well
as MgU_4O_{12} have been examined by adiabatic calorimetry from 5 to 350 K, and supplemented
by drop calorimetry to 900 K for $MgUO_4$. The work was undertaken primarily to ascertain
the thermodynamic stability of these uranium compounds, which may be important from the
point of view of reactor safety, and to provide a better understanding of the thermodynamics
of ternary oxides. Relatively uncomplicated heat capacity curves were obtained. The entropies
$(\Delta S^0/R)$ at 298.15 K of typical members are: $MgUO_4$ = 15.87, $CaUO_4$ = 14.89, α-$SrUO_4$= 18.42,
and $BaUO_4$ = 21.39.

1. INTRODUCTION

Information on alkali and on alkaline-earth uranates is important because of
the considerable likelihood that one or more of these compounds may form in
nuclear fuel elements upon reaction of the uranium oxide with fission-product
cations. Conceivably some of these uranate products may play a significant role
in the swelling of the fuel element and even cause element failure. Under reactor

---

* The work at Ann Arbor has been supported in part by the Chemical Thermodynamics
Program of the Chemistry Section of the National Science Foundation under contract GP-42525X.

** Present address: Department of Chemistry, College of Science, University of Baghdad,
Baghdad, Iraq.

emergency conditions it is even more important to have a secure knowledge of the thermodynamic stability of these relatively stable compounds.

Although by virtue of the high yield of caesium as a fission product and because of the importance of its compounds in fast-reactor technology, the study of caesium uranates is of much importance [1], the study of other alkali and alkaline-earth cation—uranium—oxygen compounds should be encouraged, not only from the point of view of safety conditions, but also for the information that they provide about the relative stabilities of these ternary oxide systems, about the general thermal and phase behaviour of such uranates and about related materials (e.g. chromates, etc.).

It was for these reasons and in the further hope of understanding the relationship between structural and thermophysical properties that the present study was undertaken. Although scientists from three widely separated laboratories are involved, the work is a truly common endeavour; the present study is a progress report covering the low-temperature studies (5 to 350 K) on $MgUO_4$, $CaUO_4$, α-$SrUO_4$ and $BaUO_4$.

The summary of earlier thermodynamic studies done to date on uranates has been tersely compiled elsewhere by Cordfunke and O'Hare [2].

2.   EXPERIMENTAL

2.1. Provenance of samples

All the samples were prepared at the Nuclear Research Center at Řež and were characterized there under the direction of one of the authors, Dusan Jakes. The preparative methods have been described in the literature [3 − 7].

As an example, $MgUO_4$ was prepared from U_3O_8 (obtained by ignition in air of ammonium polyuranate at 973 K) and MgO, mixed in stoichiometric ratio and heated for 50 h at 1110 K. All chemicals used were analytical reagent grade. The triuranate was made by reacting stoichiometric proportions of $MgCl_2$ with U_3O_8 for 6 h at 1110 K. For both preparations excess $MgCl_2$ was leached out with absolute ethanol, and magnesium was determined as the oxide after ignition of $Mg(NH_4)_2CO_3$ and uranium as U_3O_8. Both were stoichiometric within the mean deviation of the analytical methods (± 0.4%). Infra-red spectra were also taken and tabulated. Both samples were yellowish-brown powders and both were homogeneous under the microscope. Both samples are exceedingly hygroscopic and were handled only in an inert-atmosphere glove-box, in sealed glass bulbs or in vacuum lines. Tetravalent uranium was not detected. X-ray powder diffraction patterns were in good accord with published data.

TABLE I.   ANALYTICAL DATA ON MONOURANATE SAMPLES

Compound	Infra-red spectra	X-ray diffraction	Value[a] of x in MUO_{4-x}
$MgUO_4$	+	+	0.0059
$CaUO_4$	+	+	0.007
$SrUO_4$	+	+	0.030
$BaUO_4$	+	+	0.0022

[a] For samples produced at pressure of 0.021 MPa (0.2 atm) compared with preparation at 4.9 MPa (50 atm) oxygen.

Although we are confident of the stoichiometry and the reliability of the values as claimed, independently prepared samples will be studied soon in order to confirm some rather surprising results. Some of the most relevant analytical data are tersely summarized in Table I. Further details may be noted elsewhere [3 – 7].

2.2. Low-temperature heat capacity measurements

Heat capacity studies were made over a temperature range of 5 to 350 K in the University of Michigan Mark II adiabatic calorimetric cryostat by an intermittent-heating technique that permitted the achievement of thermal equilibrium after each determination. The details have been described elsewhere [8] but we note that all determinations of mass, potential, current, time and temperature were ultimately referred to calibrations made by the National Bureau of Standards, USA.

2.3. Higher temperature enthalpy increment measurements

Higher temperature enthalpy increment measurements have only been made on $MgUO_4$ thus far, but such determinations will be extended to the other compounds as soon as possible. The $MgUO_4$ measurements were all made by the method of mixtures (drop calorimetry) at Řež.

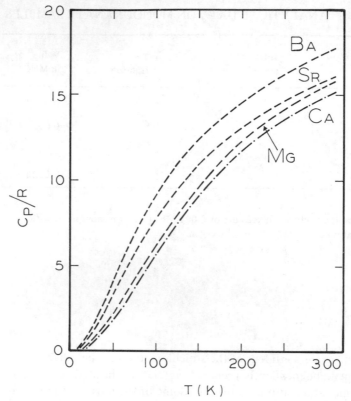

FIG.1. *The low-temperature heat capacities of alkaline-earth monouranates,* MUO_4.

3. RESULTS

3.1. Low-temperature data

The smoothed values of heat capacities given in Table III and shown graphically in Fig.1 are based on a gram formula mass as listed in Table II on the basis of the 1961 international atomic weights. The observed heat capacities were curve-fitted to polynomials in reduced temperature by the method of least-squares, and then integrated by digital computer to provide values of the thermophysical functions at selected temperature intervals. Values of the entropy and enthalpy increments below 5 K were obtained from plots of C_p/RT versus T^2, as shown in Fig.2. These extrapolations are only small fractions of the totals at 298.15 K. It was assumed that the zero-point entropy was zero and that no magnetic contribution occurred between 0 K and 5 K. The imprecisions in the thermodynamic functions in Table III are considered to be less than 0.1% from 100 to 350 K.

TABLE II. DATA ON THE MONOURANATES

Compound	Molar mass $(g \cdot mol^{-1})$	Sample mass (g)	Crystal form (space group)	Z	Lattice parameters (Å)	Refs
$MgUO_4$	326.34	23.02	orthorhombic (Pbcm)	4	a = 6.520 b = 6.595 c = 6.924	[4] (also [17])
$CaUO_4$	342.11	104.00	rhombohedral ($R\bar{3}m$)	1	a = 6.2683±0.0006 α = 36.040±0.002	[22]
α-$SrUO_4$	389.65	94.97	orthorhombic (Pbcm)	4	a = 5.4896±0.0007 b = 7.9779±0.0009 c = 8.1297±0.0012	[22]
$BaUO_4$	439.39	171.36	orthorhombic (Pbcm)	4	a = 5.7553±0.0010 b = 8.1411±0.0021 c = 8.2335±0.0011	[22]

3.2. High-temperature data

The high-temperature enthalpy data for similarly prepared samples are in good accord with the low-temperature data. They will not be discussed further in this presentation.

4. DISCUSSION

No anomalies have been noted in the heat capacities of these compounds — and indeed the Debye-theta (Θ_D) plot of Fig.3 provides additional confirmation of an absence of other contributions. In Figs 1 and 3 and Table III, it is evident that the heat capacity of $CaUO_4$ is less over the entire region studied than is that of $MgUO_4$, which has a less massive cation. How can this sequence — which applied also to the entropies — be reconciled?

TABLE III. THERMOPHYSICAL DATA ON ALKALINE EARTH MONOURANATES

(in energy dimensionless units)

T (K)	Mg	Ca	Sr	Ba
C_p/R				
25	0.61	0.37	0.88	1.27
50	2.26	1.84	3.34	4.19
100	6.27	5.70	7.77	9.16
200	12.30	11.72	13.08	14.50
300	15.44	14.90	15.75	17.35
S^0/R				
25	0.23	0.14	0.33	0.52
50	1.13	0.80	1.67	2.28
100	3.92	3.25	5.42	6.82
200	10.31	9.22	12.66	15.05
300	15.96	14.64	18.52	21.52
$(H^0-H_0^0)/R$ (K)				
25	4	3	6	9
50	39	28	58	77
100	252	313	342	419
200	1209	1210	1416	1637
300	2613	2701	2871	3240
$-(G^0-H_0^0)/RT$				
25	0.06	0.04	0.09	0.14
50	0.35	0.23	0.51	0.74
100	0.70	3.12	2.01	2.63
200	4.27	6.01	5.58	6.87
300	7.25	16.02	8.95	10.71

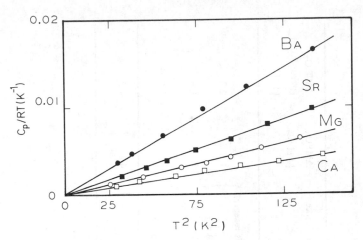

FIG.2.   Fit of the very low temperature data for extrapolation to the rectified relation
$C = aT + bT^3$ for the monouranates, MUO_4.

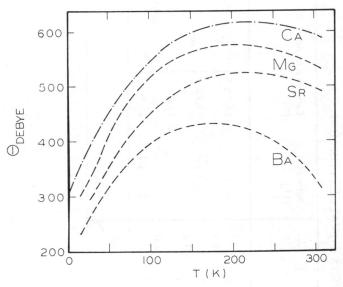

FIG.3.   Arbitrarily devised Debye-theta trends for the alkaline-earth monouranates, MUO_4.

Correlators have frequently had occasion to estimate entropies of members
of the monouranate series studies. Some data are presented in Table IV. These
include values from several tabulations, application of the Latimer scheme [9, 10],
and two other ternary oxide systems [11, 12]. The Latimer scheme for the
estimation of entropies for homologous series of compounds is the one most
widely used for this purpose, although there are relatively few entropy values on

TABLE IV. $\Delta S^0/R$ (298.15 K) VALUES FOR MUO_4

Compound	This research	Δ	Barin et al. [11]	Latimer scheme [9]	Δ	Cordfunke, O'Hare [2]	ΔMSO_4 [11]	$\Delta MTiO_3$ (perovskite) [13]
$MgUO_4$	15.87 ± 0.02	−1.3	--	16.5 ± 1	0.8	This research		
$CaUO_4$	14.56 ± 0.02	3.9	17.2 ± 1.4	17.3 ± 1	1.3	17.3 ± 1	1.8	2.3
$\alpha\text{-}SrUO_4$	18.42 ± 0.03	3.0	--	18.6 ± 1	0.9	--	1.3	1.8
$BaUO_4$	21.39 ± 0.04		20.3 ± 1.4	19.5 ± 1		20.9 ± 1	1.7	−0.1
$\Sigma\Delta =$		5.6			3.0		4.8	4.0

TABLE V. MOLAR VOLUMES (298.15 K) OF MUO_4 FROM
CRYSTALLOGRAPHIC DATA

Compound	Atomic mass of M	$V(cm^3 \cdot mol^{-1})$	$\Delta V(cm^3 \cdot mol^{-1})$
$MgUO_4$	24.3	44.65	
			0.9
$CaUO_4$	40.1	45.55	
			7.9
α-$SrUO_4$	87.6	53.37	
			4.2
$BaUO_4$	137.3	57.57	

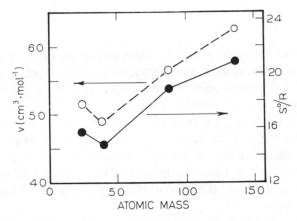

FIG.4. The trends of the molar volumes and the S^0/R values for the alkaline-earth
monouranates, MUO_4.

ternary oxides which may be taken into account for comparison. The alkaline-
earth sulphates seem to be the best example thus far and indicated a trend [10]
rather different from that of the Latimer curve itself. It is immediately evident
that the increments between adjacent members are not uniform and that the
$\Sigma \Delta$ values vary by nearly a factor of two.

Caution should, therefore, be observed in applying the Latimer scheme
uncritically. For example, studies by Grønvold and Westrum [13] on the application of the Latimer scheme to binary chalcogenides in the transition element

TABLE VI. C_p^0/R AT 298.15 K FOR MUO$_4$

Compound	This work	From high-temperature enthalpy increments	
MgUO$_4$	16.05	14.9 [4]	14.7 [19]
CaUO$_4$	14.89	14.9 [20]	15.5 [21]
α-SrUO$_4$	15.71		
BaUO$_4$	17.30	16.0 [20]	

region yield results indicating a virtually constant contribution for the 3-d, 4-d and 5-d elements [14], and more recent studies [15, 16] confirm a trend over several series of lanthanide compounds perpendicular to the trend of the Latimer curve. In fact, a far greater dependence on volume than on mass seems to obtain [16]. The molar volumes of the four compounds have been presented in Table V and graphically in Fig.4. Both slopes and increments readily attest to the unusually small volume of the calcium monouranate. If account is taken of the importance of the molar volumes for the derived thermodynamic functions [16] for isostructural compounds, the trend of the S^0/R plot at 298.15 K (see Fig.4) seems less surprising.

However, the fact that the calcium monouranate has a structure significantly different to that of the adjacent uranates interposes a parametric change beyond that contemplated by Latimer and may provide additional argument for the unexpectedly low value for the S^0/R of the calcium compound as compared with that for the related compounds. Zachariasen [17] has indicated that in calcium monouranate two of the oxygens are close to uranium atoms (1.91 Å) and six are more remote (2.33 Å), in contrast to the work of Samson and Sillén [18] which shows the oxygens to be surrounded by four oxygens in a plane at 2.20 Å.

Although the present authors are confident of the validity of their interpretation, they have considered how these conclusions might be further tested. New samples of the two lowest members have been prepared independently and will be measured. Moreover, samples of mixed magnesium calcium monouranates, differing somewhat in composition, will permit control of the crystal structure of the product and may yield some interesting information.

Finally, if β-SrUO$_4$ can be prepared in high purity, its rhombohedral structure – comparable to that of CaUO$_4$ — will provide a means of direct comparison with data on the α-SrUO$_4$ values of the present endeavour.

At present, data on the phase diagrams and thermochemistry of these compounds are relatively rudimentary. Values which do exist may be combined with the present data to produce some striking trends. It is, of course, interesting to ask to what extent the literature is able to provide confirmation of the heat-capacity values of this work. These have been summarized in Table VI. Since all values are derived from high-temperature enthalpy increment determinations by the method of mixtures, it should be appreciated that the region near 300 K is the least accurate region of such measurements. Confirmation such as is evident in Table VI is all that can be anticipated.

Samples of HgUO$_4$ and Ag$_2$UO$_4$ are also to be measured. These transition element uranates may prove to be especially interesting.

ACKNOWLEDGEMENTS

The authors are appreciative of the assistance of H. Magadanz and D. Weinstock in the cryogenic calorimetry and of Laurel Harmon, John Suter and W.G. Lyon in the evaluation of the data.

REFERENCES

[1] CORDFUNKE, E.H.P., WESTRUM Jr., E.F., Investigation on caesium uranates: VII. Thermochemical properties of Cs$_2$U$_4$O$_{12}$, these Proceedings, Vol.2, paper IAEA-SM-236/34.

[2] CORDFUNKE, E.H.P., O'HARE, P.A.G., Miscellaneous Actinide Compounds, Part III of The Chemical Thermodynamics of Actinide Elements and Compounds (MEDVEDEV, V., RAND, M.H., WESTRUM Jr., E.F., Eds; OETTING, F.L., Exec. Ed.), IAEA, Vienna (1978).

[3] JAKES, D., SCHAUER, V., Proc. Br. Ceram. Soc. 8 (1967) 123.

[4] KLIMA, J., JAKES, D., MORAVEC, J., J. Inorg. Nucl. Chem. 28 (1966) 1861.

[5] JAKES, D., KRIVY, I., J. Inorg. Nucl. Chem. 36 (1974) 3885.

[6] JAKES, D., SEDLAKOVA, L.N., MORAVAC, J., GERMANIC, J., J. Inorg. Nucl. Chem. 30 (1968) 525.

[7] JAKES, D., KRIVY, I., J. Inorg. Nucl. Chem. 36 (1974) 385.

[8] WESTRUM Jr., E.F., FURUKAWA, G.T., McCULLOUGH, J.P., "Adiabatic low-temperature calorimetry", in Experimental Thermodynamics (McCULLOUGH, J.P., SCOTT, D.W., Eds) Vol.I, Butterworth's, London (1968) 133–214.

[9] LATIMER, W.M., J. Am. Chem. Soc. 43 (1921) 818; ibid. 73 (1951) 1480.

[10] LATIMER, W.M., The Oxidation States of the Elements and their Potentials in Aqueous Solution, Prentice-Hall, New York (1952).

[11] BARIN, I., KNACKE, O., KUBASCHEWSKI, O., Thermochemical Properties of Inorganic Substances, Supplement I, Springer Verlag, Berlin (1977).

[12] KELLEY, K.K., KING, E.G., Contributions to the Data on Theoretical Metallurgy, XIV. Entropies of the Elements and Organic Compounds, Bulletin 592, Bureau of Mines, Washington, DC (1961).

[13] GRØNVOLD, F., WESTRUM Jr., E.F., Inorg. Chem. 1 (1962) 36.

[14] WESTRUM Jr., E.F., Exciting developments in the thermophysics of the pnictides and chalcogenides of the transition elements, Usp. Khim. (1979, in press).

[15] WESTRUM Jr., E.F., "Developments in chemical thermodynamics of the lanthanides", in Lanthanide/Actinide Chemistry (GOULD, R.F., Ed.), American Chemical Society, Washington, DC (1967) 25–50.

[16] CHIRICO, R.D., WESTRUM Jr., E.F., GRUBER, J.B., WARMKESSEL, J., Low-temperature heat capacities, thermophysical properties, optical spectra, and analysis of Schottky contributions to $Pr(OH)_3$, J. Chem. Thermodyn. (1979, in press).

[17] ZACHARIASEN, W.H., Acta Crystallogr. 1 (1948) 281.

[18] SAMSON, S., SILLÉN, L.G., Ark. Kemi Min. Geol. 25 21 (1947).

[19] O'HARE, P.A.G., BOERIO, J., FREDRICKSON, D.R., HOEKSTRA, H.R., J. Chem. Thermodyn. 8 (1976) 845.

[20] LEONIDOV, V.Ya., REZUKHINA, T.N., BEREZNIKOVA, I.A., Zh. Fiz. Khim. 34 (1960) 1862.

[21] FAUSTOVA, D.G., IPPOLITOVA, E.A., SPITSYN, V.I., Investigations in the Field of Uranium Chemistry, Argonne National Laboratory Rep. ANL-TRANS-33 (1961).

[22] LOOPSTRA, B.O., RIETVELD, H.M., Acta Crystallogr., B25 (1969) 787.

DISCUSSION

J. FUGER: Although all orthorhombic, $MgUO_4$ on the one hand and $SrUO_4$ and $BaUO_4$ on the other probably belong to different Laue groups, a, b and c being close to each other only in $MgUO_4$. Therefore this compound seems to exhibit a distortion in a cubic phase. Could you please comment?

E.F. WESTRUM Jr.: I agree that, although I have dealt primarily with the major contrasts in structure between rhombohedral $CaUO_4$ and the other three monouranates, the orthorhombic compounds do indeed show interesting changes in structure, which are hinted at by the space groups in our Table II.

Note added in proof: Recent studies by P.A.G. O'Hare and H.E. Flotow at Argonne National Laboratory have raised doubts concerning the integrity of the $BaUO_4$ sample.

E.F. Westrum Jr.,
1980-09-09

ЭНТАЛЬПИЯ ОБРАЗОВАНИЯ $YH_{1,994}$ и $YH_{2,792}$

В.В.АХАЧИНСКИЙ

Всесоюзный научно-исследовательский институт
неорганических материалов
Государственного комитета
по использованию атомной энергии СССР,
Москва,
Союз Советских Социалистических Республик

Abstract—Аннотация

ENTHALPY OF FORMATION OF $YH_{1.994}$ AND $YH_{2.792}$.

Solution calorimetry has been used to determine the standard enthalpy of formation, $\Delta H^0_{f,298}$, of two hydrides of yttrium, the dihydride and a composition close to the trihydride. Yttrium and yttrium hydride were dissolved in 1.5M and 0.3M hydrochloric acid in a semi-microcalorimeter with an isothermal jacket. The composition of the dihydride was determined by the increase in the weight of the yttrium as a result of hydriding and the composition of the trihydride from the amount of hydrogen evolved during the dissolution. It was found that:

$$\Delta H^0_{f,298}(YH_{1.994 \pm 0.005}) = -52.3 \pm 0.30 \text{ kcal} \cdot \text{mol}^{-1}$$

$$\Delta H^0_{f,298}(YH_{2.792 \pm 0.011}) = -59.08 \pm 0.41 \text{ kcal} \cdot \text{mol}^{-1}$$

ЭНТАЛЬПИЯ ОБРАЗОВАНИЯ $YH_{1,994}$ и $YH_{2,792}$.

Методом калориметрии растворения определена стандартная энтальпия образования, $\Delta H^0_{f,298}$, гидридов иттрия двух составов – дигидрида и близкого к тригидриду. Иттрий и гидрид иттрия растворяли в 1,5$\underline{M}$ и 0,3$\underline{M}$ соляной кислоте в полумикрокалориметре с изотермической оболочкой. Состав дигидрида определили по увеличению веса иттрия в результате гидрирования, а состав тригидрида – по количеству водорода, выделяющегося при растворении. Найдено, что $\Delta H^0_{f,298}(YH_{1,994 \pm 0,005}) = -52,3 \pm 0,30$ ккал·моль$^{-1}$; $\Delta H^0_{f,298}(YH_{2,792 \pm 0,011}) = -59,08 \pm 0,41$ ккал·моль$^{-1}$.

1. ВВЕДЕНИЕ

Гидрид иттрия и композиции на его основе представляют интерес как замедлители нейтронов в активной зоне жидкометаллических или газоохлаждаемых реакторов на промежуточных и тепловых нейтронах. Несмотря на сравнительно хорошую изученность системы иттрий – водород, энтальпия образования гидридов иттрия методами калориметрии не определялась, а была вычислена на основании результатов измерения давления диссоциации. В связи с присущей этому способу определения

ΔH_f не очень высокой точностью, большим расхождением результатов разных исследователей и трудностью пересчета к стандартной температуре (298 К), целесообразно было провести новое определение энтальпии образования гидридов иттрия при комнатной температуре другим, более точным методом.

Был использован метод калориметрии растворения, который мог обеспечить определение энтальпии образования гидрида иттрия любого состава с абсолютной ошибкой не более ± 0,5 ккал · моль$^{-1}$.

2.   ЭКСПЕРИМЕНТАЛЬНАЯ ЧАСТЬ

2.1. Приготовление гидридов

Для синтеза гидридов иттрия были использованы водород, получаемый при разложении гидрида урана, и дистиллированный иттрий высокой чистоты, содержащий следующие примеси (в вес%): Ca < 0,02; Fe − 0,04; Cu < 0,005; Ta < 0,03; O − 0,08. Редкоземельных элементов, кроме Sm (< 0,1%), не обнаружено.

Для получения дигидрида было взято 0,84304 г иттрия. Гидрирование начинали при температуре 1273 К с последующим ее снижением. Вес полученного гидрида составлял 0,86210 г. Его состав, вычисленный по изменению веса, выражался формулой $YH_{1,994 \pm 0,005}$.

Тригидрид иттрия получали при давлении водорода 760 мм рт. ст. и температуре 520-620 К. Точно взвешенные кусочки иттрия помещали каждый в отдельную кварцевую микроколбочку и гидрировали одновременно. Достаточное время гидрирования и малый размер образцов обеспечивали получение однородного продукта, тем более, что образцы рассыпались в порошок. По окончании синтеза закрытую ампулу, в которой проводили синтез, переносили в герметичную камеру, заполненную аргоном, где колбочки извлекали из ампулы, закрывали специальными пробками, герметизировали расплавленным воском, после чего извлекали из камеры и использовали для калориметрических измерений. Состав каждого образца определяли по количеству выделившегося при его растворении водорода, поскольку вес иттрия, взятого для гидрирования, был известен.

2.2. Калориметрические измерения

Энтальпию образования полученных гидридов определяли по разности теплот растворения иттрия и этих гидридов в соляной кислоте состава HCl · 38,5 H_2O (5%-ной, или ≃ 1,5$\underline{M}$). Количество наливаемой в калориметр кислоты составляло 22,5 мл. Средняя температура растворителя во время опыта равнялась 298 К. Калориметр и методика работы подробно описаны в [1].

Результаты калориметрических измерений приведены в табл. I.

На основании измерения количества водорода, выделяющегося при растворении иттрия и $YH_{1,994}$, было принято, что при растворении тригидрида водород насыщается парами кислоты на $30 \pm 10\%$ от полного насыщения (в случае иттрия насыщение составляло 20%, а в случае $YH_{1,994} - 30\%$). При таком достаточно обоснованном допущении ошибка в определении количества водорода составляла не более $\pm 0,05$ см^3. Поскольку из общего количества водорода, выделяющегося при растворении образца, объем водорода, содержавшегося в гидриде, составлял около 7 см^3, максимальная ошибка в его определении не должна была превышать $\pm 0,7\%$. Хорошая сходимость результатов при измерении водорода свидетельствовала о том, что все образцы каждого гидрида имели одинаковый состав и для расчетов можно было брать среднее значение состава, полученное из всех опытов. Рассчитанный таким образом состав растворенного нами тригидрида иттрия выражался формулой $YH_{2,792 \pm 0,011}$.

Приведенные в табл. I значения теплоты растворения мольных количеств дигидрида и тригидрида даны в расчете на г-формулы $YH_{1,994 \pm 0,005}$ и $YH_{2,792 \pm 0,011}$. При вычислении поправки на теплоту испарения кислоты принимали, что она равна 0,32 ккал на 1 моль выделяющегося водорода при 100%-ном насыщении его парами кислоты.

При расчетах принимали, что 1 кал = 4,184 джоуля. В качестве показателя точности приводится средняя квадратичная ошибка результата.

3. ВЫЧИСЛЕНИЕ ЭНТАЛЬПИИ ОБРАЗОВАНИЯ $YH_{2,994 \pm 0,005}$ и $YH_{2,792 \pm 0,011}$

На основании найденных теплот растворения энтальпия образования $YH_{1,994}$ и $YH_{2,792}$ может быть вычислена из следующих уравнений, при составлении которых не учитывалась некоторая разница в конечной концентрации растворов YCl_3, т.к. растворы были очень разбавленными и даже двукратное уменьшение количества растворяемого иттрия (опыты 5 и 6) и пятикратное уменьшение концентрации кислоты (опыты 15 и 16) не сопровождалось ощутимым тепловым эффектом.

$$Y \text{ (кр.)} + 140\,HCl \cdot 5390\,H_2O = YCl_3 \text{ (р – р, } 137\,HCl \cdot 5390\,H_2O) + 1{,}5\,H_2 \text{ (г.)}$$
$$\Delta H_1 = -173{,}46 \pm 0{,}18 \text{ ккал} \tag{1}$$

$$YH_{1,994} \text{ (кр.)} + 140\,HCl \cdot 5390\,H_2O = YCl_3 \text{ (р – р, } 137\,HCl \cdot 5390\,H_2O) + 2{,}497\,H_2 \text{ (г.)}$$
$$\Delta H_2 = -121{,}16 \pm 0{,}24 \text{ ккал} \tag{2}$$

$$YH_{2,792} \text{ (кр.)} + 140\,HCl \cdot 5390\,H_2O = YCl_3 \text{ (р – р, } 137\,HCl \cdot 5390\,H_2O) + 2{,}896\,H_2 \text{ (г.)}$$
$$\Delta H_3 = -114{,}38 \pm 0{,}37 \text{ ккал} \tag{3}$$

$$Y \text{ (кр.)} + 0{,}997\,H_2 \text{ (г.)} = YH_{1,994} \text{ (кр.)}$$
$$\Delta H_4 = \Delta H_1 - \Delta H_2 = -52{,}30 \pm 0{,}30 \text{ ккал} \tag{4}$$

ТАБЛИЦА I. РЕЗУЛЬТАТЫ КАЛОРИМЕТРИЧЕСКИХ ИЗМЕРЕНИЙ

Номер опыта	Вещество	Навеска (мг)	Выделилось тепла (кал)	Выделилось газа (при норм. усл.) (см³)	Должно выделиться водорода (при норм. усл.) (см³)	Выделилось водорода с поправкой на P_{HCl} (при норм. усл.) (см³)	Выделилось тепла на 1 г вещества (кал)	Выделилось тепла на 1 моль вещества [г] (ккал)
1	2	3	4	5	6	7	8	9
1	Y	20,583	40,139	7,90	7,79	7,85[а]	1950,0	173,37
2	Y	20,056	39,071	7,63	7,59	7,59[а]	1948,0	173,19
3	Y	19,540	37,985	7,45	7,39	7,38[а]	1943,9	172,83
4	Y	10,139	19,842	3,89	3,84	3,86[а]	1957,0	173,98
5	Y	9,949	19,379	3,78	3,76	3,75[а]	1947,8	173,17
6	Y	9,890	19,321	3,72	3,74	3,69[а]	1953,6	173,68
						Среднее:	1950,0 ± 1,9	173,36 ± 0,17
					Среднее с учетом теплового испарения HCl :		1951 ± 2,0	173,46 ± 0,18
7	YH$_{1,994}$	19,726	26,344	12,23	12,15	12,13[б]	1344,9	121,42
8	YH$_{1,994}$	19,797	26,369	12,31	12,19	12,20[б]	1332,0	121,10
9	YH$_{1,994}$	20,704	27,403	12,90	12,75	12,80[б]	1323,6	120,34
10	YH$_{1,994}$	20,438	27,174	12,71	12,59	12,59[б]	1329,6	120,89
						Среднее:	1300,0 ± 2,4	120,92 ± 0,23
					Среднее с учетом теплового испарения HCl :		1302,6 ± 2,5	121,16 ± 0,24
11	YH$_{2,792}$	20,293[д]	24,269	14,890	—	14,750[б]	1195,9	109,69
12	YH$_{2,792}$	20,211[д]	24,499	14,990	—	14,860[б]	1212,1	111,17
13	YH$_{2,792}$	19,940[д]	24,088	14,259	—	14,531[б]	1208,0	110,80
14	YH$_{2,792}$	20,433[д]	25,010	15,139	—	15,005[б]	1224,0	112,26
15	YH$_{2,792}$	20,204[д]	24,475	14,878	—	14,746[б]	1211,4	111,08[в]
16	YH$_{2,792}$	19,858[д]	23,994	14,604	—	14,474[б]	1208,3	110,82[б]
						Среднее:	1209,9 ± 3,7	110,97 ± 0,34
					Среднее с учетом теплового испарения HCl :		1214,3 ± 4,0	114,38 ± 0,37

[а] При вычислении количества водорода и поправки на теплоту испарения растворителя в случае растворения иттрия принято, что водород насыщается парами HCl на 20%.

[б] В случае растворения YH$_{1,994}$ и YH$_{2,792}$ принято 30%-ное насыщение водорода парами.

[в] В опытах 15 и 16 для растворения применяли 1%-ную соляную кислоту.

[г] При вычислении мольной теплоты растворения гидридов учтена ошибка в определении их состава.

[д] Указан вес иттрия, взятого для гидрирования.

$$Y \text{ (кр.)} + 1{,}396 \, H_2 \text{ (г.)} = YH_{2,792} \text{ (кр.)}$$
$$\Delta H_5 = \Delta H_1 - \Delta H_3 = -59{,}08 \pm 0{,}41 \text{ ккал} \tag{5}$$

Таким образом

$$\Delta H^0_{f,298} (YH_{1,994 \pm 0,005}) = -52{,}30 \pm 0{,}30 \text{ ккал} \cdot \text{моль}^{-1} \tag{6}$$

$$\Delta H^0_{f,298} (YH_{2,792 \pm 0,011}) = -59{,}08 \pm 0{,}41 \text{ ккал} \cdot \text{моль}^{-1} \tag{7}$$

Из этих данных следует, что энтальпия образования тригидрида $YH_{3,00}$ из дигидрида $YH_{2,00}$ равна $-8{,}50$ ккал $\cdot$ моль$^{-1}$, т.е.

$$\Delta H^0_{f,298}(YH_{3,00}) = -61{,}00 \pm 0{,}55 \text{ ккал} \cdot \text{моль}^{-1}$$

4. ОБСУЖДЕНИЕ РЕЗУЛЬТАТОВ

Найденные нами значения энтальпии образования дигидрида и тригидрида иттрия и вытекающие из них значения энтальпии образования тригидрида из дигидрида не противоречат значениям, полученным методом определения зависимости равновесного давления диссоциации от температуры в работах Лундина и др. [2], Яннопулоса и др. [3] и Флотова и др. [4], но все же значительно от них отличаются.

Блекледж, экстраполируя полученные в [2] данные для $YH_{1,75}$, оценил энтальпию образования стехиометрического $YH_{2,00}$, как равную $-56{,}2$ ккал $\cdot$ моль$^{-1}$ [5]. Разница в 4 ккал по сравнению с величиной, полученной нами, обусловлена, по-видимому, тремя одновременно действующими причинами — ошибкой в определении $\Delta H^0_f (YH_{1,75})$, ошибкой в экстраполяции и зависимостью от температуры. Точный учет последнего фактора невозможен, т.к. имеющиеся данные по высокотемпературной теплоемкости дигидрида иттрия ($YH_{1,90}$) недостаточно надежны [5]. Из вполне надежных данных Флотова и др. [4] по низкотемпературной теплоемкости $YH_{2,0}$ следует, что для реакции

$$Y \text{ (кр.)} + H_2 \text{ (г.)} = YH_2 \text{ (кр.)}$$
$$\Delta C_{p,298} = -5{,}0 \text{ кал} \cdot \text{моль}^{-1} \cdot \text{K}^{-1} \tag{8}$$

Если бы эта разность теплоемкости сохранялась вплоть до 1100 K, то из высокотемпературных измерений Яннопулоса [3] и оценки Блекледжа [5] следовало, что $\Delta H^0_{f,298}(YH_{2,0}) = -52{,}2$ ккал $\cdot$ моль$^{-1}$, что полностью совпадает с величиной, найденной нами. Однако теплоемкость YH_2 быстро растет с ростом температуры и ΔC_p должна становиться менее отрицательной. Если все же воспользоваться данными о высокотемпературной теплоемкости $YH_{1,90}$ [5], то оказывается, что при 1033 K

$\Delta C_p = 6,8$ кал $\cdot$ моль$^{-1}\cdot$ К$^{-1}$. Таким образом, может оказаться, что в области температур порядка 1000 К энтальпия образования должна быть менее отрицательной, чем при 298 К, и близка к $-50,0$ ккал $\cdot$ моль$^{-1}$. Если эта оценка верна, то данные Лундина и др. [2] по давлению диссоциации более точны, чем данные Яннопулоса и др. [3], поскольку из [2] следует, что ΔH_f^0 (YH$_2$) в интервале 1173-1623 К равна $-48,2$ ккал $\cdot$ моль$^{-1}$.

По данным работы [3] относительная парциальная энтальпия водорода в гидриде иттрия при изменении состава гидрида от YH$_{2,0}$ до YH$_{2,6}$ изменяется от $-9,86$ до $-11,5$ ккал на г-атом Н. Из результатов измерения давления диссоциации над стехиометрическим YH$_{3,0}$ (Флотов и др.)[4]) следует, что парциальная энтальпия водорода в YH$_{3,0}$ равна $-10,0$ ккал на г-атом Н. Следовательно, энтальпия образования тригидрида из дигидрида близка к $-10,0$ ккал $\cdot$ моль$^{-1}$ в интервале температур 320-600 К. Разница с полученной нами величиной ($-8,5 \pm 0,55$ ккал $\cdot$ моль$^{-1}$) сравнительно невелика и обусловлена теми же причинами, что и в случае ΔH_f^0 (YH$_2$).

ЛИТЕРАТУРА

[1] АХАЧИНСКИЙ, В.В., КОПЫТИН, Л.М., Ат. Энерг. 9 6 (1960) 504.
[2] LUNDIN, C.E., BLACKLEDGE, J.P., J. Electrochem. Soc. 109 9 (1965) 838.
[3] YANNOPOULOS, L.N., EDWARDS, R.K., WAHLBECK, P.G. J. Chem. Phys. 69 8 (1965) 2510.
[4] FLOTOW, H.E., OSBORNE, D.W., OTTO, K., ABRACHAM, B.M. J. Chem. Phys. 38 11 (1963) 2620.
[5] MUELLER, W.M., BLACKLEDGE, J.P., LIBOWITZ, G.G. Metal Hydrides, Acad. Press, New York and London, (1968).

О ТЕРМОДИНАМИЧЕСКОЙ НЕСТАБИЛЬНОСТИ ТВЕРДОГО РАСТВОРА АЛЮМИНИЯ В ПЛУТОНИИ (δ-ФАЗЫ) ПРИ КОМНАТНОЙ ТЕМПЕРАТУРЕ

В.В. АХАЧИНСКИЙ, Л.Ф. ТИМОФЕЕВА
Всесоюзный научно-исследовательский
институт неорганических материалов
Государственного комитета по использованию
атомной энергии СССР,
Москва,
Союз Советских Социалистических Республик

Abstract—Аннотация

THERMODYNAMIC INSTABILITY OF A SOLID SOLUTION OF ALUMINIUM IN
PLUTONIUM (δ-PHASE) AT ROOM TEMPERATURE.

Using solution calorimetry, the authors determine the standard enthalpies of formation
of solid solutions of aluminium in plutonium (δ-phase) with compositions $Pu_{0.9401}$ $Al_{0.0599 \pm 0.0001}$
and $Pu_{0.9103}$ $Al_{0.0897 \pm 0.0001}$. It was found that:

$$\Delta H^0_{f,298}(Pu_{0.9401} Al_{0.0599 \pm 0.0001}) = -160 \pm 180 \ cal \cdot mol^{-1}$$

$$\Delta H^0_{f,298}(Pu_{0.9103} Al_{0.0897 \pm 0.0001}) = -390 \pm 190 \ cal \cdot mol^{-1}$$

On the basis of data yielded by this research and of published data, the authors evaluate the
variation in the free energy of reaction involving the decomposition of the δ-phase, and draw
the conclusion that the δ-phase is thermodynamically unstable at room temperature. The
lower stability boundary for the δ-phase is calculated as 383 K (110°C).

О ТЕРМОДИНАМИЧЕСКОЙ НЕСТАБИЛЬНОСТИ ТВЕРДОГО РАСТВОРА АЛЮМИНИЯ
В ПЛУТОНИИ (δ-ФАЗЫ) ПРИ КОМНАТНОЙ ТЕМПЕРАТУРЕ.

Методом калориметрии растворения определены стандартные энтальпии образования твер-
дых растворов алюминия в плутонии (δ-фазы) состава $Pu_{0,9401}$ $Al_{0,0599 \pm 0,0001}$ и $Pu_{0,9103}$ $Al_{0,0897 \pm 0,0001}$.
Найдено, что $\Delta H^0_{f,298}$ $(Pu_{0,9401} Al_{0,0599 \pm 0,0001}) = -160 \pm 180$ кал · моль$^{-1}$, $\Delta H^0_{f,298}$ $(Pu_{0,9103} Al_{0,0897 \pm 0,0001})$
$= -390 \pm 190$ кал · моль$^{-1}$. На основании результатов работы и литературных данных произве-
дена оценка изменения свободной энергии реакции распада δ-фазы и сделан вывод о термодинамической нестабильности δ-фазы при комнатной температуре. Нижняя граница стабильности δ-фазы оценена в 383 К (110°С).

ТАБЛИЦА I.   РЕЖИМЫ ОТЖИГА И РЕЗУЛЬТАТЫ АНАЛИЗА СПЛАВОВ

Состав сплава	Режим отжига		Параметры решетки δ-фазы (Å)	Микротвердость (20 г) (кг/мм²)
	Температура (°C)	Время (ч)		
1	2	3	4	5
Pu + 6 ат % Al	450 470	350 120	4,595 + 0,001	56
Pu + 9 ат % Al	450 470	350 100	4,577 + 0,001	62

1. ВВЕДЕНИЕ

В настоящее время нет единого мнения о температурной границе существования δ-фазы в системе плутоний—аллюминий. По данным А.А. Бочвара и др. [1] твердый раствор на основе δ-плутония, легко фиксирующийся при комнатной температуре, метастабилен ниже 175°C. На диаграммах, представленных А.С. Кофинберри и др. [2] и М.Б. Уолдроном и др. [3], δ-фаза стабильна до комнатной температуры.

Данная работа посвящена решению вопроса о стабильности δ-фазы в системе плутоний—аллюминий при комнатной температуре путем определения энтальпии образования δ-фазы при комнатной температуре методом калориметрии растворения с последующей оценкой изменения энтальпии и свободной энергии при распаде δ-фазы на α-плутоний и интерметаллид Pu_3Al.

2. ЭКСПЕРИМЕНТАЛЬНАЯ ЧАСТЬ

2.1. Получение сплавов

Для получения сплавов использовался алюминий марки АВ 000 (99,997%-ной чистоты) и плутоний, содержащий не более 0,2 вес % примесей.

Сплавление металлов производили путем индукционного нагрева шихты в тигле из окиси бериллия в вакууме с разрежением 10^{-5} мм рт. ст. Было получено два сплава. Шихта первого состояла из 465,930 ± 0,010 мг плутония и 3,352 ± 0,005 мг алюминия, шихта второго — из 302,920 ± 0,010 мг плутония и 3,369 ± 0,005 мг алюминия.

Вес корольков сплавов был равен весу исходной шихты с точностью до $5 \cdot 10^{-6}$ г. Поверхность сплавов была светлой, неокисленной. С целью гомогенизации сплавы подвергали длительному отжигу. Закалку осуществляли путем сбрасывания сплавов в масло комнатной температуры. О достижении гомогенности судили по результатам рентгенографического и металлографического изучений шлифов, сделанных на двух противоположных участках королька сплава. Режимы отжига и результаты анализа шлифов приведены в табл. I.

2.2. Калориметрические измерения

Для определения энтальпии образования полученных сплавов достаточно было определить теплоту их растворения в 6М соляной кислоте, т.к. теплоты растворения алюминия и плутония были определены ранее [4-6].

В работе использовали калориметр, подробно описанный в [4]. В качестве растворителя применяли свободную от кислорода, насыщенную водородом 6М соляную кислоту. Средняя температура кислоты во время опыта составляла $25,2^0$ С, а количество кислоты во всех опытах было равным 22,5 ± 0,05 мг.

Результаты калориметрических измерений приведены в табл. II.

В четвертом столбце табл. II в скобках приведены значения поправки на теплообмен. Как видно, она была невелика и составляла 6-8 % от исправленного подъема температуры.

Воспроизводимость результатов определения теплового значения была обычной для данного калориметра и вполне удовлетворительной — отклонения от среднего из двух определений в каждом опыте в большинстве случаев не превышали ±0,1 %.

Приведенные в двух последних столбцах таблицы теплоты растворения даны без учета поправки на теплоту испарения соляной кислоты.

Специальное изучение вопроса о степени насыщения парами соляной кислоты выделяющегося при растворении водорода и накопленный при этом материал позволяют утверждать, что полного насыщения не происходит. С достаточной точностью можно принять, что водород насыщается парами кислоты на 50 ± 25 %. Поскольку при полном насыщении поправка составляет 0,206 ккал на моль выделившегося водорода [4], то для растворявшихся нами сплавов поправку следует принять равной 0,15 ± 0,07 ккал на моль сплава.

Таким образом, теплота растворения сплава $Pu_{0,9401} Al_{0,0599 \pm 0,0001}$ равна 140,09 ± 0,15 ккал·моль$^{-1}$, а сплава $Pu_{0,9103} Al_{0,0897 \pm 0,0001}$ — 139,41 ± 0,17 ккал·моль$^{-1}$

Небольшая ошибка, которая может быть допущена при введении поправки на теплоту испарения кислоты, в основном элиминируется при вычислении энтальпии образования сплава, т.к. кинетика растворения сплава и чистого плутония очень похожа, а следовательно, близка и степень насыщения водорода парами кислоты.

ТАБЛИЦА II.   РЕЗУЛЬТАТЫ ИЗМЕРЕНИЯ ТЕПЛОТЫ РАСТВОРЕНИЯ СПЛАВОВ

№ опыта	Состав сплава	Масса растворявшей-ся навески (мг)	Исправленный подъем температуры (Ом)	Тепловое значение калориметра (кал/Ом)	Выделилось тепла (кал/г)	Выделилось тепла (ккал/моль)
1	2	3	4	5	6	7
1	$Pu_{0,9401} Al_{0,0599} \pm 0,0001$	27,20	0,91397 (0,07732)	18,37 18,34	616,77	139,65
2	$Pu_{0,9401} Al_{0,5599} \pm 0001$	48,60	1,63278 (0,11392)	18,37 18,40; 18,45	618,45	140,03
3	$Pu_{0,9401} Al_{0,0599} \pm 0001$	26,05	0,87505 (0,6713)	18,43 18,42	618,90	140,13
				Среднее	618,04±0,66	139,94±0,15*
4	$Pu_{0,9103} Al_{0,0897} \pm 0,0001$	30,80	1,05367 (0,11096)	18,50 18,41	631,32	138,95
5	$Pu_{0,9103} Al_{0,0897} \pm 0,0001$	40,16	1,48094 (0,09677)	18,41 18,43	634,74	139,71
6	$Pu_{0,9103} Al_{0,0897} \pm 0,0001$	43,62	1,49876 (0,07008)	18,36 18,41	631,79	139,06
7	$Pu_{0,9103} Al_{0,00897} \pm 0,0001$	46,06	1,58084 (0,08783)	18,41 18,48	633,05	139,33
				Среднее	632,73 ± 0,76	139,26±0,17*

* Среднеквадратичная ошибка

3.   ВЫЧИСЛЕНИЕ ЭНТАЛЬПИИ ОБРАЗОВАНИЯ СПЛАВОВ

Энтальпия образования сплава равна разности теплот растворения сплава и образующих его компонентов.

Теплота растворения алюминия и плутония в данной работе не определялась. Для расчета были использованы величины, полученные ранее [4-6] при растворении плутония и алюминия той же чистоты, что и использованных металлов для получения сплавов. В результате анализа приведенных в [4-6]

данных принято, что Q_{298}^0 (раствор. Al) = 127,46 ± 0,30 ккал · моль$^{-1}$ и
Q_{298}^0 (раствор. Pu) = 141,00 ± 0,10 ккал·моль$^{-1}$.   Эти значения получены в предположении о 50%-ном насыщении водорода парами кислоты.

Конечная концентрация Pu^{3+} в растворе как при растворении сплава, так и при растворении чистого плутония была практически одинаковой и равной примерно $1,7 \cdot 10^{-4}\underline{M}$, и лишь в опытах 1 и 3 была почти вдвое меньшей.   В литературе нет данных о теплоте разбавления растворов Pu^{3+} в 6 $\underline{M}$ соляной кислоте, чтобы можно было ввести поправку.   Однако, на основании данных работ [7-10] по теплоте растворения плутония и хлорида плутония при других концентрациях плутония и кислоты, а также по аналогии с теплотой разведения растворов хлоридов лантаноидов [11], в качестве оценки можно принять теплоту разведения раствора PuCl$_3$ в 6 $\underline{M}$ соляной кислоте от $1,7 \cdot 10^{-4}\underline{M}$ до $0,8 \cdot 10^{-4}\underline{M}$, равной 0,1 ± 0,05 ккал·моль$^{-1}$ и уменьшить теплоту растворения, полученную в опытах 1, 3 и 4 на 0,1; 0,1 и 0,08 ккал·моль$^{-1}$, соответственно, приведя ее к конечной концентрации $1,7 \cdot 10^{-7}\underline{M}$.

Конечная концентрация Al^{3+} при растворении сплавов составляла в среднем $(1,0 \div 1,5) \cdot 10^{-5} \underline{M}$, а при растворении чистого алюминия $- 2 \cdot 10^{-4}\underline{M}$. В литературе также нет данных, чтобы достаточно точно ввести поправку на разведение.   Из приведенных в [11] данных о теплоте разведения AlCl$_3$ в менее концентрированной кислоте и воде можно сделать вывод, что разница в теплоте растворения при конечных концентрациях Al^{3+}, равных $2 \cdot 10^{-4}\underline{M}$ и $1 \cdot 10^{-5}\underline{M}$, не будет больше 0,7 ккал.   Однако, нет возможности оценить более точно не только величину поправки, но даже ее знак.   Поэтому поправку на теплоту разведения раствора AlCl$_3$ не вводили.   Могущая возникнуть из-за этого ошибка при вычислении энтальпии образования сплава будет невелика, т.к. содержание алюминия в сплаве мало.

После приведения теплот растворения к условиям одинаковой конечной концентрации можно вычислить энтальпии образования сплавов

$$\Delta H_{f,298}^0 \,(\text{Pu}_{0,9401}\,\text{Al}_{0,0599\pm0,0001}) = (140,02 \pm 0,15) - 0,9401 \cdot (141,00 \pm 0,10)$$
$$- (0,0599 \pm 0,0001) \cdot (127,46 \pm 0,30) = -0,16 \pm 0,18 \text{ ккал·моль}^{-1} \tag{1}$$

$$\Delta H_{f,298}^0 \,(\text{Pu}_{0,9103}\,\text{Al}_{0,897\pm0,0001}) = (139,39 \pm 0,17) - 0,9103 \cdot (141,00 \pm 0,10)$$
$$- (0,0897 \pm 0,0001) \cdot (127,46 \pm 0,30) = -0,39 \pm 0,19 \text{ ккал·моль}^{-1}. \tag{2}$$

4. ОБСУЖДЕНИЕ РЕЗУЛЬТАТОВ

4.1. Оценка изменения энтальпии и свободной энергии при реакции распада δ-фазы.

Удовлетворительная точность полученных значений энтальпии образования δ-фазы двух составов позволяет сделать оценку термодинамической стабильности δ-фазы при комнатной температуре.

Для реакций распада δ-фазы

$$Pu_{0,9401}Al_{0,0599} = 0,7604\,\alpha\text{-}Pu + 0,0599\,Pu_3Al$$
$$\Delta H^0_{298} = -0,86 \pm 0,19 \text{ ккал}\cdot\text{моль}^{-1*} \tag{3}$$

$$Pu_{0,9103}Al_{0,0897} = 0,6412\,\alpha\text{-}Pu + 0,897\,Pu_3Al$$
$$\Delta H^0_{298} = -1,13 \pm 0,21 \text{ ккал}\cdot\text{моль}^{-1} \tag{4}$$

Как видно, реакции распада протекают с уменьшением энтальпии, что термодинамически выгодно. Однако решающим фактором, определяющим направление реакции, является изменение свободной энергии (изобарно-изотермического потенциала) ΔG^0, которое нами экспериментально не определялось.

Имеющиеся в литературе сведения позволяют произвести оценку ΔG^0_{298} реакций (3) и (4).

По данным работы [12] S^0_{298} δ-фазы сплава Pu – Al, содержащего 8 ат%Al, равна $17,52 \pm 0,11$ ккал·моль$^{-1}$·K^{-1}. Однако Рэнд считает эту величину ошибочной [13]. Действительно, Тейлор и др. [14] измерили низкотемпературную теплоемкость сплавов (δ-фазы), содержащих 4,92 ат% алюминия и 2,6; 3,2 и 6,1 ат% галлия. Вычисленная на основании их данных энтропия этих сплавов (S^0_{298}), оказалась равной 15,7; 15,9; 16,2 и 16,5 кал·моль$^{-1}$·K^{-1}, соответственно. По-видимому, можно принять, что

$$S^0_{298}(Pu_{0,94}Al_{0,06}) = 15,8 \pm 0,3 \text{ кал}\cdot\text{моль}^{-1}\cdot\text{K}^{-1} \tag{5}$$

$$S^0_{298}(Pu_{0,91}Al_{0,09}) = 16,1 \pm 0,3 \text{ кал}\cdot\text{моль}^{-1}\cdot\text{K}^{-1} \tag{6}$$

Энтропия α-плутония $S^0_{298} = 13,42 \pm 0,10$ кал·моль$^{-1}$·K^{-1} [15].

Энтропия Pu_3Al никем не определялась. По-видимому, инкремент энтропии плутония в этом соединении должен быть выше, чем энтропия чистого α-плутония и близок к энтропии δ-фазы плутония, стабилизированной алюминием или галлием. Дополнительным основанием для такого предположения является то обстоятельство, что инкремент энтропии Pu в соединении Pu_6Fe равен 15,87 кал·моль$^{-1}$·K^{-1} (вычислено из данных работы [16]). Приняв эту величину и для Pu_3Al, получим (по аддитивности), что $S^0_{298}(Pu_3Al) = 54,4 \pm 1,0$ кал·моль$^{-1}$·K^{-1}.

Поскольку

$$\Delta G^0_{298} = \Delta H^0_{298} - 298\,\Delta S^0_{298} \tag{7}$$

---

* Необходимая для расчета величина $\Delta H^0_{f,\,298}(Pu_3Al)$ взята из работы [5] и равна 17 ± 1 ккал·моль$^{-1}$.

то для реакции (3)

$$\Delta G^0_{298} = (-860 \pm 190) - 298 \cdot [0,7604(13,42 \pm 0,10)$$
$$+ 0,0599(54,4 \pm 1,0) - (15,8 \pm 0,3)] = -252 \pm 212 \text{ кал·моль}^{-1} \cdot K^{-1} \tag{8}$$

Для реакции (4)

$$\Delta G^0_{298} = (-1130 \pm 210) - 298 \cdot [0,6412(13,42 \pm 0,10)$$
$$+ 0,0897(54,4 \pm 1,0) - (16,1 \pm 0,3)] = -349 \pm 230 \text{ кал·моль}^{-1} \cdot K^{-1} \tag{9}$$

Таким образом, распад твердого раствора алюминия в плутонии с содержанием от 6 до 9 ат% алюминия на α-плутоний и Pu_3Al при комнатной температуре должен сопровождаться убылью свободной энергии, и, следовательно, δ-фаза такого состава при комнатной температуре находится в метастабильном состоянии. Однако "движущая сила" распада, т.е. уменьшение свободной энергии, очень мала, и процесс распада самопроизвольно практически не идет.

Представляет интерес оценить температуру, выше которой δ-фаза становится термодинамически стабильной. Для этого нужно знать зависимость теплоемкости от температуры всех компонентов реакций (3) и (4) с тем, чтобы воспользоваться уравнением

$$\Delta G^0_T = H^0_{298} + \int\limits_{298}^{T} \Delta C_p \, dT - T\Delta S^0_{298} - T \int\limits_{298}^{T} \frac{\Delta C_p}{T} \, dT \tag{10}$$

(без учета фазового превращения $\alpha \rightarrow \beta$ в плутонии, если $T < 395 K$).

По данным Роуза и др. [17], теплоемкость сплава $Pu + 3,3$ ат% Ga (δ-фаза) изменяется с температурой по уравнению

$$C_p (\text{кал·моль}^{-1} \cdot K^{-1}) = 17 \cdot 563 - 3,502 \cdot 10^{-2} T + 3,68 \cdot 10^{-5} T^2$$
$$- 2,461 \cdot 10^5 T^{-2} \quad (298 - 773 K) \tag{11}$$

Можно принять, что по такому же закону изменяется теплоемкость δ-фазы $Pu - Al$, только формулу (11) нужно изменить так, чтобы при $298 K$ C_p равнялась не 7,57, как у сплава $Pu + 3,3$ ат% Ga, а 7,62, как найдено Тейлором для сплава $Pu + 4,9$ ат% Al [14], и принять также, что теплоемкость сплавов с 4,9 и 6,0 ат% Al одинакова. Тогда

$$C_p (Pu_{0,9401} Al_{0,0599}) = 17,613 - 3,502 \cdot 10^{-2} T$$
$$+ 3,618 \cdot 10^{-5} T^2 - 2,461 \cdot 10^5 T^{-2} \quad (298 - 773 K) \tag{12}$$

Теплоемкость α-плутония изменяется в интервале 298-395 K линейно по уравнению

$$C_p = 3,56 + 0,0144 T \quad [15] \tag{13}$$

Можно принять, что

$$C_p(Pu_3Al) = 3C_p(Pu) + C_p(Al) \tag{14}$$

$$C_p(Al) = 4,94 + 2,96 \cdot 10^{-3} T \quad (298 - 993K) \tag{15}$$

Из уравнений (12) и (15) можно оценить температурную зависимость изменения инкремента теплоемкости плутония в Pu_3Al, приняв ее равной теплоемкости Pu в δ-фазе.

$$C_p(Pu \text{ в } Pu_3Al) = [C_p(Pu_{0,94}Al_{0,06}) - 0,06\,C_p(Al)] : 0,94$$
$$= 18,422 - 3,744 \cdot 10^{-2} T + 3,849 \cdot 10^{-5} T^2 - 2,621 \cdot 10^5 T^{-2} \text{ кал·моль}^{-1} \tag{16}$$

Из (14)-(16) получаем

$$C_p(Pu_3Al) = 60,206 - 10,936 \cdot 10^{-2} T + 11,547 \cdot 10^{-5} T^2$$
$$- 7,863 \cdot 10^5 T^{-2} \text{ кал·моль}^{-1} \cdot K^{-1} (298 - 700K) \tag{17}$$

Для реакции (3)

$$\Delta C_p = 0,7604\,C_p(\alpha\text{-}Pu) + 0,0599\,C_p(Pu_3Al)$$
$$- (Pu_{0,9401}Al_{0,0599}) = 11,295 + 3,940 \cdot 10^{-2} T$$
$$- 2,925 \cdot 10^{-5} T^2 + 11,989 \cdot 10^5 T^{-2} \quad (298 - 395K) \tag{18}$$

Подставив (18) в (10) и учитывая, что $\Delta H^0_{298} = -860$ кал·моль$^{-1}$ и $\Delta S^0_{298} = 2,04$ кал·моль$^{-1} \cdot K^{-1}$, после проведения соответствующих расчетов находим, что

$$\Delta G^0 = 0 \text{ при } T = 383K (110^0C) \tag{19}$$

Если не учитывать зависимости C_p и S^0 от температуры, то

$$\Delta G^0_T = \Delta H^0_{298} - T\Delta S^0_{298} \tag{20}$$

и

$$\Delta G^0 = 0 \text{ при } 421,5K (148,5^0C) \tag{21}$$

Подобный приближенный расчет для реакции (4) показывает, что

$$\Delta G^0 = 0 \text{ при } 431K (158^0C). \tag{22}$$

Таким образом, вычисленная на основании термодинамических данных температурная граница стабильности δ-фазы (Pu, содержащий 6-9 ат%Al) хотя и находится несколько ниже приведенной А.А. Бочваром и др. [1], но качественно с ней согласуется и, во всяком случае, лежит явно выше комнатной температуры.

ЛИТЕРАТУРА

[1] БОЧВАР, А.А., КОНОБЕЕВСКИЙ, С.Т., КУТАЙЦЕВ, В.И., МЕНЬШИКОВА, Т.С., ЧЕБОТАРЕВ, Н.Т., Труды Второй международной конференции по мирному использованию атомной энергии, 3, М., Атомиздат, 1959.
[2] КОФИНБЕРРИ, А.С. и др., Труды Второй международной конференции по мирному использованию атомной энергии, 6, М., Атомиздат, 1953.

[3]   УОЛДРОН, М.Б. и др., Труды Второй международной конференции по мирному исполь-
      зованию атомной энергии, 6, М., Атомиздат, 1953.

[4]   АХАЧИНСКИЙ, В.В., КОПЫТИН, Л.М., Ат. Энерг. 9 (1960) 504.

[5]   АХАЧИНСКИЙ, В.В., КОПЫТИН, Л.М., ИВАНОВ, М.И., ПОДОЛЬСКАЯ, Н.С.,
      Thermodynamics of Nuclear Materials (Proc. Symp. Vienna, 1961), IAEA, Vienna (1962) 309.

[6]   АХАЧИНСКИЙ, В.В., Thermodynamics (Proc. Symp. Vienna, 1965) Vol. 2, IAEA, Vienna (1966) 561.

[7]   WESTRUM Jr., E.F., ROBINSON, H.P., The Transuranium Elements, McGraw Hill, New York,
      14B (1949) 914.

[8]   ROBINSON, H.P., WESTRUM Jr., E.F., The Transuranium Elements, McGraw Hill, New York,
      14B (1949) 922.

[9]   FUGER, J., CUNNINGHAM, B.B., J. Inorg. Nucl. Chem. 25 (1963) 1423.

[10] HINCHEY, R.J., COBBLE, J.W., Inorg. Chem. 9 (1970) 922.

[11] Термические Константы Веществ (Выпуск VIII, часть I), под ред. Глушко, В.П.,
      Медведева, В.А. и др., М., изд. АН СССР, 1978.

[12] SANDENAW, T.A., J. Phys. Chem. 16 (1960) 329.

[13] RAND, M.H., et al., At. Energy Rev. 4, Special Issue № 1, IAEA, Vienna (1966) 39.

[14] TAYLOR, J.C., LINFORD, P.F.T., DEAN, D.J., J. Inst. Met. 96 (1968) 178.

[15] OETTING, F.L., RAND, M.H., ACKERMANN, R.J., The Chemical Thermodynamics of Actinide
      Elements, Part 1. The Actinide Elements, IAEA, Vienna (1976).

[16] SANDENAW, T.A., HARBUR, D.R., J. Phys. Chem. Solids 34 (1973) 1487.

[17] ROSE, R.L., ROBINS, J.L., MASSALSKI, T.B., J. Nucl. Mater. 36 (1970) 99.

DISCUSSION

M.H. RAND: There are similar measurements of the low-temperature heat capacities and thus entropies for the Pu—Ga system. Do you plan any measurements on the enthalpies of formation of these phases in order to complement phase diagram information?

V.V. AKHACHINSKIJ: Yes, we are going to perform such measurements.

ТЕРМОДИНАМИКА И КИНЕТИКА ВЗАИМОДЕЙСТВИЯ МЕТАЛЛОВ С МОНОКАРБИДОМ, МОНОСУЛЬФИДОМ И КАРБОСУЛЬФИДОМ УРАНА

В.Н.ЗАГРЯЗКИН, А.С.ПАНОВ, Е.В.ФИВЕЙСКИЙ
Институт металлургии им. А.А.Байкова АН СССР,
Москва,
Союз Советских Социалистических Республик

Abstract—Аннотация

THERMODYNAMICS AND KINETICS OF INTERACTIONS OF METALS WITH URANIUM MONOCARBIDE, MONOSULPHIDE AND CARBOSULPHIDE.
    The authors consider the thermodynamics of uranium carbide, uranium sulphide and uranium carbosulphide and perform a thermodynamic analysis of their interactions with transition metals. Theoretical calculations and experimental results show that at $1750-1950^{\circ}C$ carbosulphide solid solutions based on uranium monosulphide do not carburize tungsten and molybdenum, but do carburize niobium and tantalum, at appreciably lower rates than UC does. Uranium carbosulphide in solid solution with uranium monocarbide carburized all metals.

ТЕРМОДИНАМИКА И КИНЕТИКА ВЗАИМОДЕЙСТВИЯ МЕТАЛЛОВ С МОНОКАРБИДОМ, МОНОСУЛЬФИДОМ И КАРБОСУЛЬФИДОМ УРАНА.
    Рассмотрена термодинамика карбида, сульфида и карбосульфида урана и проведен термодинамический анализ процессов их взаимодействия с переходными металлами. Теоретические расчеты и результаты экспериментов показали, что при $1750-1950^{\circ}C$ карбосульфидные твердые растворы на основе моносульфида урана не карбидизируют W и Mo и с заметно меньшими скоростями, чем UC, карбидизируют Nb и Ta. Карбосульфидные твердые растворы на основе монокарбида урана карбидизировали все металлы.

Повышение требований к различным ядерно-энергетическим установкам выдвигает необходимость разработки новых, более эффективно работающих топливных материалов. Помимо широко применяющейся двуокиси урана, внимание многих исследователей привлекли монокарбид и, особенно в последнее время, моносульфид урана, имеющие определённые преимущества перед двуокисью урана. В то же время как UC, так и US имеют серьезные недостатки, сдерживающие их применение. Причём, если какое-либо свойство характеризуется как неудовлетворительное у монокарбида урана(например,плохая совместимость с материалами оболочек), то это же свойство у моносульфида урана выглядит как вполне приемлемое, и наоборот, неудовлетворительность какого-либо свойства у моносульфида урана (например, низкое процентное содержание урана в единице объёма) отсутствует у монокарбида урана. Поэтому интересно комбинированное карбосульфидное топливо, объединяющее достоинства как монокарбида, так и моносульфида урана.

Для суждения о возможности применения карбосульфидного топлива необходимы сведения о его свойствах. В настоящей работе решено рассмотреть взаимодействие карбосульфидного топлива с рядом металлов, которые могут использоваться в качестве оболочек или диффузионных барьеров, Помимо карбосульфида урана, рассматривалось для сравнения взаимодействие металлов с монокарбидом и моносульфидом урана.

Экспериментальным работам предшествовали теоретические оценки возможности протекания реакций в рассматриваемых системах.

I. Термодинамика монокарбида, моносульфида и карбосульфида урана.

Для оценки возможностей протекания реакций взаимодействия монокарбида, моносульфида и карбосульфида урана с различными металлами нам потребуются сведения о парциальных и интегральных термодинамических свойствах как исходных веществ, так и продуктов реакций. Для расчёта парциальных термодинамических свойств рассматриваемых моносоединений урана воспользуемся подходом, развитым нами ранее [I-3] для нестехиометрических моносоединений $Me_{1-x}C_x$, имеющих структуру типа $NaCl$. Парциальные термодинамические свойства компонент можно представить в виде следующих уравнений:

$$\overline{F}_1 = E_1 + 0.5E_{11} - 0.5\left(\frac{x}{1-x}\right)^2 E_{22} - T\left[R\ln\frac{1-x}{1-2x} + S_1 + 0.5S_{11} - 0.5\left(\frac{x}{1-x}\right)^2 S_{22}\right], \quad (1)$$

$$\overline{F}_2 = E_2 + E_{12} + \left(\frac{x}{1-x}\right)E_{22} - T\left[R\ln\frac{1-2x}{x} + S_2 + S_{12} + \left(\frac{x}{1-x}\right)S_{22}\right], \quad (2)$$

где индекс I относится к металлу, а индекс 2 – к металлоиду.

Применяя уравнения (I) и (2) к анализу известных эксперименальных данных [4,5] по активностям урана и углерода в нестехиометрическом монокарбиде урана в области составов $x < 0.5$, получим в итоге следующие уравнения:

$$\overline{F}_U^{UC} = -37000 + 15.8T + 50000\left(\frac{x}{1-x}\right)^2 - 19.1\left(\frac{x}{1-x}\right)^2 T + RT\ln\frac{1-2x}{1-x}, \quad (3)$$

$$\overline{F}_C^{UC} = 64000 - 35.9T - 100000\left(\frac{x}{1-x}\right) + 38.2\left(\frac{x}{1-x}\right)T + RT\ln\frac{x}{1-2x}, \quad (4)$$

Термодинамические свойства моносульфида урана изучались в работах [6-10]. С использованием калориметрической методики уточнена [7] энтальпия образования моносульфида урана, близкого по составу к стехиометрическому. Оказалось, что в отличие от оценок авторов работы [8], энтальпия образования $US_{1.01}$ равнялась $\Delta H_f = -73.2$ ккал/г-формулу [7]. Этот результат подтверждается в работе [9]. Энтропия образования моносульфида $US_{1.01}$ была равна $\Delta S_f = -1.10$ кал/ K г-формулу [7].

В работе [6] измерены парциальные давления компонент над моносульфидом урана переменного состава. Было установлено, что парциальная теплота испарения урана меняется в пределах от 120 до 131,6 ккал/г-атом при изменении состава от нижней границы области гомогенности до состава, лежащего вблизи верхней границы области гомогенности. По данным работы [II] нижняя граница области гомогенности моносульфида урана приходится на состав $US_{0.96}$.

Используя вышеупомянутые результаты, можно получить аналитическое выражение для парциальных молярных свободных энергий урана и серы в моносульфиде урана переменного состава. Уравнения для $\bar{F}_U^{US}$ и $\bar{F}_S^{US}$ запишутся в виде:

$$\bar{F}_U^{US} = 175500 - 196000 \left(\frac{x}{1-x}\right)^2 + RT \ln \frac{1-2x}{1-x} -$$
$$- T\left[113 - 130,5\left(\frac{x}{1-x}\right)^2\right], \text{кал}/\text{г-атом} \qquad (5)$$

$$\bar{F}_S^{US} = -444700 + 392000 \left(\frac{x}{1-x}\right) + RT \ln \frac{x}{1-2x} -$$
$$- T\left[-244,6 + 261\left(\frac{x}{1-x}\right)\right], \text{кал}/\text{г-атом} \qquad (6)$$

Для расчёта термодинамических свойств карбосульфидного уранового топлива воспользуемся методом, согласно которому отдельные двойные соединения формально рассматриваются как элементы, входящие в состав многокомпонентного сплава, обладающего свойствами регулярного раствора [12]. Рассмотрим термодинамические свойства трёхкомпонентного твёрдого раствора уран-углерод-сера, образованного по реакции:

$$y\, U_{1-x} C_x + (1-y) U_{1-x} S_x \longrightarrow U_{1-x} C_{yx} S_{(1-y)x}, \qquad (7)$$

Парциальные молярные свободные энергии компонент этого регулярного раствора могут быть представлены в виде следующих уравнений [12]:

$$\bar{F}_U = y\, \bar{F}_U^{UC} + (1-y)\, \bar{F}_U^{US}, \qquad (8)$$

$$\bar{F}_C = \bar{F}_C^{UC} + \frac{(1-y)(1-x)}{x}\left(\bar{F}_U^{UC} - \bar{F}_U^{US}\right) + W_{CS}(1-y)^2 + RT \ln y, \qquad (9)$$

$$\bar{F}_S = \bar{F}_S^{US} + \frac{y(1-x)}{x}\left(\bar{F}_U^{US} - \bar{F}_U^{UC}\right) + W_{CS}\, y^2 + RT \ln(1-y), \qquad (10)$$

где $\bar{F}_U^{UC}$; $\bar{F}_C^{UC}$; $\bar{F}_U^{US}$ и $\bar{F}_S^{US}$ — парциальные молярные свободные энергии урана, углерода и серы в монокарбиде и моносульфиде урана;

W_{CS} — энергия взаимодействия между атомами углерода и серы.

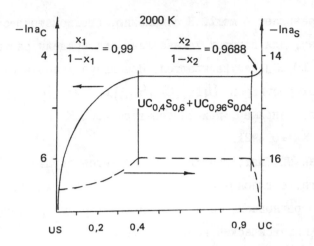

Рис.1. Зависимость термодинамических активностей углерода и серы в твердых растворах систем $UC - US$ при температуре 2000 К.

Энергия взаимодействия W_{CS}, найденная из уравнения (9), равна в твёрдом растворе $UC - US$ величине 9,5$\pm$2,5 ккал/г-атом. Неопределённость в значении W_{CS} связана с соответствующей неопределённостью в температуре и стехиометрических коэффициентах сосуществующих фаз, при которых указанные [10,13] пределы взаимной растворимости реализуются. Рассчитанные значения W_{CS} охватывают интервал температур 1500-2500°К и отношений $x/_{1-x}$ =0,96÷0,99.

Результаты расчётов активностей металлоидных компонент в карбосульфиде урана, выполненных с учётом диаграмм фазовых равновесий для температуры 2000°К, представлены на рис.I.

Из уравнения (8) для металлической подрешётки, содержащей только атомы урана, следует, что ограниченные твёрдые растворы в системе $UC - US$ должны иметь в условиях термодинамического равновесия различные стехиометрические составы. Из работ [10,13] известно, что UC растворяется в моносульфиде урана в количествах до 40 мол%, а предельная растворимость US в монокарбиде урана не превышает 4 мол%. К сожалению, стехиометрические коэффициенты фаз, для которых установлены эти пределы растворимости, в работах [10,13] не указываются. Используя условия равновесия двух твёрдых растворов $U_{1-x_1}(C_{0,4}S_{0,6})_{x_1}$ и $U_{1-x_2}(C_{0,96}S_{0,04})_{x_2}$ найдём, что, например, при температуре 2000 К и $x_1 = 0,4975$, величина $x_2 = 0,4921$.

Отмеченный факт, по-видимому, является отражением общей закономерности, согласно которой при разрыве смесимости твёрдых растворов в равновесии могут находиться твёрдые растворы, имеющие разные стехиометрические коэффициенты.

2. Термодинамический анализ процессов взаимодействия металлов с монокарбидом, моносульфидом и карбосульфидом урана

В рассматриваемых системах $Me-US$, $Me-UC$ и $Me-U_{1-x}$ в принципе могут образовываться карбиды и сульфиды различных металлов. Степень и возможность их образования у различных металлов разные.

Сульфиды ниобия, тантала, молибдена и вольфрама устойчивы по имеющимся данным [14-16] в области температур 1000-1600 К. Поэтому они могут явиться продуктами реакций взаимодействия ниобия, тантала, молибдена и вольфрама с моносульфидом и карбосульфидом урана. Так как моносульфид урана является более прочным с термодинамической точки зрения соединением, чем сульфиды ниобия, тантала, молибдена и вольфрама, то реакции взаимодействия в данном случае могут проходить по реакциям типа:

$$(1-x_2)U_{1-x_1}S_{x_1} + (x_1-x_2)Me \rightarrow (1-x_1)U_{1-x_2}S_{x_2} + (x_1-x_2)MeS \quad (II)$$

Определить пределы по составам и температурам и оценить выход реакций типа (II) применительно к образованию сульфидов ниобия, тантала, молибдена и вольфрама в данный момент затруднительно, ввиду недостаточности сведений о термодинамических свойствах этих соединений.

Карбиды молибдена и вольфрама устойчивы при высоких температурах. Поэтому они могут быть продуктами реакций взаимодействия молибдена и вольфрама с монокарбидом и карбосульфидом урана. При оценках возможности образования карбидных фаз молибдена и вольфрама в рассматриваемых системах встретились трудности, связанные с немногочисленностью и противоречивостью литературных сведений о термодинамических свойствах низших карбидов молибдена и вольфрама. Так, например, теплота образования Mo_2C по данным [17] составляет – 11,5 ккал/моль, а по сведениям работы [4] всего – 5,5 ккал/моль. Теплота образования карбида вольфрама $\overline{W_2C}$ по данным [17] составляет – 11 ккал/моль и всего – 6,3 ккал/моль по данным работы [4]. Последний результат согласуется с работой [18], в которой определялись термодинамические свойства карбида W_2C, находящегося в равновесии с вольфрамом, что позволяет считать результаты этой работы достоверными. Подобной определённости по термодинамическим свойствам карбидов молибдена в целом нет, однако в области температур 1000–1400°К результаты, приводимые в работах [4,17], согласуются и поэтому могут быть использованы при анализе процессов взаимодействия топлива с молибденом.

Так как карбиды молибдена и вольфрама являются менее стойкими соединениями, чем монокарбид урана, то в системах $UC - Me$ будут превалировать реакции типа (11). Что касается системы $Me - U_{1-x}C_{yx}S_{(1-y)x}$, то в ней следует рассмотреть отдельно два случая. Первый случай – это взаимодействие молибдена и вольфрама с твёрдым карбосульфидным раствором на основе моносульфида урана. Сравнивая в этом случае $\overline{F}_c^{U(c,s)}$ и ΔF^{Me_2C} и пользуясь при этом уравнением (9), можно показать, что образование карбидных фаз молибдена и вольфрама в зоне контакта $U_{1-x}C_{yx}S_{(1-y)x} - Mo(W)$

не должно проходить при температурах, по крайней мере, до 2000 К в широком интервале составов. Во втором случае рассмотрим карбо-сульфидный твёрдый раствор на основе монокарбида урана. В этом случае изменение $\bar{F}_c$ $U(C,S)$ при введении в твёрдый раствор серы невелико (предельная растворимость $\leqslant$ 4 мол%). Поэтому взаимо-действие карбосульфидов на основе монокарбида урана практически не будет отличаться от взаимодействия монокарбида. То есть при взаимодействиии твёрдых карбосульфидных растворов на основе моно-карбида урана с молибденом и вольфрамом возможно образование кар-бидных фаз на этих металлах.

Выполненный аналогичным способом термодинамический анализ процессов контактного взаимодействия карбосульфидов урана с нио-бием и танталом позволяет ожидать образования карбидных фаз в ши-роком интервале температур и составов карбосульфидного топлива.

3. Кинетика взаимодействия металлов с монокарбидом, моносульфидом и карбосульфидом урана

Экспериментально взаимодействие монокарбида урана с различ-ными металлами изучалось в работах [13,19-23] . Показано, что образование карбидных фаз в зоне контакта $UC-Nb(Ta, Mo, W)$ заметно происходит, начиная с 1000°С. Причём скорости карбидизации наибольшие отмечались в системах с участием ниобия и тантала и на-именьшие в системах с молибденом и особенно с вольфрамом. Карбид-ные слои росли, как правило, по параболическому закону [19,20] , что свидетельствовало о лимитировании процесса диффузионным пере-носом углерода через образующийся карбид.

В настоящей работе были проведены некоторые уточняющие экспе-рименты. Объектами исследования служили образцы монокарбида урана, имеющие 4,8 вес%С, и монокристаллы ниобия (99,99%) и молибдена (99,9%). Было установлено, что при 1000-1800°С в зоне контакта $UC-Nb$ и $UC-Mo$ образовывались карбидные слои (Nb_2C + $+ NbC$) и Mo_2C.

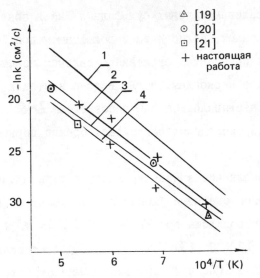

Рис. 2. Температурная зависимость константы скорости роста карбидных слоёв (K = δ²/t) на молибдене и ниобии: 1 – Mo – C [24]; 2 – Nb – UC; 3 – Nb–C [26]; 4 – Mo – UC.

Результаты экспериментов вместе с литературными данными показаны на рис. 2. Там же для сравнения приведены скорости роста карбидных слоёв на ниобии и молибдене при взаимодействии их с графитом [24-26]. Видно, что карбидные слои на молибдене в системе UC – Mo растут медленнее, чем в случае взаимодействия молибдена с графитом [24]. Отмеченный факт закономерен и связан с меньшим значением термодинамической активности углерода в монокарбиде урана (уравнение 4) по сравнению с графитом, где $a_c \simeq 1$.

Что касается кинетики роста карбидных слоёв на ниобии, то оказалось, что скорости роста карбидов ниобия в системах UC – Nb и графит-ниобий [25,26] близки (и даже несколько больше в системе UC – Nb) (рис. 2). По-видимому, это связано с тем фактом, что процессу образования карбидов ниобия в системе UC – Nb сопутствует появление фазы жидкого урана.

В известных экспериментальных работах по взаимодействию моносульфида урана с различными металлами [27-31] сообщается об

отсутствии образования новых сульфидных фаз в зоне контакта US
с ниобием, танталом, молибденом и вольфрамом при 1650-2500°С. В
тех же работах приводятся сведения о сильном взаимодействии моносульфида урана с цирконием, когда в зоне контакта $US-Zr$ уже
при 800°С образовывался за 340 часов слой ZrS толщиной 700мкм,
а при 1825°С происходило полное растворение циркония в моносульфиде урана.

Нами в дополнение к известным литературным данным проводились
исследования взаимодействия моносульфида урана с рядом металлов.
Эксперименты проводилсь при 1700-1900°С (Mo,W), 1200-1700°С
(Nb,Ta) и 1200°С (Zr,Ti,V). Продолжительность экспериментов составляла 1000 ч при всех температурах, кроме 1900°С,
при которой время проведения экспериментов не превышало 340 часов.

С помощью выполненных экспериментов было показано отсутствие
появления каких-либо новых сульфидных соединений на границе контакта $US-Me$ для всех рассмотренных случаев. Всё взаимодействие
сводилось лишь к диффузии урана и серы из моносульфида в контак-
тирующий с ним металл и образованию твёрдых растворов. Такой характер взаимодействия объясняется соотношением свободных энергий образования моносульфида и сульфидов переходных металлов (см.раз-
дел I). Образование сульфидов переходных металлов по реакции (II)
за счёт наличия области гомогенности у моносульфида урана в условиях наших экспериментов не проходило по той причине, что моносульфид урана имел состав, располагающийся вблизи нижней границы области гомогенности.

Так как коэффициенты диффузии серы в Nb,Ta,Zr и V [32,33]
(а по аналогии с ними, по-видимому, и в Ti) относительно велики,
то во всех случаях в этих металлах образовывались насыщенные твёрдые растворы, которые при охлаждении распадались, что приводило
к выделению по всему объёму точечных мелкодисперсных включений
сульфидных фаз.

Насыщение молибдена и вольфрама ураном и серой происходило более медленно по сравнению с Nb, Ta, Zr, Ti и V. Так, например, концентрация урана в вольфраме и молибдене после испытаний при 1700°С в течение 1000 часов не превышала 10^{-2} ат% на расстоянии 4 микрон от границы контакта и уменьшалась по мере дальнейшего удаления от границы. Небольшие скорости насыщения молибдена и особенно вольфрама ураном и серой объясняются малыми значениями коэффициентов диффузии U и S в этих металлах [34–36].

Так как скорости диффузии урана и серы в металлах, как правило, отличаются, то при взаимодействии моносульфида урана с различными металлами будет происходить изменение состава моносульфида урана. Ранее нами была показана [36] возможность изменения состава двуокиси и мононитрида урана при их взаимодействии с молибденом и вольфрамом и были приведены значения стабильных составов, при которых скорости выноса урана и металлоидной компоненты были пропорциональны отношению их мольных долей.

В настоящей работе с использованием полученных ранее данных по термодинамике моносульфида урана (уравнения 5,6) и результатов работ [34–36] была оценена температурная зависимость стабильных составов моносульфида урана для случаев контактирования его с Mo и W. Было установлено, что стабильные составы для систем US_{1-z}-Mo и US_{1-z}-W располагались вблизи стехиометрического коэффициента и описывались следующими уравнениями:

$$Z = 1{,}6 \cdot 10^{-1} exp\left(-37500/RT\right),\qquad (12)$$

$$Z = 2{,}5 \cdot 10^{4} exp\left(-65900/RT\right).\qquad (13)$$

Помимо относительно чистого моносульфида урана ($0 \leqslant 0{,}07\%$ и $C \leqslant 0{,}09$), нами рассматривался моносульфид, в котором имелись повышенные количества кислорода (0,14–1,0%) и углерода (0,18–0,60%).

Повышение концентрации кислорода приводило к появлению легкоплавкой фазы эвтектического типа, что согласовывалось с результатами работ [37,38] , в которых подчёркивалось, что моносульфид урана растворяет очень небольшие количества кислорода и при высоких температурах легко образуется оксисульфид UOS , который в свою очередь образует легкоплавкие эвтектики с US или UO_2. По-видимому, предельная растворимость кислорода в моносульфиде находится в интервале 0,07 – 0,14 %.

Наличие в моносульфиде 0,18-0,60% углерода ни в коей мере не повлияло на процессы контактного взаимодействия его с молибденом и вольфрамом, но значительно изменяло характер его взаимодействия с ниобием и танталом. Так, например, при 1700°C на ниобии вырастал карбидный слой NbC , насыщенный серой. При продолжительности эксперимента 1000 часов толщина его составляла 25 мкм.

Сопоставляя скорости карбидизации ниобия в системах с моносульфидом, содержащим углерод, и монокарбидом, видно, что в последнем случае скорость карбидизации ($K_{1600°}$ =1,1·10^{-9} см^2/с) значительно выше, чем в первом ($K_{1700°}$ =1,7·10^{-12} см^2/с). Причём, если в системе UC-Nb наблюдалась при 1700°C жидкая фаза, то в системе $\left[US + (0,6\% C) \right] - Nb$ жидкой фазы при этой температуре не обнаруживалось.

Увеличение содержания углерода в моносульфиде урана до 1,9÷2,0% не изменяло общей картины взаимодействия. Так, например, молибден и вольфрам не карбидизировались при взаимодействии с $US_{0,55}C_{0,45}$ при 1750 и 1950°C в течение 25 часов. Никаких жидких фаз в зоне контакта также не наблюдалось. Уместно напомнить, что температура плавления перитектики в системе UC-Mo по данным [39,40] находится вблизи 1850°C, а скорость карбидизации молибдена в системе UC-Mo уже при 1400°C равна 2,5·10^{-11} см^2/с .

Ниобий и тантал при взаимодействии с карбосульфидом $US_{0,55}C_{0,45}$ при 1750 и 1950°C образовывали насыщенные твёрдые растворы углерода и серы и карбидные фазы. Однако скорости их образования были значительно меньше, чем в системах с монокарбидом урана.

Полученные результаты находятся в согласии с данными работы [13], носящей качественный характер, и с результатами термодинамических оценок, выполненных в настоящей работе.

Были проведены также эксперименты по взаимодействию вольфрама и молибдена с монокарбидом урана, содержащим ∼ 0,01вес% S. Не было отмечено какой-либо разницы в поведении чистого монокарбида урана и монокарбида, содержащего серу в качестве примеси.

ЗАКЛЮЧЕНИЕ

Привлечение термодинамики фаз переменного состава позволило предсказать поведение карбосульфида урана по отношению ко многим металлам. С помощью теоретических оценок и экспериментальных данных было показано, что карбосульфид урана на основе моносульфида является более инертным по отношению к различным металлам, чем монокарбид урана. Поэтому US-UC может рассматриваться как ядерное топливо, конкурентно способное по отношению к монокарбиду (с точки зрения совместимости с материалами оболочек) и по отношению к моносульфиду урана (с точки зрения удельного содержания урана).

ЛИТЕРАТУРА

[1] ЗАГРЯЗКИН, В.Н., ПАНОВ, А.С., УСАЧЕВА, М.И., ФИВЕЙСКИЙ, Е.В., Ж. Физ. Хим., 47 (1973) 1946.
[2] ЗАГРЯЗКИН, В.Н., ПАНОВ, А.С., ФИВЕЙСКИЙ, Е.В., Ж. Физ. Хим., 47 (1973) 1951.
[3] ЗАГРЯЗКИН, В.Н., ПАНОВ, А.С., ФИВЕЙСКИЙ, Е.В., Thermodynamics of Nuclear Materials, (Proc. Symp. Vienna, 1974) Vol. 2, IAEA, Vienna (1975) 431.
[4] СТОРМС, Э., Тугоплавкие Карбиды, М., Атомиздат, 1970.
[5] STORMS, E.K., Thermodynamics (Proc. Symp. Vienna, 1965) Vol. 1, IAEA, Vienna (1966) 309.
[6] ACKERMANN, R.J., RAUH, E.J., J. Phys. Chem. 73 (1969) 769.
[7] O'HARE, P.A.G., et al., Thermodynamics of Nuclear Materials (Proc. Symp. Vienna, 1967),IAEA, Vienna (1968) 265.

[8] RAND, M.H., KUBASCHEWSKI, O., The Thermochemical Properties of Uranium Compounds, Oliver and Boyd, Edinburgh (1963).

[9] BASKIN, Y., SMITH, S.D., J. Nucl. Mater. 37 (1970) 209.

[10] IMOTO, S., NIIHARA, K., STÖCKER, H.-J., Thermodynamics of Nuclear Materials (Proc. Symp. Vienna, 1967), IAEA, Vienna (1968) 371.

[11] SHALEK, P.D., J. Am. Ceram. Soc. 46 (1963) 155.

[12] ЗАГРЯЗКИН, В.Н., Доклад на Международной конференции по реакторному материаловедению, г. Алушта, май 1978 г.

[13] SHALEK, P.D., WHITE, G.D., Carbides in Nuclear Energy, v.I, London, (1964) p. 266.

[14] HAGER, J.P., ELLIOTT, J.F., Trans. Metall. Soc. AIME 239 (1967) 513.

[15] LARSON, H.R., ELLIOTT, J.F., Trans. Metall. Soc. AIME 239 (1967) 1713.

[16] FREDRICKSON, D.K., CHASANOV, M.G., J. Chem. Thermodyn. 3 (1971) 693.

[17] SCHICK, H.L., Thermodynamics of Certain Refractory Compounds, Vols 1, 2, New York (1966).

[18] GUPTA, D.K., SEIGLE, L.L., Metall. Trans. 6A (1975) 1939.

[19] ЕРЕМЕЕВ, В.С., ПАНОВ, А.С., УШАКОВ, Б.Ф., ФИВЕЙСКИЙ, Е.В., Thermodynamics of Nuclear Materials (Proc. Symp. Vienna, 1965) Vol. 2, IAEA, Vienna (1966) 161.

[20] WEINBERG, A.F., YANG, L., Advanced Energy Conversion, New York (1963) p. 101.

[21] COEN, V., et al., Ceramic Nuclear Fuels, New York (1963) p. 147.

[22] CHUBB, W., DICKERSON, R.L., Am. Ceram. Soc. Bull. 9 (1962) 564.

[23] KATZ, S., J. Nucl. Mater. 6 (1962) 172.

[24] ЕРЕМЕЕВ, В.С., ПАНОВ, А.С., Известия АН СССР, сер. неорг. матер. 6 9 (1968) 1507.

[25] TOBIN, J.M., et al., Nuclear Applications of Nonfissionable Ceramics, H., (1966) p. 257.

[26] FUJIKAWA, Y., et al., J. Jap. Inst. Metals 34 (1970) 1259.

[27] SCHUMAR, J.F., USAEC Report TID-7687, (1964).

[28] SHALEK, P.D., HANDWERK, J.H., USAEC Report ANL-FGF-397, (1962).

[29] EDWARDS, R.K., USAEC Report ANL-6900, (1962).

[30] React. Mater. 6 2 (1963).

[31] ROBINSON, L.E., et al., Nucl. Appl. 1 (1965) 168.

[32] ВАНДЫШЕВ, Б.А., ПАНОВ, А.С., Известия АН СССР, сер. металлы 1 (1968) 206; 1 (1969) 244.

[33] ВАНДЫШЕВ, Б.А., ГРУЗИН, П.Л., ПАНОВ, А.С., Физ. Мет. Металловед. 23 (1967) 908.

[34] ИОВКОВ, В.П., ПАНОВ, А.С., РЯБЕНКО, А.В., Физ. Мет. Металловед. 34 (1972) 1322.

[35] ВАНДЫШЕВ, Б.А., ПАНОВ, А.С., Физ. Мет. Металловед. 25 (1968) 321.

[36] ЗАГРЯЗКИН, В.Н., ПАНОВ, А.С. и др. Thermodynamics of Nuclear Materials (Proc. Symp. Vienna, 1974) Vol. 2, IAEA, Vienna (1975) 193.

[37] PETITJEAN, G., ACCARY, A., J. Nucl. Mater. 34 (1970) 59.

[38] SHALEK, P.D., Dissertation Abstracts 29 8 (1969) 2861 B.

[39] CHUBB, W., J. Nucl. Mater. 23 (1967) 336.

[40] АЛЕКСЕЕВА, З.М., ИВАНОВ, О.С., Физико-химия Сплавов и Тугоплавких Соединений с Торием и Ураном, М., Наука, 1968, стр. 136.

DISCUSSION

P.E. POTTER: Have you taken the solubility limits of UC in US and of US in UC from already published data or do you have some more information?

A.S. PANOV: The data on the solubility limits of UC in US and of US in UC used in the calculations are taken from Shalek's work. We have no experimental data of our own on this problem.

INVESTIGATIONS ON SILICOTHERMIC
REDUCTION OF URANIUM FLUORIDES

J.M. JUNEJA, S.P. GARG, RAJENDRA PRASAD,
D.D. SOOD
Bhabha Atomic Research Centre,
Trombay, Bombay,
India

Abstract

INVESTIGATIONS ON SILICOTHERMIC REDUCTION OF URANIUM FLUORIDES.

The reactions of silicon with uranium trifluoride and uranium tetrafluoride have been investigated thermogravimetrically with a view to optimizing parameters for preparation of uranium-silicon intermetallic compounds. It was observed that the reaction proceeds in a stepwise sequence:

$$UF_4 \rightarrow UF_3 \rightarrow USi_3 \rightarrow USi_{1.88} \rightarrow U_3Si_5$$

The thermodynamic properties of the USi_3 phase have been determined over the temperature range 870–1040 K by measurement of the partial pressure of SiF_4 in equilibrium with $UF_3 + USi_3 + Si$. These data can be represented by the equation:

$$\log p_{SiF_4} = (9.02 \pm 0.20) - (8272 \pm 200) \, T^{-1}$$

where p is in kilopascals and T in kelvin. The data have been used to obtain the free energy of formation of USi_3 as given by the relation:

$$\Delta G_f^0 = -133.7 - 0.3 \times 10^{-3} \, T$$

where ΔG_f^0 is in $kJ \cdot mol^{-1}$.

1. INTRODUCTION

Uranium and silicon form six intermetallic compounds: USi_3, $USi_{1.88}$, U_3Si_5, USi, U_3Si_2 and U_3Si [1]. Uranium–silicon alloys with compositions close to U_3Si have been considered for use as nuclear fuels [2]. Some of the methods suggested for the preparation of these silicides are: (i) arc melting of constituent elements; (ii) co-reduction of UF_4 and SiO_2 with magnesium or calcium; and (iii) silicothermic reduction of uranium fluorides [1]. The third method is very simple and has the potential of giving essentially pure silicides, but details regarding this are not available in the literature. The reactions of silicon with UF_4 and UF_3 have therefore been studied thermogravimetrically in the present investigation.

The thermodynamics of the U–Si system have been investigated only by Alcock and Grieveson [3], using the Knudsen effusion method over the temperature range 1680–1790 K. In the present investigation the thermodynamic properties of the USi_3 phase have been determined over the temperature range 870–1040 K by measurement of the partial pressure of SiF_4 in equilibrium with a mixture of UF_3, USi_3 and Si, using the static method of vapour pressure measurement.

2.   EXPERIMENTAL

2.1. Materials

Uranium tetrafluoride was prepared by reducing high purity UO_3 with H_2 to UO_2, followed by hydrofluorination with anhydrous HF at 850 K. The conversion of UO_2 to UF_4 was checked by the analysis of water in the effluent and by pyrohydrolysis of the product. Uranium trifluoride was prepared by hydrogen reduction of UF_4 at 1100 K. Silicon powder (−200 mesh) of >99.99% purity was obtained by grinding semiconductor grade silicon rod in a dry box under an argon atmosphere.

2.2. Procedure

2.2.1.   Thermogravimetric studies

The reaction of UF_4/UF_3 with silicon has been investigated in a Mettler vacuum thermobalance. Mixtures of UF_4+Si or UF_3+Si (~1.5 g charge), having known compositions, were heated in a tantalum crucible under a dynamic vacuum (10 to 0.1 Pa). The experiments were carried out either by heating the sample at a constant heating rate over the range 2–10 K/min up to 1230 K or by heating the charge at a predetermined temperature for 2 to 3 hours. The progress of the reaction was monitored by recording the weight change (TG) and the rate of weight change (DTG) together with the temperature and pressure of the system. After each experiment the charge was cooled and examined by X-ray diffraction analysis. Investigations were limited to temperatures below 1230 K because of the substantial weight loss due to volatilization of UF_3 above this temperature.

2.2.2.   Thermodynamic studies

The pressures of SiF_4 in equilibrium with UF_3, USi_3 and Si were measured using the static method. A mixture of UF_3, Si and USi_3 was put into a tantalum boat and placed in a nickel tube closed at one end. The open end of the tube was

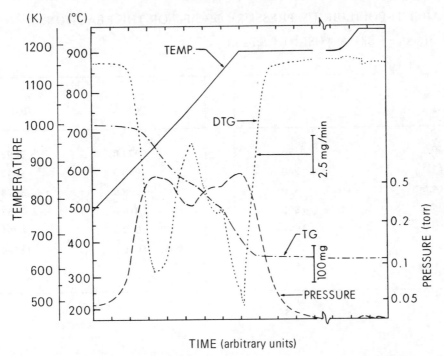

FIG.1. *Thermogram for the reaction of UF$_4$ with Si (heating rate 10 K/min; UF$_4$ = 1.291 g; Si = 0.264 g).*

connected to a manometer and a vacuum manifold. The fluorolube oil (density = 1.908 g·cm^{-3} at 301 K) used as the manometric fluid was non-reactive to SiF$_4$. After loading the charge, the system was checked for vacuum tightness and degassed at 600 K before isolating it from the vacuum manifold and heating it to the desired temperature. The temperature was measured with a calibrated chromel/alumel thermocouple and was controlled to within ±1 K. The uncertainty in pressure measurement was ±10 Pa.

3. RESULTS AND DISCUSSION

3.1. Thermogravimetric studies

The thermogravimetric studies on the reactions of silicon with UF$_4$, at various heating rates ranging from 2 to 10 K per minute, revealed that the reaction proceeds up to the formation of the U$_3$Si$_5$ phase only. A typical thermogram is given in Fig.1. It is seen that the DTG curve has three prominent peaks with maxima at 933 K, 1113 K and 1177 K. The pressure record also shows peaks corresponding

TABLE I. EQUILIBRIUM PRESSURE OF SiF$_4$ FOR THE REACTION:

$$\tfrac{4}{3} UF_3(s) + 5 Si(s) \rightleftharpoons \tfrac{4}{3} USi_3(s) + SiF_4(g)$$

T (K)	p (Pa)	T (K)	p (Pa)
869.5	376	977.0	3592
912.0	940	998.0	5642
914.0	830	998.0	5078
945.5	1805	1016.0	7767
950.0	1994	1017.5	8754
956.0	2144	1035.0	11528
976.0	3310		

TABLE II. FREE ENERGY DATA FOR SOME REACTIONS REPRESENTED BY:

$$\Delta G^0 = A + BT^{-1}$$

where ΔG^0 is in kilojoules and T is in kelvin.

Reaction	Reference and temp. range	A	B
$\tfrac{4}{3} UF_3(s) + 5 Si(s) \rightarrow \tfrac{4}{3} USi_3(s) + SiF_4(g)$	Present work (870 – 1040 K)	(158.3±3.8)	$-(134.3\pm4.0)\times10^{-3}$
$U(s) + \tfrac{3}{2} F_2(g) \rightarrow UF_3(s)$	Long and Blankenship [5] (800 – 1300 K)	−1464.4	209.2×10^{-3}
$Si(s) + 2 F_2(g) \rightarrow SiF_4(g)$	JANAF tables [6] (800 – 1100)	−1615.9	145.1×10^{-3}
$U(s) + 3 Si(s) \rightarrow USi_3(s)$	Present work (870 – 1040 K)	−133.7	-0.3×10^{-3}
$U(s) + 3 Si(s) \rightarrow USi_3(s)$	Alcock and Grieveson [3] (1680 – 1790 K)	−135.4	4.2×10^{-3}

to the DTG curve, probably resulting from the rapid evolution of SiF_4 gas. The most likely reactions in this system are as follows:

$$4\,UF_4(s) + Si(s) \rightarrow 4\,UF_3(s) + SiF_4(g) \tag{I}$$

$$4\,UF_3(s) + 15\,Si(s) \rightarrow 4\,USi_3(s) + 3\,SiF_4(g) \tag{II}$$

$$4\,UF_3(s) + 9.4\,USi_3(s) \rightarrow 13.4\,USi_{1.88}(s) + 3\,SiF_4(g) \tag{III}$$

$$4\,UF_3(s) + 42.4\,USi_{1.88}(s) \rightarrow 15.47\,U_3Si_5(s) + 3\,SiF_4(g) \tag{IV}$$

The shapes of the DTG and TG curves suggest that these reactions proceed in a sequential manner, but quantitative analysis of the data have revealed that the steps overlap with one another. Preparation of individual product phases was investigated during experiments in which mixtures containing stoichiometric quantities of UF_4 and Si were heated under vacuum at a fixed temperature for 2 to 3 hours. It was observed that essentially pure UF_3, USi_3, $USi_{1.88}$ and U_3Si_5 phases could be prepared by carrying out the reactions at 900, 1085, 1210 and 1223 K, respectively. The weight loss during these experiments corresponded to more than 99% of the value expected from the reaction equations (I)–(IV). The reaction product was analysed by X-ray diffraction and the patterns obtained agreed with those reported in the literature [4] in each case. Lines corresponding to uranium fluorides, silicon or other silicides were not detected in this analysis. In some of the experiments UF_3 was used as a starting material instead of UF_4, but this did not alter the course of the reactions.

Attempts to prepare lower silicides, viz. USi, U_3Si_2 and U_3Si, by this method were not successful because up to 1230 K the reaction did not proceed beyond the formation of the U_3Si_5 phase. Even at 1350 K, where the volatilization of UF_3 becomes significant, the product did not reveal the presence of lower silicides. Silicothermic reduction of fluorides can thus be used for the preparation of fluoride-free higher silicides, but the preparation of USi and lower silicides by this method does not appear to be feasible.

3.2. Thermodynamic study of the USi_3 phase

The thermodynamic properties of the USi_3 phase have been determined over the temperature range 870–1040 K by measurement of the equilibrium pressures of SiF_4 for reaction (II). The equilibrium, which could be attained in less than 24 hours, was found to be reversible. The experimental data are given in Table I and can be represented by the following linear relation:

$$\log_{10}p_{SiF_4} = (9.02 \pm 0.20) - (8272 \pm 200)T^{-1}$$

where p is in kilopascals and T in kelvin. The reaction mixture containing UF_3, Si and USi$_3$ phases was analysed by X-ray diffraction before and after the equilibrium-pressure measurements. It was observed that the equilibrium studies did not lead to the formation of any new phase and that there was no shift in the d-spacings for any of the phases, thereby indicating the absence of any solid-solution formation in the mixture. Thus if all phases are present in their pure states the equilibrium constant (K_{II}) for reaction (II) can be given by:

$$K_{II} = p_{SiF_4}^3 \qquad (1)$$

The free energy change for reaction (II) is given by:

$$\Delta G_2^0 = -3RT \ln p_{SiF_4} = 4\Delta G_{USi_3}^0 + 3\Delta G_{SiF_4}^0 - 4\Delta G_{UF_3}^0 \qquad (2)$$

The free energy change for this reaction, per mole of SiF$_4$ produced, is given by the expression (Table II):

$$\Delta G_{SiF_4}^0 = A + BT^{-1} \qquad (3)$$

where ΔG^0 is in kJ/mol and T is in kelvin. The value of the free energy change at 1000 K is (24.0 ±0.2) kJ. The standard free energy of formation of the USi$_3$ phase is given in Table II. It was obtained by combining the present data with the free energies of formation of UF$_3$ as reported by Long and Blankenship [5] and of SiF$_4$ as reported in the JANAF tables [6]. The table also includes the free energy data for the USi$_3$ phase reported by Alcock and Grieveson [3]. The value of $\Delta G_{USi_3}^0$ at 1000 K calculated from the present data and that from the data of Alcock and Grieveson are 134.2 kJ·mol^{-1} (870–1040 K) and 131.2 kJ·mol^{-1} (1680–1790 K), respectively. It is seen that the agreement is fairly good, despite the great difference in the temperature ranges of the investigations.

ACKNOWLEDGEMENTS

The authors thank Mr. R. Venkataramani and Dr. K.N. Roy for their help during the investigations. They are also grateful to Shri C.V. Sundaram, Head, Metallurgy Division, and Dr. M.V. Ramaniah, Head, Radiochemistry Division, BARC, for their interest in this work.

REFERENCES

[1]   WILKINSON, W.D., Uranium Metallurgy, Vol. 2, Interscience Publishers (1962) 1109.
[2]   CHALDER, G.H., BOURNS, W.T., FERADAY, M.A., VEEDER, J., Atomic Energy of
       Canada Ltd Rep. AECL-2874 (1967).
[3]   ALCOCK, C.B., GRIEVESON, P., J. Inst. Metals **90** (1962) 304.
[4]   POTTER, P.E., "Carbides, silicides and borides", Comprehensive Inorganic Chemistry,
       Vol. 5, Actinides (BAILAR, J.C., et al., Eds) Pergamon Press, Oxford (1973).
[5]   LONG, G., BLANKENSHIP, F.F., USAEC Rep. ORNL-TM-2065 (1969).
[6]   JANAF Thermochemical Tables, The Dow Chemical Company, Michigan (1971).

DISCUSSION

M.H. RAND: Can the reaction of UF_4 and silicon be used as a clean process
for the preparation of UF_3 which is employed in solid-state electrolyte studies for
uranium compounds? Many people have found difficulty in preparing UF_3 by
traditional methods.

D.D. SOOD: Yes, it is quite possible to use the reaction of UF_4 and silicon
to prepare pure UF_3; however our other investigations, which will be reported
shortly, reveal that it is equally easy to prepare UF_3 by reduction of UF_4 with H_2
at 1100 K.

Section J

PHASE DIAGRAMS

ADVANCED FUELS FOR FAST BREEDER REACTORS
A critical assessment of some phase equilibria

P.E. POTTER
Chemistry Division,
Atomic Energy Research Establishment,
Harwell, Didcot, Oxfordshire,
United Kingdom

K.E. SPEAR
Materials Research Laboratory,
Pennsylvania State University,
University Park, Pennsylvania,
United States of America

Abstract

ADVANCED FUELS FOR FAST BREEDER REACTORS: A CRITICAL ASSESSMENT OF
SOME PHASE EQUILIBRIA.
    In this paper are assessed critically six ternary systems of great significance to the
preparation, fabrication and performance of advanced fuels for use in fast breeder nuclear reactors.
The systems which have been considered are: U—C—O, Pu—C—O, U—C—N, Pu—C—N, U—N—O,
Pu—N—O. All the systems are characterized by partial or complete solid solutions, and a major
task of this assessment has been to develop simple models for these solutions which allow of
consistency between the known thermodynamic and phase equilibrium data of the binary systems
and the known condensed and gaseous phase equilibria of the ternary systems. Either ideal or
regular solution models have been employed to describe the behaviour of the various solutions.

1. INTRODUCTION

In this paper we review critically many of the published experimental data on
the phase equilibria of six ternary systems of great significance to the prediction and
to the understanding of the behaviour of the 'so-called' advanced nuclear fuels for
fast breeder reactors. The fuels are solid solutions of uranium and plutonium
carbides or nitrides. The systems which we have chosen to examine here are:
U—C—O, Pu—C—O, U—C—N, Pu—C—N, U—N—O, Pu—N—O. These studies form
part of the IAEA co-ordinated project on the critical assessment of the thermodynamic properties of the actinide elements and their compounds [1].

The uranium-plutonium carbide fuels would most likely be prepared by the carbothermic reduction of the oxides. The nitrides could also be prepared by such a method in the presence of nitrogen gas. With such preparative methods it is obvious that the carbides would contain some oxygen, and that the nitrides would contain both carbon and oxygen. Both uranium-plutonium oxycarbides and carbonitrides have also been considered as potential fuels for fast breeder reactors.

For the successful exploitation of all these potential fuel materials, that is, to optimize the conditions for their preparation and fabrication as well as to understand and to predict their behaviour during operation, it would be desirable to have a detailed knowledge of the appropriate regions of the five-component system, U—Pu—C—N—O. Before an accurate description of the equilibria in such a complex system can be given it is essential to have detailed descriptions of the systems of lower order.

The phase diagrams and thermodynamic data for the binary uranium systems are now quite well established although there remain some uncertainties in the sesquinitride region of the uranium—nitrogen system. For the phase diagrams and thermodynamic data of the systems and compounds containing plutonium some further experimental studies and evaluations are required. Before describing the major conclusions of our critical analysis of the six ternary systems we shall briefly review the phase diagrams of the relevant binary systems, namely U—C, Pu—C, U—N, Pu—N, U—O and Pu—O. The Gibbs energies of formation of the binary compounds which we have used in the analysis are also included.

2. THE BINARY SYSTEMS

2.1. Uranium—carbon

The phase diagram which describes the essential features of this system is that determined by Benz, Hoffman and Rupert [2]. The system is characterized by three compounds, the monocarbide, the sesquicarbide and the dicarbide. Uranium monocarbide has a face-centred cubic rock-salt structure and exists over a range of composition which increases with increase in temperature. Uranium sesquicarbide has a Pu_2C_3-type body-centred cubic structure. Uranium dicarbide exists in two forms, the low temperature α-body-centred tetragonal CaC_2-type structure, which transforms to the β-face-centred cubic KCN-type structure at higher temperatures (>2000 K). The monocarbide and the dicarbide melt congruently at 2780 K and 2720 K respectively. The Gibbs energies of formation for these compounds have been taken from the assessment of Storms [3], and these data, together with those for the other binary compounds required in this assessment, are given in Table I.

TABLE I. GIBBS ENERGIES OF FORMATION OF VARIOUS URANIUM
OR PLUTONIUM CARBIDES, NITRIDES AND OXIDES

Compound	Gibbs energy of formation, ΔG_f^0 (cal·mol^{-1})		Temperature (K)	Reference
UC	−26311	+0.28 T	1400–3000	[3]
U_2C_3	−46910	−3.28 T	1400–2100	[3]
$UC_{1.9}$	−17810	−4.71 T	1400–2800	[3]
$PuC_{0.88}$	−12402	−0.68 T	1000–1800	[5]
$PuC_{1.00}$	−13860	−0.18 T	1000–1800	this work
Pu_2C_3	−39272	−1.32 T	1000–2300	[5]
PuC_2	−8465	−6.41 T	1000–2500	[5]
UN	−71070	+20.60 T	298–3000	[19]
$UN_{1.50}(\beta)$	−87793	+31.80 T	298–2500	[19]
$UN_{1.54}(\alpha)$	−88930	+32.03 T	298–2500	[19]
PuN	−72377	+23.25 T	298–2500	[19]
UO_2	−258658	+40.84 T	1000–3000	[24]
Pu_2O_3	−399558	+63.81 T	1000–2150	[28]
$PuO_{1.61}$	−210623	+33.27 T	1000–2150	[28]

2.2. Plutonium—carbon

The phase diagram for the plutonium—carbon system is essentially based on
the work of Mulford et al. [4]. There are four compounds in the system, Pu_3C_2,
the monocarbide ('PuC'), the sesquicarbide and the dicarbide. Pu_3C_2 is of unkown
structure and decomposes peritectoidally at ca. 850 K into ε-Pu and 'PuC'. 'PuC',
which is isostructural with uranium monocarbide and also exists over a range of
composition, is always hypostoichiometric with respect to carbon. The
sesquicarbide is, of course, isomorphous with U_2C_3 but unlike the latter compound
exists over a range of composition. The dicarbide has the same structure as β-UC_2
and only exists at high temperatures. The form of the phase diagram is very
different from that for the U—C system; all the compounds, with the possible
exception of the dicarbide, decompose by either peritectoid or peritectic reactions.
The data for the Gibbs energies of formation of the plutonium carbides have
been taken from a recent assessment by Rand [5]. The values given in Table I for
the Gibbs energy of formation of the hypothetical stoichiometric monocarbide have

been obtained by an extrapolation of the enthalpy of formation/composition relationship for the hypostoichiometric monocarbide, together with a modification to the entropy term; there would be no stabilization due to the randomization of the vacancies in the carbon lattice of the stoichiometric compound.

2.3. Uranium—nitrogen

There are four compounds of this system, the mononitride, the α and β sesquinitrides and the nitrogen-deficient dinitride. Uranium mononitride has a face-centred cubic rock-salt structure and exists over a small range of composition (N:U ratio, 0.990 ± 0.006 → ca. 1.000) in the temperature range 1873—3073 K [6, 7]. It has also been suggested [8] that the mononitride is an hypostoichiometric compound in the nitrogen-rich region below ca. 1100 K (N:U ratio, 0.995—0.997). The quoted ranges of existence for this compound are much smaller than originally suggested [9]. The mononitride melts congruently at ca. 3123 K under a nitrogen pressure of ca. 2.5 atm [10].

The β-uranium sesquinitride is isomorphous with La_2O_3 and Pu_2O_3 (A-type rare earth structure): α-uranium sesquinitride is isomorphous with the body-centred cubic Mn_2O_3 structure. If the N:U ratio is greater than 1.70—1.75, then the face-centred cubic phase with the CaF_2 structure is formed [11].

There are many discrepancies in the literature regarding the exact compositions of the α-sesquinitride at the lower phase boundary and that of the β-sesquinitride. For β-$UN_{1.5}$, compositions which vary between $UN_{1.34}$ and $UN_{1.51}$ have been reported [8, 9, 12—15]; by analogy with the A-type rare earth oxides we have taken the composition of this compound to be $UN_{1.50}$. For the α-$UN_{1.5}$, the reported compositions at the lower phase boundary of this phase vary between 1.54 and 1.595 [14, 16—18]. We have taken the composition of this compound in equilibrium with β-$UN_{1.50}$ to be $UN_{1.54}$. The data for the Gibbs energies of formation of UN, $UN_{1.50}$ and $UN_{1.54}$ given in Table I are from a preliminary assessment [19].

2.4. Plutonium—nitrogen

Unlike the uranium—nitrogen system, this system possesses only one compound, the mononitride. Plutonium mononitride is isomorphous with uranium mononitride and thus has a face-centred cubic rock-salt structure; it probably has only a very narrow range of existence. Under 1 atm of nitrogen, PuN decomposes on heating at 2843±30 K to give Pu(ℓ) and N_2(g) [20, 21]. Even at 25 atm N_2, the melting behaviour was still incongruent [22]; Spear and Leitnaker [23] estimated that PuN melts congruently at 3103±50 K at a pressure of 50±20 atm of N_2 gas. The Gibbs energy of formation of this compound given in Table I is taken from a preliminary assessment [19].

2.5. Uranium—oxygen

For the purposes of this paper we shall be concerned only with the fluorite structured uranium dioxide phase. This compound exists over a very wide range of composition, and even at room temperature there is a small homogeneity range. At temperatures around 600 K, the upper phase boundary of urania deviates sharply from the stoichiometric composition, rising to a composition of $UO_{2.24}$ at 1400 K, at which temperature the ordered U_4O_{9-y} and U_3O_{8-z} are in equilibrium. Above 1400 K, UO_{2+x} is in equilibrium with U_3O_{8-z}, and then x increases only slowly with temperature; $x \approx 0.27$ at 2000 K. The value of x in UO_{2-x} at the lower phase boundary is close to zero up to temperatures of ca. 1500 K, but then increases to 0.33 at the temperature (2700 K) of the monotectic reaction

$$\text{Liquid } 1 + UO_{2-x} \rightleftharpoons \text{Liquid } 2$$

Above the monotectic temperature the lower phase boundary moves back sharply to higher O/U ratios until the maximum temperature is reached, the congruent melting point (3120 K), at a composition close to $UO_{2.00}$. The above information about the phase diagram for this region of the system is taken from a recent assessment of the thermodynamic properties and phase equilibria of this binary system [24]. The Gibbs energy of formation of UO_2 which is given in Table I is also taken from the assessment.

2.6. Plutonium—oxygen

There are still uncertainties in the nature of the phase diagram of this system. There are four compounds of the system, Pu_2O_3 (which has the hexagonal A-type rare earth oxide structure), $PuO_{1.52}$ and $PuO_{1.61}$ (which both have the body-centred cubic Mn_2O_3 structure, and finally PuO_2 (which has the face-centred cubic fluorite structure). $PuO_{1.61}$, which does not exist at room temperature, can accommodate additional oxygen atoms within its lattice, and PuO_2 can be oxygen deficient with many anion vacancies within its lattice. The major uncertainties in the phase relationships are those involving the $PuO_{1.61+x}$ and PuO_{2-y} phases [25], and the Pu_2O_3—$PuO_{1.61}$ two-phase region. For this latter region it is not clear whether the position of the upper phase boundary moves closer or whether it continues parallel to the lower phase boundary with increasing temperature [26]. We shall only be concerned with the Pu_2O_3—$PuO_{1.61}$ region of this system. The estimated melting point of Pu_2O_3 is 2358 K [27]. The Gibbs energies of formation for these two compounds given in Table I are based on the assessment of Ackermann and Chandrasekhariah [28].

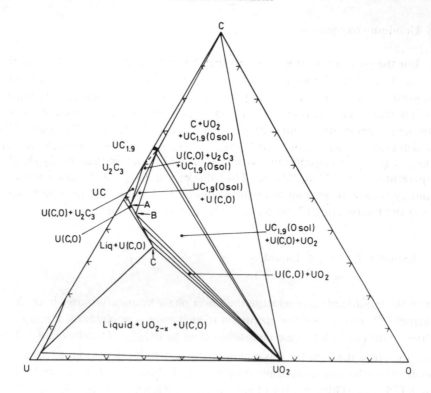

FIG.1.   A phase diagram of the U−C−O system at ca. 2000 K (see § 3.1).

3.   THE TERNARY SYSTEMS

The uranium–carbon–oxygen and plutonium–carbon–oxygen systems are characterized by the stabilization of the hypothetical uranium and plutonium monoxide phases by their dissolution in the respective monocarbides [29]. In the uranium–carbon–nitrogen and plutonium–carbon–nitrogen systems there is complete solubility between the respective monocarbides and mononitrides [30], but in the uranium–nitrogen–oxygen and plutonium–nitrogen–oxygen systems there is hardly any solubility of the hypothetical monoxide in the mononitride phases [31].

The main task of this presentation is to determine the nature of the solid solutions described above which together with the thermodynamic data for the relevant binary compounds reproduce the experimentally determined phase relationships. In addition to a successful description of the phase relationships the experimentally determined vapour pressures must be reproduced. We shall see that uranium dicarbide and probably plutonium dicarbide dissolve both oxygen and

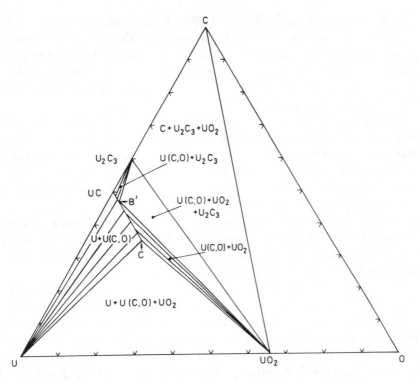

FIG.2. A phase diagram of the U–C–O ternary system when U_2C_3 is the only higher carbide present (see § 3.1).

nitrogen but that there is very little quantitative information available. In any calculations, we shall simply lower the thermodynamic activity of the dicarbide to describe these solubilities. Neither uranium nor plutonium sesquicarbide dissolve significant amounts of oxygen or nitrogen.

3.1. Uranium–carbon–oxygen

The form of the phase diagram for the conditions where uranium dicarbide is a phase of the system is shown in Fig.1. This is essentially the phase diagram presented by Henry et al. [32] for a temperature of 1973 K. Because of the increase in the range of composition of uranium monocarbide with temperature, the single phase uranium monoxycarbide region [U(C,O)] will exist over an area which increases with increase in temperature. In this part of the assessment, however, we shall be mainly concerned with the phase relationships at temperatures for which the area of the U(C,O) phase field will be quite small and thus deviations from the stoichiometric composition [(C+O):U = 1] will be rather small.

In the region of temperature 1800–2040 K, the range of existence of the α-UC_2 phase increases with increase in temperature. The influence of dissolved oxygen in the dicarbide phase on the equilibria in this region of the phase diagram together with the ternary system above the temperature of the peritectoid decomposition of U_2C_3 at 2100 K have received some attention [33–37] but will not be discussed in detail here.

At somewhat lower temperatures, for which the α-UC_2 phase may not be stabilized by oxygen solubility with respect to decomposition to U_2C_3, UO_2 and carbon, a phase diagram of the type shown in Fig.2 would describe the equilibria within the system. At temperatures below ca. 1500 K the only phase in the U–C system which is thermodynamically stable will be UC; the solubility of the hypothetical compound UO in UC will stabilize the monoxycarbide phase to higher temperatures. There will be a two-phase region of U(C,O)+C, and a three-phase region of U(C,O)+UO_2+C. These low-temperature phase relationships, which will not be changed significantly by the more recent thermodynamic data for the binary compounds, have been discussed in some detail [38].

As previously mentioned, the major feature of this ternary system is the stabilization of the compound UO by dissolution in UC, and much effort has been given to the determination of the variation of the lattice parameter of this face-centred cubic rock-salt structured solid solution with composition, as well as to the limit of solubility of UO in UC. Although there is considerable scatter in the experimental data for the variation of lattice parameter with composition [29], it is apparent that there is an appreciable decrease in lattice parameter with increase in oxygen concentration in the solid solution; for example, for UC, $a_0 = 4.9605$ Å whilst for $UC_{0.65}O_{0.35}$, $a_0 = 4.949$ Å. The majority of the experimental studies [22, 32, 39–46] indicate that the maximum solubility of UO in UC is between 32 and 37 mol%. Some higher values of 80 and 88 mol% [47, 48] for the maximum solubility are inexplicable, particularly as the lattice parameter given for $UC_{0.2}O_{0.8}$ is close to that given for the solid solution with 32–37 mol% UO substitution in UC. Some apparently low values of 25 mol% UO for the maximum solubility [37, 49] might be due simply to the fact that the three-phase univariant phase field [U(C,O)+UO_2+U] might not have been reached in the reduction experiments – indeed the CO pressures reported are higher than those determined by Henry et al. [32] and the earlier calculated values [29].

An essential requirement for the self-consistency of these phase diagrams is that any model which we choose to describe the solid solutions, here the uranium monoxycarbide phase [U(C,O)], together with the thermodynamic data for the binary compounds, should reproduce the experimentally determined positions of the invariant points which are points A and B in Fig.1 and point B' in Fig.2. The position of the invariant point C in both figures is known already. Another requirement is that the measured pressures of the gas phase species should also be reproduced. Carbon monoxide is the major gas phase species above the regions of this ternary system with which we are concerned: only at temperatures above 2000 K will the pressures of the uranium-containing gaseous species become significant in particular regions of the phase diagram [29, 50]. This is the reason for the ease with which uranium monocarbide containing only very small quantities of oxygen can be prepared by the carbothermic reduction of urania.

In order to calculate the compositions of the invariant points A, B and B′, an estimate of the Gibbs energy of formation of the uranium monoxide phase will be required. In order to obtain this we shall require the value of the maximum solubility of UO in UC, which we have taken as 35 mol%, and a model for the solid solution. For the model of the solid solution we have taken that of a regular solution which becomes ideal when the interaction parameter, E, is zero. At equilibrium:[1]

$$\langle UO_2 \rangle + \{U\} = 2[UO]_{UC} \quad (T > 1405 \text{ K}) \tag{I}$$

and thus

$$\Delta G^0_{f, \langle UO_2 \rangle} = 2\Delta \bar{G}_{[UO]} \tag{1}$$

where $\Delta G^0_{f, \langle UO_2 \rangle}$ and $\Delta G_{[UO]}$ are the Gibbs energy of formation and the partial molal Gibbs energy of the two phases, respectively, and:

$$\Delta \bar{G}_{[UO]} = \Delta G^0_{f, \langle UO \rangle} + RT \ln x + (1-x)^2 E \tag{2}$$

$\Delta G^0_{f, \langle UO \rangle}$ is the Gibbs energy of formation of pure UO, x is the mole fraction of UO dissolved in UC, and T is the temperature. As x = 0.35, $\Delta G^0_{f, \langle UO \rangle}$ can be calculated from the relationship:

$$\Delta G^0_{f, \langle UO \rangle} = 0.5 \, \Delta G^0_{f, \langle UO_2 \rangle} + 2.08 \, T - 0.42 \, E \tag{3}$$

We now examine the parameters required to reproduce the experimentally determined invariant points given in Table II. In order to calculate these points we have equated the chemical potentials of one component of the system for the neighbouring phase fields. For example at point A in Fig.1 the uranium chemical potential of the U(C,O)+UC$_{1.9}$ phase field will equal that for the U(C,O)+U$_2$C$_3$ phase field. At point B of Fig.1, the uranium chemical potential for the U(C,O)+UC$_{1.9}$ phase field will equal that for the U(C,O)+UO$_2$ phase field. (One could alternatively equate the chemical potentials of either of the other two components, namely carbon or oxygen, or indeed the pressures of one of the gas phase species, for example CO.)

---

[1] The following notation is used throughout the paper: $\langle \ \rangle$ pure element or compound in solid state; [ ] solid solution (suffix denotes solvent); { } liquid state; ( ) gaseous state.

TABLE II. THE EXPERIMENTALLY DETERMINED COMPOSITION OF
THE URANIUM MONOXYCARBIDE PHASE AT THE INVARIANT POINTS A
AND B (FIG.1) AND B′ (FIG.2)

Reference	Temperature (K)	UO concentration at point A (mol%)	UO concentration at point B or B′ (mol%)
Alcock et al. [42]	1573	7	11
Anselin et al. [22]	1673	–	5 (B′)
Brett et al. [40]	1373–1773	–	5 (B′)
Henry et al. [32]	1973	7	10
Magnier et al. [39]	1773	–	~6 (B′)
Pialoux, Dodé [35]	2043	–	10

For the phase field $U(C,O)+UC_{1.9}$, the chemical potential of uranium $(\Delta \overline{G}_U)$ is given by:

$$\Delta \overline{G}_U = 2.11 \, \Delta \overline{G}_{[UC]} - 1.11 \, \Delta \overline{G}_{[UC_{1.9}]} \qquad (4)$$

where $\Delta \overline{G}_{[UC]}$ and $\Delta \overline{G}_{[UC_{1.9}]}$ are the partial molal Gibbs energies of UC and $UC_{1.9}$ respectively.

$$\Delta \overline{G}_{[UC]} = \Delta G^0_{f,\langle UC \rangle} + RT \ln (1-x) + x^2 E \qquad (5)$$

where $\Delta G^0_{f,\langle UC \rangle}$ is the Gibbs energy of formation of UC, and x is the mole fraction of UC in the monoxycarbide phase.

$$\Delta \overline{G}_{[UC_{1.9}]} = \Delta G^0_{f,\langle UC_{1.9} \rangle} + RT \ln a \qquad (6)$$

where $\Delta G^0_{f,\langle UC_{1.9} \rangle}$ is the Gibbs energy of formation of $UC_{1.9}$, and a is its activity — which is possibly reduced below unity by dissolution of oxygen.

For the phase field $U(C,O)+UO_2$, the chemical potential of uranium $(\Delta \overline{G}_U)$ is given by:

$$\Delta \overline{G}_U = 2 \, \Delta \overline{G}_{[UO]} - \Delta G^0_{f,\langle UO_2 \rangle} \qquad (7)$$

where $\Delta\bar{G}_{[UO]}$ is the partial molal Gibbs energy of UO dissolved in UC, and $\Delta G^0_{f,\langle UO_2\rangle}$ is the Gibbs energy of formation of UO_2, and (cf. Eq.(2)):

$$\Delta\bar{G}_{[UO]} = \Delta G^0_{f,\langle UO\rangle} + RT \ln x + (1\text{-}x)^2 E$$

Thus in order to obtain the composition, x, of the U(C,O) phase at point B (Fig.1) for a given temperature, T, and value of the interaction parameter, E, the two expressions for the chemical potential of uranium are equated. In order to obtain the composition of the U(C,O) phase at point B' (Fig.2), the chemical potentials for the phase fields U(C,O)+U_2C_3 and U(C,O)+UO_2 are equated.

Some calculated compositions of the U(C,O) phase at the invariant points for different values of T, E and a, the activity of $UC_{1.9}$, are given in Table III. These calculated values indicate that the dicarbide phases will be stabilized to lower temperatures by increasing oxygen dissolution; i.e. a phase diagram of the type shown in Fig.1 will exist down to lower temperatures the lower the activity of the dicarbide. The concentration of UO in the solid solution at points B or B' increases with decreasing values of the interaction parameter, E. To reproduce the experimentally observed compositions at the invariant points given in Table II for the highest temperatures (1973 K and 2043 K), an interaction parameter for the regular solution of ca. −20 000 cal would be required. Whilst to reproduce the diagrams of the type given in Fig.2, where the composition of the solid solution at the invariant point B' is 5−6 mol% UO over the temperature range 1373−1773 K, an interaction parameter somewhat less negative of ca. − 12 000 cal would be required. It is, however, not possible to reproduce the values of the invariant points given by Alcock et al. at 1573 K; for the dicarbide to be a phase of the system with the given compositions of the U(C,O) phase at the invariant points, considerable amounts of oxygen would have to dissolve in the dicarbide in order to lower its activity sufficiently.

Alcock and colleagues [42−45] have determined experimentally the activities of uranium and oxygen for the uranium−uranium monoxycarbide, and uranium dioxide−uranium monoxycarbide regions of the phase diagram at temperatures of 1473, 1573 and 1648 K. These measurements could be interpreted in terms of a regular solution for UC−UO with an interaction parameter of −19 000 cal. It seems reasonable to give more weight to the experimental determinations of the invariant points at the highest temperatures, where there would be a better chance of attaining equilibrium. Thus with the direct measurements by Alcock et al. of the activities in regions of the phase diagram where the higher carbide phases were absent, we accept a value of ca. −20 000 cal for the regular solution interaction parameter. It is clear that further studies on the detailed phase relationships in the regions where the sesquicarbide and dicarbide phases are present would be desirable.

TABLE III. THE DEPENDENCE OF THE COMPOSITION OF THE U(C,O) PHASE WITH TEMPERATURE, T, INTERACTION PARAMETER, E, AND ACTIVITY OF URANIUM DICARBIDE, $a_{UC_{1.9}}$, AT THE INVARIANT POINTS

Temperature (K)	E (cal)	$a_{UC_{1.9}}$	Composition at invariant points		
			A	B	B'
			(x=mol. fraction UO)		
2100	−20000	1.0	0.035	0.106	−
	−20000	0.9	*	0.109	−
2000	−20000	1.0	0.067	0.100	−
	−20000	0.9	*	0.103	−
	−10000	1.0	−	−	0.053
	−10000	0.9	*	0.055	−
	0	1.0	−	−	0.017
	0	0.9	*	0.018	−
1900	−20000	1.0	−	−	0.094
	−20000	0.9	0.015	0.097	−
	−10000	1.0	−	−	0.047
	−10000	0.9	0.015	0.048	−
	0	1.0	−	−	0.014
	0	0.9	−	−	0.014
1800	−20000	1.0	−	−	0.090
	−20000	0.9	0.058	0.090	−
	−10000	1.0	−	−	0.042
	−10000	0.9	−	−	0.042
	0	1.0	−	−	0.011
	0	0.9	−	−	0.011
1700	−20000	1.0	−	−	0.084
	−10000	1.0	−	−	0.036
	0	1.0	−	−	0.009

* For these conditions either U_2C_3 is not a phase of the system or the UO concentration in the solid solution is negligible.'

Much attention has been paid to the determination of the variation of CO pressures with temperature for the univariant phase fields containing the dicarbide. These equilibria are, for the phase field $UC_{1.9}$, UO_2 and graphite:

$$\langle UO_2 \rangle + 3.9\ \langle C \rangle = [UC_{1.9}] + 2\ (CO) \tag{II}$$

and for the phase field $UC_{1.9}$, U(C,O) and UO_2:

$$0.9\ \langle UO_2 \rangle + 3\ [UC_{1.9}] = 3.9\ [UC]_{UO} + 1.8\ (CO) \tag{III}$$

The studies of these equilibria have already been summarized [29] and only a limited number of additional measurements [33–37] have been made since that summary. It should be remembered that, as the temperature increases, the values of the stoichiometric coefficients in the equations will change to compensate for the changes in composition of the phases. There is good agreement between the measured values and calculated values of the CO pressures in the temperature range 1500–2200 K which would preclude further measurements; the value of the CO pressure at a given temperature is little dependent on the value of the activity of $UC_{1.9}$ and, for the second equilibrium reaction, little dependent on the value of the interaction parameters for the monoxycarbide solid solution. By means of the measurement of the dependence of CO pressures within a carbon tube furnace on the extent of reaction between mixtures of graphite and UO_2 of different proportions at different temperatures, Dodé and coworkers have attempted to determine the position of the phase boundaries in this ternary system. Particular emphasis was placed on the compositions of the solid phases in the univariant phase fields. However, the values given for the O:U ratios of UO_{2-x} in equilibrium with $UC_{1.9}$ (oxygen saturated), namely 1.95 at 2043 K [35] and 1.97 at 2023 and 2038 K [37] require that these results and those for the compositions of the other phases should be critically examined. Within this phase field the activity of uranium must be the same in all the phases. For the two-phase region $UC_{1.9}$ + graphite, the uranium activity will decrease with decrease in $UC_{1.9}$ activity. At 2000 K, when $a_{UC_{1.9}} = 1$, log $a_U = -2.92$, and when $a_{UC_{1.9}} = 0.9$, log $a_U = -2.99$. The data of Tetenbaum and Hunt [51] suggest that the composition of UO_{2-x} with these uranium activities would have O:U ratios greater than 1.99. This assessment agrees with the experimental determinations of Wheeler [52]. For $UO_{1.97}$, log $a_U = -1.85$; it thus seems unlikely that the amount of reduction suggested by Dodé and coworkers occurs within this phase field. The composition of urania in equilibrium with uranium is $UO_{1.93}$ at 2000 K.

Measurements which have been made of the CO pressure in the univariant phase field UO_{2-x}, U, U(C,O) [34, 37, 53], for which the equilibrium reaction is:

$$\langle UO_{2-x} \rangle + (2-x)\ [UC]_{UO} = (3-x)\ \{U\} + (2-x)(CO) \quad (T > 1405\ K) \tag{IV}$$

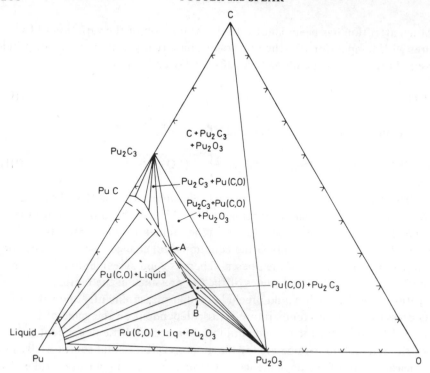

FIG.3. *A phase diagram of the Pu–C–O ternary system (see § 3.2).*

give pressures which are higher than the expected values, most probably because the conditions under which the experiments were carried out did not allow a sufficiently reducing environment for this reaction to occur. Only the single measurement of Henry et al. [32] at 1973 K, log $p_{CO} = -5.85$, is in reasonable agreement with the calculated value, log $p_{CO} = -5.5$, obtained with an interaction parameter, E, of $-20\,000$ cal. A more positive interaction parameter would increase the CO pressure, but only very slightly.

Whilst the essential features of the phase diagram are known there remain uncertainties, particularly in the relationships between the higher carbides in the presence of oxygen at different temperatures.

3.2. Plutonium–carbon–oxygen

The phase equilibria for this system were reviewed in 1971 [29], and some further assessments on this system have been done recently [54]. In this section, however, we present the calculations of the phase equilibria using the recently assessed data for the plutonium carbides given in Table I. The form of the phase diagram is given in Fig.3 and is quite similar to that for the U–C–O system.

The hypothetical compound plutonium monoxide is stabilized by dissolution in the monocarbide. There are three determinations of the maximum solubility [55–57] of the monoxide in the monocarbide in the temperature range 1273–1673 K which give between 65 and 78 mol%. All these monoxycarbide phases which have a face-centred cubic rock-salt structure and which are in equilibrium with Pu_2O_3 and $Pu(\ell)$ have nearly the same lattice parameter ($a_0 = 4.956$–4.958 Å). We have taken a value of 67 mol% PuO for the maximum solubility of this phase in the monocarbide (point B, Fig.3). The lattice parameter of $PuC_{0.88}$ increases with increase of oxygen concentration up to the composition $PuC_{0.9}O_{0.1}$, when it is assumed that all the vacancies in the carbon lattice of the monocarbide have been filled; further oxygen substitution results in a decrease of the lattice parameter of the solid solution [56].

The PuO concentration in the monoxycarbide solid solution at the other invariant point (point A, Fig.3) is 36–38 mol% [55, 56] for temperatures in the range 1273–1673 K.

We have calculated the change in position of the invariant point with temperature by a method similar to that described for the calculation of the invariant points in the U–C–O system.

The Gibbs energy of formation of the hypothetical compound plutonium monoxide is calculated from the equilibrium:

$$\{Pu\} + \langle Pu_2O_3 \rangle = 3\,[PuO]_{PuC} \tag{V}$$

At equilibrium:

$$\Delta\bar{G}_{[PuO]} = 0.33\,\Delta G^0_{f,\,\langle Pu_2O_3 \rangle} \tag{8}$$

where $\Delta\bar{G}_{[PuO]}$ and $\Delta G^0_{f,\,\langle Pu_2O_3 \rangle}$ are the partial molal Gibbs energy of PuO dissolved in PuC and the Gibbs energy of formation of Pu_2O_3, respectively. For a regular solution of PuO in PuC:

$$\Delta\bar{G}_{[PuO]} = \Delta G^0_{f,\,\langle PuO \rangle} + (1-x)^2 E + RT \ln x \tag{9}$$

and thus:

$$\Delta G^0_{f,\,\langle PuO \rangle} = 0.33\,\Delta G^0_{f,\,\langle Pu_2O_3 \rangle} - (1-x)^2 E - RT \ln x \tag{10}$$

and here $x = 0.67$.

The invariant point was then calculated by equating the chemical potentials of one of the three components for the phase fields $Pu_2C_3 + Pu$ monoxycarbide and $Pu_2O_3 + Pu$ monoxycarbide. In this calculation we used the Gibbs energy of formation of the hypothetical compound $PuC_{1.00}$, because the composition of the solid solution at the invariant point would be close to stoichiometric $((C+O)/Pu = 1)$.

The calculated values of the PuO concentration in the solid solution varied from 36.1 mol% at 1273 K to 38.0 mol% at 1673 K, with an interaction parameter, E, of zero; an ideal solid solution.

It is not possible to prepare the monocarbide by the carbothermic reduction of the plutonium oxides. The pressures of plutonium gas become comparable with those of CO in the phase fields Pu_2C_3+Pu monoxycarbide and Pu_2O_3+Pu monoxycarbide, and attempted reductions result in considerable loss of plutonium [56, 57]. The pressures of all the gas phase species above this system have been calculated previously [29, 54] and assessments using the more recent thermodynamic data for all the condensed and gaseous species will make little difference to the overall picture. There are only very few measurements of gaseous pressures above this system. Some measurements of the CO pressures above the phase field Pu_2O_3+Pu_2C_3+graphite [58, 59] give inexplicably low values for the Gibbs energy of formation of Pu_2C_3. Although there may be some slight solubility of oxygen in this phase, such solubility could not account for a stabilization of some 20 000 cal·mol^{-1} Pu_2C_3. The observation [56] of nearly congruent vaporization in a free evaporation experiment of a monoxycarbide of composition close to $PuC_{0.5}O_{0.5}$ in the temperature range 1600–1900 K would be expected from the calculations of the pressures of the main gaseous species, plutonium atoms and CO molecules.

The regions of the phase diagram where plutonium dicarbide could be a phase of the system have yet to be considered.

3.3. Uranium–carbon–nitrogen

The tenary U–C–N system has been studied quite extensively [22, 60–77]. The phase diagram of the system at 1773 K is shown in Fig.4. No ternary compounds exist, but the rock-salt structured UC and UN phases are completely miscible. The following discussion will be limited mostly to the thermodynamic characterization of this solid solution. There are three invariant compositions of the U(C,N) solid solution at points A, B and C of Fig.4. These invariant points define the three univariant phase fields U(C,N), U_2C_3 and $UC_{1.9}$, U(C,N), $UC_{1.9}$ and carbon, and finally U(C,N), β-$UN_{1.50}$ and carbon. At temperatures sufficiently low that $UC_{1.9}$ is no longer a phase of the system, the points A and B are replaced by a invariant point B' which defines the univariant phase field U(C,N), U_2C_3 and carbon.

Thermodynamic measurements on the UC–UN solution require a knowledge of the composition of the solution. Direct determination of the elemental composition by the techniques of analytical chemistry can be quite difficult, especially compared to measurements of lattice constants of the solution. Since the (C+N)/U ratio remains essentially constant at 1.00 in the temperature range with which we will be concerned [60, 70, 77], we have chosen to make use of a

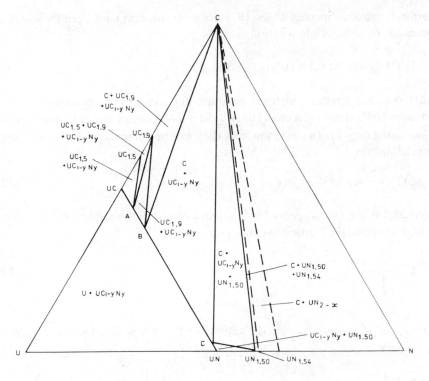

FIG.4.   A phase diagram of the U–C–N ternary system (see § 3.3).

recently assessed lattice constant/composition relationship [78] in order to
recalculate the compositions from the given values of lattice constants of the
solutions for the published equilibrium results.

Both Leitnaker et al. [70] and Cordfunke [77] have made accurate measurements of lattice constants for UC–UN solutions which have been analysed
chemically very carefully. When these data are corrected for dissolved oxygen and
surface layers of uranium dioxide [78], there is extremely good agreement between
the two sets of values, and we have fitted the following relationships to the data.

$$N/(C+N) = 1.000 - 11.2977\,(Ba_0) - 36.5833\,(Ba_0)^2$$

$$+\ 1.36532 \times 10^3\,(Ba_0)^3 - 1.92034 \times 10^4\,(Ba_0)^4 \tag{11}$$

where $Ba_0 = (a_0 - 4.8886)$, a_0 is the lattice constant in ångstroms.

Thermodynamic properties of the UC−UN solid solution have been measured by studies of the equilibrium reaction:

$$[UC]_{UN} + 0.5 (N_2) = [UN]_{UC} + \langle C \rangle_{graphite} \tag{VI}$$

at different temperatures, nitrogen pressures and solution compositions. Leitnaker [64], Naoumidis et al. [63, 68, 71], and Katsura and Nomura [74] have all presented data for this reaction, and they can be compared directly by using the relationship:

$$\Delta G_T^0 = - RT \ln K \tag{12}$$

where ΔG_T^0 is the Gibbs energy for the above equilibrium reaction and K is the equilibrium constant for the reaction, i.e.:

$$K = \frac{a_C \, a_{UN}}{a_{N_2}^{\frac{1}{2}} \, a_{UC}} \tag{13}$$

where a_C, a_{UN}, a_{UC} and a_{N_2} are the activities of graphite, UN, UC and nitrogen gas, respectively.

$$\left. \begin{aligned} a_C &= 1 \\[1em] a_{N_2} &= p_{N_2} \\[1em] a_{UN} &= \gamma_{UN} \, x_{UN} \\[1em] a_{UC} &= \gamma_{UC} x_{UC} \end{aligned} \right\} \tag{14}$$

where p_{N_2} is the pressure of N_2 gas, γ_{UN} and x_{UN} are activity coefficient and mole fraction of UN in the solid solution, and γ_{UC} and x_{UC} are the corresponding quantities of the UC component of the solution.

From the identities above we obtain:

$$\ln \left(\frac{\gamma_{UN}}{\gamma_{UC}} \right) = \frac{-\Delta G_T^0}{RT} - \ln \left(\frac{x_{UN}}{x_{UC}} \right) + \frac{1}{2} \ln p_{N_2} \tag{15}$$

The value of ΔG_T^0 can be obtained from assessed data [3, 19], and from the experiments at temperature T we obtain x_{UN}/x_{UC} and p_{N_2}.

From the published data of the authors mentioned above and using a calculated value of ΔG_T^0, we have calculated values of γ_{UN}/γ_{UC}. Where the reported values of x_{UN}/x_{UC} were derived from given lattice parameter measurements, we have recalculated this ratio by using our critically evaluated relationship between lattice parameter and composition of the solution [78]. Plots of the variation of the activity coefficient ratio with composition of the solution are shown in Fig.5. The horizontal line for $\gamma_{UN}/\gamma_{UC} = 1.0$ represents the case of an ideal solid solution.

Figure 5 indicates that the only set of data which exhibits good precision is that of Leitnaker [64]. Close examination of the other sets of data reveals that the scatter in the experimental determinations cannot be explained by a temperature or composition dependence of the activity coefficients. The scatter is most probably due to the fact that equilibrium was not attained in the experiments.

Values of γ_{UN}/γ_{UC} obtained from the results of Leitnaker [64] almost all fall between 0.9 and 1.1. At values of x_{UN} approaching unity, the value of γ_{UN} should also approach unity. Since the value of the ratio γ_{UN}/γ_{UC} is also close to unity at these values of x_{UN}, the value of γ_{UC} must also be close to unity. If the value of γ_{UC} is close to unity at these values of x_{UC} it is necessarily unity at $x_{UC} = 1$ and it is probably unity for all compositions of the solution. We can thus conclude that γ_{UN} must also have a value close to unity for all experimentally determined compositions since the ratio γ_{UN}/γ_{UC} has a value close to unity. These data seem not to be temperature dependent and it thus seems reasonable to assume that the solid solution U(C,N) is ideal at all temperatures and compositions.

Some calculated equilibrium pressures of N_2 gas for the U(C,N) solution in equilibrium with U_2C_3, $UC_{1.9}$ and/or graphite are shown in Fig.6. The mono-carbonitride solution was assumed to be ideal and the other phases were assumed to have unit activities.

Ikeda et al. [79] have recently measured the pressures of uranium gas and N_2 gas over UN and three compositions of the U(C,N) solution in the temperature range 1900–2300 K. The compositions of the solid solution corresponded to UN mole fractions, x_{UN}, of 0.30, 0.48 and 0.69. In order to obtain information about the ideality of the U(C,N) solution phase, the following reaction was considered:

$$[UN]_{UC} = \{U\} + 0.5(N_2) \tag{VII}$$

The reported experimental pressures were used to calculate the equilibrium constant:

$$K = \frac{p_U \, p_{N_2}^{\frac{1}{2}}}{a_{UN}} = \frac{p_U \, p_{N_2}^{\frac{1}{2}}}{\gamma_{UN} \, x_{UN}} \tag{16}$$

Providing that both the uranium and nitrogen pressures are known, it is of no significance whether the material under investigation is single-phased or contains excess of uranium or a non-metal.

The reported data for the sample of UN ($a_{UN} = 1$) were used to calculate the value of the equilibrium constant, K, and the experimental data for the three

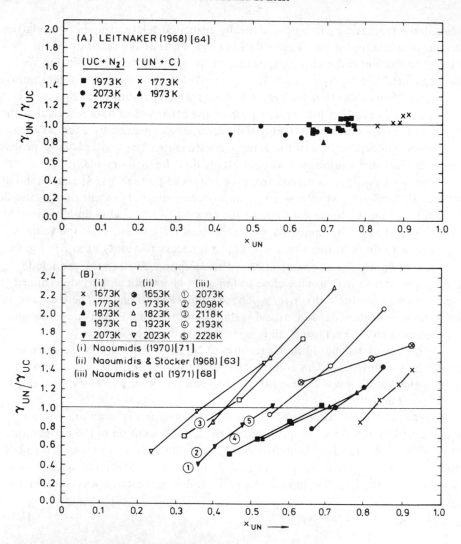

compositions of the solid solution were then used to calculate values of γ_{UN} for each composition. Possible systematic errors are cancelled out when such an analytical method is used. The calculated values of $\gamma_{U\dot{N}}$ vary from 0.47 to 0.10 or, when converted to regular solution interaction parameters, E, the variation is from −10 500 to −43 600 cal. These values are in poor agreement with the equilibrium measurements which have been previously discussed.

Pialoux [80] has recently reported phase equilibria studies in the U—C—N system for temperatures of 1073 to 2273 K. The equilibria in the system were determined using high-temperature X-ray diffractometry at constant temperatures

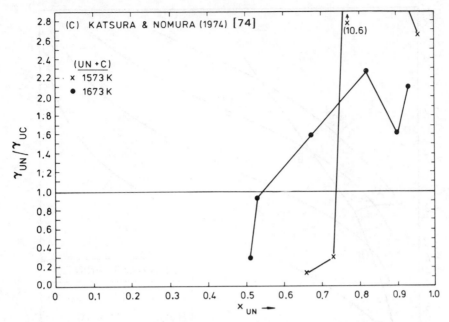

FIG.5.   *The relationship between the activity coefficient ratio (γ_{UN}/γ_{UC}) and mole fraction of UN (x_{UN}) in the UC–UN solid solution (see §3.3).*

and nitrogen pressures. The compositions of the U(C,N) solution at the invariant point B (Fig.4) together with the nitrogen pressures are in good agreement with the calculated values (Fig.6).

A number of investigators [60, 61, 68, 81, 82] have also reported the compositions of the U(C,N) phase at the invariant points A and B (Fig.4). These experimental values are scattered about the calculated values given in Fig.6.

This assessment would indicate that many of the measurements of the phase diagram in terms of the position of the invariant points are consistent with the chosen ideal solid-solution model which we have taken. Many of the experimental values of nitrogen pressures for various regions of the system are probably misleading in that equilibrium may not have been attained.

3.4. Plutonium–carbon–nitrogen

The phase diagram of the Pu–C–N system is similar in form to that of the U–C–N system (Fig.4). The complete miscibility of the monocarbide and mononitride is a major feature of the system and this solid solution is the main concern of the present discussion of the thermodynamic properties of the system.

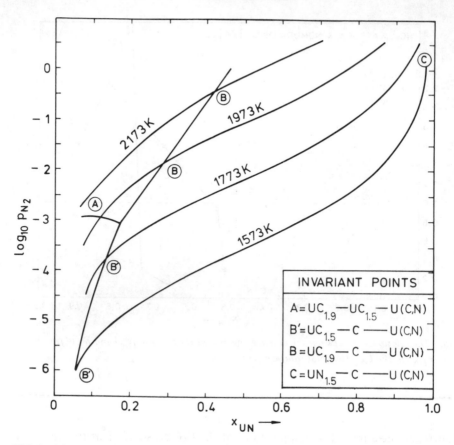

FIG.6.   The variation of nitrogen pressure (atm) with composition of the UC−UN solid solution in equilibrium with U_2C_3, $UC_{1.9}$ and/or graphite (see § 3.3).

The system, like that of U−C−N, is characterized by the invariant points A and B (Fig.4), and when Pu_2C_3 is the only carbide other than the monocarbide the invariant point is B'. Point A is the composition of the Pu(C,N) phase in equilibrium with Pu_2C_3 and PuC_2, point B is the composition of the solid solution in equilibrium with PuC_2 and graphite and point B' is the composition of the solid solution in equilibrium with Pu_2C_3 and graphite.

Lorenzelli [60] and de Franco [83] have examined the temperature−pressure−composition relationships in this system over the respective ranges of temperature of 1673−1873 K and 1823−2113 K. Lorenzelli [60] found that the solid solution at point B' contained 75 mol% PuN; there was no observed variation of this composition with temperature. de Franco [83] found that the monocarbonitride phase in equilibrium with PuC_2 and carbon at 2005 K contained 75 mol% PuN and

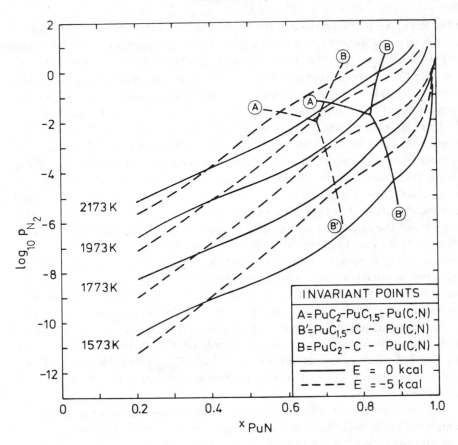

FIG.7.   *The variation of nitrogen pressure (atm) with composition of the PuC—PuN solid solution in equilibrium with Pu₂C₃, PuC₂, and/or graphite (see § 3.4).*

gave some equations for the temperature dependence of the nitrogen pressure for the three invariant phase fields defined by the invariant points A, B and B′.

Attempts have been made to develop a thermodynamic model for the Pu(C,N) solid solution phase which would reproduce the limited experimental data.

Figure 7 shows some results of our calculations of the dependence of the equilibrium pressures of nitrogen gas on temperature and composition of the solid solution for the non-metal rich regions of this ternary system. The solid lines are the results of the calculations using an ideal solid-solution model and the broken lines are the results of the calculations using a regular solution model with a negative interaction parameter, E, of − 5000 cal. The variations of the positions of the invariant points in the system are depicted by the lines A, B and B′ on Fig.7.

To the left of A—B', two-phase equilibria with PuC_2 and $Pu(C,N)$ or Pu_2C_3 and $Pu(C,N)$ exist; to the right of B—B' there are two-phase equilibria involving graphite and $Pu(C,N)$; above the A—B lines, two-phase equilibria with PuC_2 and $Pu(C,N)$ exist. For the univariant regions containing the monocarbonitride, graphite and a plutonium carbide, the calculations with an ideal solution model for $Pu(C,N)$ give compositions of this solution at the invariant points which are richer in PuN (B'—B in Fig.7) than those obtained by either Lorenzelli [60] or de Franco [83]. The solution of nitrogen in either Pu_2C_3 or PuC_2 would cause a shift in the composition of the monocarbonitride at the invariant points farther toward PuN. The use of a regular solution model with a negative interaction parameter moves the invariant points in the desired direction in order to obtain agreement with the experimental results, but then discrepancies arise between the measured nitrogen pressures and the calculated values. The nitrogen pressures measured as a function of temperature by de Franco [83] for the univariant regions are greater than those calculated using an ideal solution model and even greater than those calculated using a regular solution model with a negative interaction parameter. A positive interaction parameter is needed to obtain agreement between the calculated and measured pressures, but with such an interaction parameter the compositions of the monocarbonitride at the invariant points are not in agreement with the observed values. A positive interaction parameter, E, requires that there is an immiscibility gap in the PuC—PuN system at temperatures below $T_{critical} = (E/2R)$, where R is the ideal gas constant. No evidence for immiscibility in either $U(C,N)$ or $Pu(C,N)$ has ever been reported.

The discussion, given above, indicates the inconsistencies which arise when attempts are made to fit the experimental data for this system to a simple thermodynamic model of the $Pu(C,N)$ phase. New equilibrium measurements of the pressures in the univariant regions as well as of the positions of the invariant points are required in order to resolve these inconsistencies.

If we were to accept only the determinations of the invariant points, then the solid solution can be described by a small negative deviation from ideality.

3.5. Uranium—nitrogen—oxygen

No distinct ternary phases exist in the U—N—O system, but the binary phases form solid solutions which extend into the ternary composition regions. Uranium mononitride dissolves little oxygen; solubilities of oxygen expressed as the solubility of the hypothetical monoxide which vary from zero to 7 mol% have been reported [15, 77, 81, 84—87]. The low values of oxygen solubility may be more accurate since the prevention of surface oxidation of samples after equilibration experiments but before chemical analysis is extremely difficult even in glove boxes operating with inert atmospheres with very low impurity contents.

This low solubility of UO in UN compared with a solubility of 35 mol% UO in UC (see §3.1) allows the calculation of a regular solution interaction parameter between UO and UN. This interaction parameter must be different from that for the UC—UO solution, otherwise the solubilities of UO in both UC and UN would be the same [88]. The UO—UN solid solution will be ideal or will possess rather small deviations from ideality; the nature of the solution will be dependent on the exact value of the maximum solubility of UO in UN. The oxygen solubility in the hexagonal structured phase β-$UN_{1.50}$ has not been measured, but both Holleck and Ishii [89] and Benz et al. [15] report that it is small.

At high temperatures α-$UN_{1.54}$ and UO_2 are completely miscible with one another, but as the temperature is lowered an immiscibility gap forms. The critical temperature for this immiscibility region is taken as 1423±50 K. Holleck and Ishii [89] measured values of 1403±20 K for a sample of composition $[(UO_2)_{0.32}\ (UN_{1.54})_{0.68}]$ and 1423 ±20 K for a sample of composition $[(UO_2)_{0.54}\ (UN_{1.54})_{0.46}]$. Benz et al. [15] found that an immiscibility gap existed below 1443 K, whilst Blum et al. [90] gave a critical temperature of 1343±20 K which was obtained using thermal analysis, and Smith et al. [91] determined this temperature to be below 1473 K but above 1373 K. Studies of this solid solution are hindered because the rates of dissolution are so rapid that it has not been possible to quench to room temperature the samples which are in equilibrium at high temperature.

Martin [92] has reported that the solubility of UN in UO_2 increases from 5 to 13 mol% UN as the temperature increases from 1773 to 2273 K. Benz et al. [15] give a maximum solubility of 12 mol% UN in UO_2 at 2883 K.

With the above information and further detail from the studies of Benz et al. [15], we have constructed phase diagrams for the U—N—O ternary system (Fig.8) for two temperatures, 1350 K and 1500 K, that is just below and just above the critical temperature for the α-$UN_{1.54}$—UO_2 solid solution.

The decomposition of α-U_2N_3 dissolved in UO_2 to give UN and nitrogen gas has been investigated by both Benz et al. [15] and Holleck and Ishii [89]. Data are given for nitrogen pressures at different temperatures and compositions of the α-U_2N_3—UO_2 solution. As with the other systems which have been discussed in this paper, a regular solution model is used in an attempt to compare and reproduce these data.

The critical temperature of 1423 K was used to calculate a regular solution interaction parameter ($E = 2RT_{critical}$) of 5655 cal·mol^{-1} of mixing units. It was assumed initially that only the nitrogen and oxygen species are mixing in this solution. Thus a solution of $U_{0.5}O$ and $U_{0.65}N$ (or 0.5 UO_2 and (1/1.54)$UN_{1.54}$) has been considered here. The nitrogen gas pressures for the equilibrium reaction:

$$5.704\ [U_{0.65}N]_{U_{0.5}O} \rightleftharpoons 3.704\ \langle UN \rangle + (N_2) \tag{VIII}$$

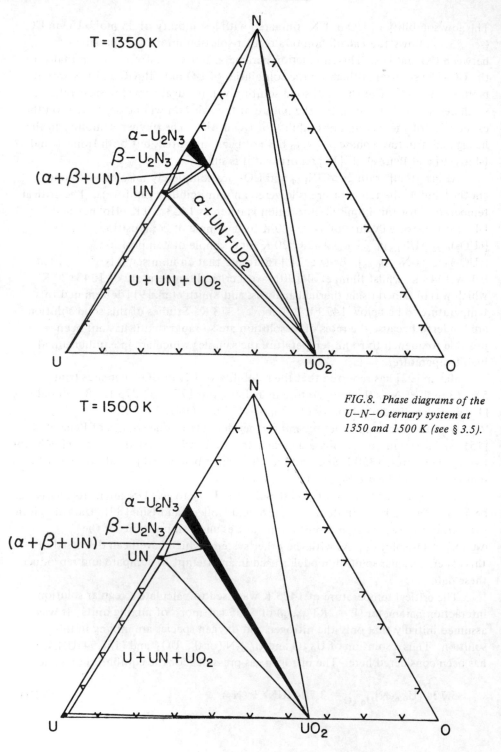

FIG.8. Phase diagrams of the U–N–O ternary system at 1350 and 1500 K (see § 3.5).

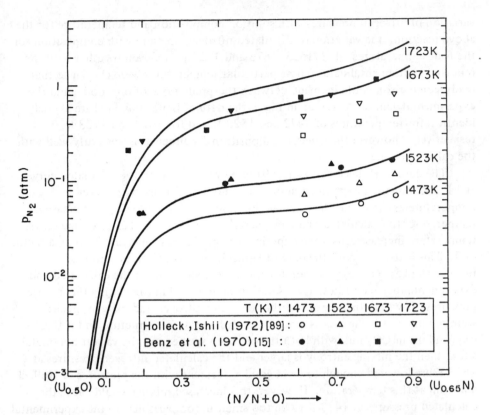

FIG.9. $\alpha\text{-}U_2N_3\text{-}UN\text{-}N_2$ equilibria in the $U\text{-}N\text{-}O$ system (see § 3.5). (Nitrogen pressure in atmospheres.)

were then calculated. The equilibrium constant, K, for this reaction is given by

$$K = \frac{p_{N_2}\, a_{UN}^{3.704}}{a_{U_{0.65}N}^{5.704}} \tag{17}$$

where p_{N_2} is the pressure of nitrogen gas and a_{UN} and $a_{U_{0.65}N}$ are the activities of UN and $U_{0.65}N$, respectively. The value of a_{UN} was taken as 0.95 ± 0.05 and $a_{U_{0.65}N}$ was represented by a regular solution relationship:

$$\ln a_{U_{0.65}N} = \ln x_{U_{0.65}N} + \frac{E}{RT}(1\text{-}x_{U_{0.65}N})^2 \tag{18}$$

where $x_{U_{0.65}N}$ is the mole fraction of $U_{0.65}N$ in the solid solution with $U_{0.5}O$, or it is the fraction [N/(N+O)] for the solution.

The Gibbs energies of formation for UN and $\alpha\text{-}UN_{1.54}$ were taken from the recent preliminary assessment [19] and are given in Table I. Figure 9 shows the variation of nitrogen pressures with solution composition and temperature for the above reaction; the variation of calculated nitrogen pressure with composition for the four temperatures 1473, 1523, 1673 and 1723 K are given together with the reported experimental data. These particular temperatures were chosen so that a ready comparison could be made between the predictions of our model and the experimental values. As is seen in Fig.9, the data of Benz et al. [15] are almost identical for temperatures of 1473 and 1523 K and for 1673 and 1723 K, respectively; however the general composition dependence agrees fairly well with the calculated data.

The data of Holleck and Ishii [89] agree very well with the calculated curve at 1473 K, but their experimental pressures do not increase with increasing temperature as rapidly as the calculated pressures. For a 200 degree increase in temperature, the calculated pressures increase by a factor of 12. A similar temperature increase causes the experimental pressures to increase by only a factor of 7. This is true for both the data of Holleck and Ishii [89] and those of Benz et al. [15]. One explanation for the difference between the calculated and experimental behaviour lies in the solution model used in calculating the nitrogen pressures. It was initially assumed that only the nitrogen and oxygen species were mixing, but if the α-U_2N_3phase is an ordered defect structure and if the oxygen ions in UO_2 mix with both the nitrogen species and the vacant non-metal sites, then the mixing entropy is larger and the calculated nitrogen pressures at 1673 K are lowered enough to be in good agreement with the pressures of Holleck and Ishii at this temperature. However, this increased mixing would cause the calculated pressures at 1473 K to be too small in comparison with the experimental data. Thus, it appears that the solution of α-U_2N_3 and UO_2 becomes more and more disordered as the temperature is increased. Clearly further experimental studies on the regions of this system containing this solid solution would be desirable.

3.6. Plutonium—nitrogen—oxygen

It has already been mentioned that the solubility of the hypothetical UO in UN is much lower than that in UC; the solubility of hypothetical PuO in PuN is also small, although one value as high as 14 mol% PuO has been reported [86]. The interaction parameter for the limited solid solution between PuN and PuO can be readily calculated from the information which we have for the PuC—PuO solid solution; PuN—PuO cannot be an ideal solution otherwise the solubilities of PuO in PuC and in PuN would be identical; thus for consistency between the two systems an interaction parameter of 8—11 kcal would be required to describe the PuN—PuO solid solution.

4. DISCUSSION

In this paper we have examined many of the features of the six ternary
equilibrium phase diagrams. Many of the data have allowed an assessment of the
likely interactions in the monoxycarbide, monocarbonitride and monoxynitride
solid solutions. There is complete miscibility in only the uranium and plutonium
monocarbonitride systems; the uranium solid solution is most probably ideal,
and that of the plutonium system may require a small negative solution interaction
parameter (E = −5000 cal) to describe it.

There is considerable solubility of the hypothetical monoxides in the monocarbides. The evidence from much of the data on the U−C−O system indicates that
the UC−UO solid solution is characterized by a quite large negative regular solution
interaction parameter (E = −20 000 cal). The PuC−PuO solid solution is however
close to ideal.

Finally, for the monoxynitride systems which possess only very limited solubility
of the hypothetical monoxides in the mononitrides, the calculations of interaction
parameters are rather uncertain because of the imprecise information on solubilities.
The UN−UO solid solution may not deviate greatly from ideality (E not greater
than 5000 cal), whilst the interaction parameter for the PuN−PuO solution will be
more positive than that for the uranium-containing solution. For the U−N−O
system there is also a solid solution between α-$UN_{1.54}$ and UO_2; these phases are
completely miscible above ca. 1423 K, and the solution has been described by a
regular solution model with an interaction parameter of ca. 5500 cal.

These assessments have indicated much of the information which is still
required in order to describe these systems in a satisfactorily consistent manner
and they should be a guide to the planning of further experimental endeavours.

REFERENCES

[1] POTTER, P.E., HOLLECK, H., SPEAR, K.E., Selected Ternary Systems, Part 11 of
    The Chemical Thermodynamics of Actinide Elements and Compounds (MEDVEDEV, V.,
    RAND, M.H., WESTRUM, E.F., Jr., Eds; OETTING, F.L., Exec. Ed.), IAEA, Vienna
    (in preparation).
[2] BENZ, R., HOFFMANN, C.G., RUPERT, G.N., High Temp. Sci. 1 (1969) 342.
[3] STORMS, E.K. in: ACKERMANN, R.J., OETTING, F.L., POTTER, P.E., RAND, M.H.,
    STORMS, E.K., The Uranium−Carbon, Plutonium−Carbon and Uranium−Plutonium−
    Carbon Systems (Report based on a Consultants' Mtg Grenoble, 1974), AERE, Harwell
    (in preparation).
[4] MULFORD, R.N.R., ELLINGER, F.H., HENDRIX, G.S., ALBRECHT, E.D., Plutonium
    1960 (Proc. 2nd Int. Conf. Grenoble, 1960: GRISON, E., LORD, W.B.H.,
    FOWLER, R.D., Eds), Cleaver Hume Press, London (1961) 301.
[5] RAND, M.H., unpublished work, AERE, Harwell, 1978 (see Ref. [3]).
[6] BENZ, R., HUTCHINSON, W.B., J. Nucl. Mater. 36 (1970) 135.
[7] ALEXANDER, C.A., OGDEN, J.S., PARDUE, W.M., J. Nucl. Mater. 31 (1969) 13.

[8] HOENIG, C.L., J. Am. Ceram. Soc. **54** (1971) 391.
[9] BENZ, R., BOWMAN, M.G., J. Am. Chem. Soc. **88** (1966) 264.
[10] OLSON, W.M., MULFORD, R.N.R., J. Phys. Chem. **67** (1963) 1952.
[11] RUNDLE, R.E., BAENZIGER, N.C., WILSON, A.S., McDONALD, R.A., J. Am. Chem. Soc. **70** (1948) 99.
[12] SASA, Y., ATODA, T., J. Am. Ceram. Soc. **53** (1970) 102.
[13] STOECKER, H.J., NAOUMIDIS, A., Ber. Dtsch. Keram. Soc. **43** (1966) 724.
[14] TAGAWA, H., J. Nucl. Mater. **41** (1971) 313.
[15] BENZ, R., BALOG, G., BACA, B.H., High Temp. Sci. **2** (1970) 221.
[16] LAPAT, P.E., HOLDEN, R.B., Compounds of Interest in Nuclear Reactor Technology (WABER, J.T., CHIOTTI, P., MINER, W.N., Eds), IMD Special Report 13, Metall. Soc. of the American Institute of Mining, Metallurgical, and Petroleum Engineers, New York, Nucl. Metall. **10** (1964) 225.
[17] BUGL, J., BAVER, A.A., Battelle Memorial Institute Rep. BMI-1692 (1964).
[18] MÜLLER, F., RAGOSS, H., Thermodynamics of Nuclear Materials, 1967 (Proc. Symp. Vienna, 1967), IAEA, Vienna (1968) 257.
[19] POTTER, P.E., *in* POTTER, P.E., MILLS, K.C., TAKAHASHI, Y., The Actinide Pnictides, Part 7 of The Chemical Thermodynamics of Actinide Elements and Compounds (MEDVEDEV, V., RAND, M.H., WESTRUM Jr., E.F., Eds.; OETTING, F.L., Exec. Ed.) IAEA, Vienna (in preparation).
[20] OLSON, W.M., MULFORD, R.N.R., J. Phys. Chem. **68** (1964) 1048.
[21] CAROLL, D.F., General Electric Co., Richland, Washington Rep. HW-SA-3370 (1964).
[22] ANSELIN, F., DEAN, G., LORENZELLI, R., PASCARD, R., Carbides in Nuclear Energy (RUSSELL, L.E., et al. Eds), Macmillan, London (1964) 131.
[23] SPEAR, K.E., LEITNAKER, J.M., Oak Ridge National Laboratory Rep. ORNL-TM-2106 (1968).
[24] RAND, M.H., ACKERMANN, R.J., GRØNVOLD, F., OETTING, F.L., PATTORET, A., paper presented at CNRS–IUPAC Conference on Refractory Oxides for High Temperature Power Sources (Odeillo, France, 1977), Rev. Int. Hautes Temp. Refract. **15** (1978) 355.
[25] SARI, C., BENEDICT, U., BLANK, H., Thermodynamics of Nuclear Materials, 1967 (Proc. Symp. Vienna, 1967), IAEA, Vienna (1968) 587.
[26] MARCON, J.P., POITREAU, J., ROULLET, G., Plutonium 1970 and Other Actinides (Proc. 4th Int. Conf. Santa Fe, New Mexico, 1970: MINER, W.N., Ed.) Metall. Soc. of the American Institute of Mining, Metallurgical, and Petroleum Engineers, New York, Nucl. Metall. **17** Part 2 (1970) 799.
[27] RILEY, B., Sci. Ceram. **5** (1970) 83.
[28] ACKERMANN, R.J., CHANDRASEKHARIAH, M.S., Thermodynamics of Nuclear Materials 1974 (Proc. Symp. Vienna, 1974) Vol.2, IAEA, Vienna (1975) 3.
[29] POTTER, P.E., J. Nucl. Mater. **42** (1972) 1.
[30] POTTER, P.E., Plutonium 1975 and Other Actinides (Proc. 5th Int. Conf. Baden-Baden, 1975; BLANK, H., LINDNER, R., Eds), Elsevier, New York *and* North Holland Publ., Amsterdam (1976) 211.
[31] POTTER, P.E., paper Conf. Reactor Materials Science, Alushta, USSR, 1978; to be published in Proceedings.
[32] HENRY, J.L., PAULSON, D.L., BLICKENSDERFER, R., KELLY, H.J., US Dept. of Interior Rep. BM-Rl-6963 (1967).
[33] PIALOUX, A., DODÉ, M., Rev. Int. Hautes Temp. Refract. **2** (1971) 154.
[34] PIALOUX, A., Rev. Int. Hautes Temp. Refract. **11** (1974) 147.
[35] PIALOUX, A., DODÉ, M., J. Nucl. Mater. **56** (1975) 221.

[36] PIALOUX, A., DODÉ, M., Colloque Int. du Centre national de la recherche scientifique
     (CNRS), Paris, No.205 (1971) 415.
[37] HEISS, A., J. Nucl. Mater. 55 (1975) 207.
[38] POTTER, P.E., Colloque Int. du Centre national de la recherche scientifique (CNRS),
     Paris, No.201 (1972) 249.
[39] MAGNIER, P., TROOVÉ, J., ACCARY, A., Carbides in Nuclear Energy (RUSSELL, L.E.,
     et al. Eds), Macmillan, London (1964) 95.
[40] BRETT, N.H., HARPER, E.A., HEDGER, H.J., POTTINGER, J.S., Carbides in Nuclear
     Energy (RUSSELL, L.E., et al., Eds), Macmillan, London (1964) 162.
[41] BESSON, J., BLUM, P.L., MORLEVAT, J.P., C.R. Hebd. Seances Acad. Sci., Ser.C 258
     (1964) 151; 260 (1965) 390: see also MORLEVAT, J.P., Commissariat à l'énergie
     atomique Rep. CEA R-2857 (1966).
[42] ALCOCK, C.B., JAVED, N.A., STEELE, B.C.H., Bull. Soc. Fr. Ceram. No.77 (1967) 99.
[43] JAVED, N.A., Ph.D. thesis, University of London, 1968.
[44] STEELE, B.C.H., JAVED, N.A., ALCOCK, C.B., J. Nucl. Mater. 35 (1970) 1.
[45] JAVED, N.A., J. Nucl. Mater. 37 (1970) 353.
[46] BONCOEUR, M., ACCARY, A., Proc. Brit. Ceram. Soc. 8 (1967) 175.
[47] SANO, T., IMOTO, S., NAMBA, S., KATSURA, M., New Nuclear Materials Including
     Non-Metallic Fuels (Proc. Conf. Prague, 1963), IAEA, Vienna (1963) 429: see also Nippon
     Genshiryoku Gakkai-Shi (J. At. Energy Soc. Japan) 3 (1961) 457 and 6 (1964) 441.
[48] CHIOTTI, P., ROBINSON, W.C., KANNO, M., J. Less-Common Met. 10 (1966) 273.
[49] STOOPS, R.F., HAMME, J.V., J. Am. Ceram. Soc. 47 (1964) 59.
[50] CHILTON, G.R., High Temperature Chemistry of Inorganic and Ceramic Materials
     (GLASSER, F.P., POTTER, P.E., Eds) Chemical Society, London, Special Publication
     No.30 (1977) 80.
[51] TETENBAUM, M., HUNT, P.D., J. Chem. Phys. 49 (1968) 4739.
[52] WHEELER, V.J., J. Nucl. Mater. 39 (1971) 315.
[53] STOOPS, R.F., HAMME, J.V., J. Am. Ceram. Soc. 47 (1964) 59.
[54] BESMANN, T., LINDEMER, T.M., J. Nucl. Mater. 67 (1977) 77.
[55] MULFORD, R.N.R., ELLINGER, F.H., JOHNSON, J.A., J. Nucl. Mater. 17 (1965) 324.
[56] POTTER, P.E., Thermodynamics of Nuclear Materials, 1967 (Proc. Symp. Vienna, 1967),
     IAEA, Vienna (1968) 337.
[57] SKAVDAHL, R.E., USAEC Rep. HW 77906 (1964).
[58] PASCARD, R., ROULLET, G., CEA, Fontenay-aux-Roses, unpublished work: see also
     Ref. [26].
[59] PICKLES, S., UKAEA Report TRG 1393 (D) (1967).
[60] LORENZELLI, R., Commissariat à l'énergie atomique Rep. R-3536 (1968).
[61] AUSTIN, A.E., GERDS, A.F., Battelle Memorial Institute Rep. BMI—1272 (1958).
[62] WILLIAMS, J., SAMBELL, R.A., J. Less-Common Met. 1 (1959) 217.
[63] NAOUMIDIS, A., STÖCKER, H.-J., Thermodynamics of Nuclear Materials, 1967 (Proc.
     Symp. Vienna, 1967), IAEA, Vienna (1968) 287.
[64] LEITNAKER, J.M., Thermodynamics of Nuclear Materials, 1967 (Proc. Symp. Vienna,
     1967), IAEA, Vienna (1968) 317.
[65] SANO, T., KATSURA, M., KAI, H., Thermodynamics of Nuclear Materials, 1967 (Proc.
     Symp. Vienna, 1967), IAEA, Vienna (1968) 301.
[66] MÜLLER, F., in Discussion, Thermodynamics of Nuclear Materials, 1967 (Proc. Symp.
     Vienna 1967), IAEA, Vienna (1968) 331.
[67] NAOUMIDIS, A., working paper submitted to an IAEA Panel on Thermodynamic
     Properties of Uranium and Plutonium Carbides, Vienna, 1968 (copies may be obtained
     from the author).

[68] NAOUMIDIS, A., KATSURA, M., NICKEL, H., Colloque Int. du Centre national de la recherche scientifique (CNRS), Paris, No. 205 (1971) 265.
[69] IKEDA, Y., TAMAKI, M., MATSUMOTO, G., J. Nucl. Mater. 59 (1976) 103.
[70] LEITNAKER, J.M., POTTER, R.A., SPEAR, K.A., LAING, W.R., High Temp. Sci. 1 (1969) 389.
[71] NAOUMIDIS, A., J. Nucl. Mater. 34 (1970) 230.
[72] KATSURA, M., NAOUMIDIS, A., NICKEL, H., J. Nucl. Mater. 36 (1970) 169.
[73] NOMURA, T., KATSURA, M., SANO, T., KAI, H., J. Nucl. Mater. 43 (1972) 234.
[74] KATSURA, M., NOMURA, T., J. Nucl. Mater. 51 (1974) 63.
[75] KATSURA, M., SHOHOJI, M., YATO, T., NOMURA, T., SANO, T., Thermodynamics of Nuclear Materials 1974 (Proc. Symp. Vienna, 1974) Vol.2, IAEA, Vienna (1975) 347.
[76] BLANK, H., working paper submitted to an IAEA Panel on Thermodynamic Properties of Uranium and Plutonium Carbides, Vienna, 1968 (copies may be obtained from the author).
[77] CORDFUNKE, E.H.P., J. Nucl. Mater. 56 (1975) 319.
[78] SPEAR, K.E., POTTER, P.E., to be published.
[79] IKEDA, Y., TAMAKI, M., MATSUMOTO, G., J. Nucl. Mater. 59 (1976) 103.
[80] PIALOUX, A., J. Nucl. Mater. 74 (1978) 328.
[81] HENRY, J.L., BLICKENSDERFER, R., J. Am. Ceram. Soc. 52 (1969) 534.
[82] HENNEY, J., HILL, N.A., LIVEY, D.T., Trans. Br. Ceram. Soc. 60 (1963) 955.
[83] de FRANCO, M., Commissariat à l'énergie atomique Rep. R-4573 (1974).
[84] JAVED, N.A., J. Less-Common Met. 29 (1972) 155.
[85] ANSELIN, F., Commissariat à l'énergie atomique Rep. R-2988(1966).
[86] LORENZELLI, R., DELAROCHE, M., HOUSSEAU, M., PETIT, P., Plutonium 1970 and other Actinides (Proc. 4th Int. Conf. Santa Fe, New Mexico: MINER, W.N., Ed.), Metall. Soc. of the American Institute of Mining, Metallurgical, and Petroleum Engineers, New York, Nucl. Metall. 17 Part 2 (1970) 818.
[87] BLUM, P.L., LAUGIER, J., MARTIN, J.M., MORLEVAT, J.P., C.R. Hebd. Seances Acad. Sci., Ser. C 266 (1968) 1456.
[88] IMOTO, S., STÖCKLER, H.J., Thermodynamics (Proc. Symp. Vienna, 1965) Vol.2, IAEA, Vienna (1966) 533.
[89] HOLLECK, H., ISHII, T., Thermal Analysis (Proc. 3rd Int. Conf. Davos, 1971) Vol.2, Birkhauser Verlag, Basel (1971) 137.
[90] BLUM, P.L., LAUGIER, J., MARTIN, J.M., C.R. Hebd. Seances Acad. Sci., Ser. C 268 (1969) 148.
[91] SMITH, R.A., OGDEN, J.S., ALEXANDER, C.A., Battelle Memorial Institute Rep. BMI-1872 (1969).
[92] MARTIN, J.M., J. Nucl. Mater. 34 (1970) 81.

DISCUSSION

R. BENZ: In the analysis of mono U(C,N) solid solutions as a pseudo binary mixture of UC and UN using measurements of p_{N_2} over fcc U(C,N) coexisting with higher uranium carbides, graphite or β-U_2N_3 at temperatures below about 1700°C, it is reasonable that, along a line between the terminal compounds UC and UN, the ternary equation:

$$x_N \, d\mu_N + 0.5 \, d\mu_U + x_C \, d\mu_C = 0$$

may be written for the pseudo binary system as:

$$N_{UN} \, d\mu_{UN} + N_{UC} \, d\mu_{UC} = 0.$$

The carbon activity along the non-metal rich side of U(C,N) must, however, pass through a relative maximum where graphite coexists with U(C,N), and it appears that one could assume that the uranium chemical potential, μ_U, along this boundary is approximately monotonic between UC and UN; hence it seems that UC activities must also pass through a relative maximum simultaneously with that of carbon. If this is true, then one would expect difficulties in interpreting U(C,N) as pseudo binary mixtures of UC and UN with models having mirror-symmetrical activities or potentials of UC and UN. In your calculations with existing thermodynamic data, have you noted that this peculiarity in activities does in fact exist?

P.E. POTTER: If we consider the phase field U(C,N) + UC$_{1.9}$ as the nitrogen concentration increases, the carbon activity also increases until we reach the invariant point B in Fig.4 of our paper. The carbon potential can be represented by the equilibria:

$$[UC]_{UN} + 0.9[C] = \langle UC_{1.9} \rangle$$

where [C] represents carbon at an activity less than unity. If we assume the activity of UC$_{1.9}$ to be close to unity, we can write the equilibrium constant, K_{eq}, of the above reaction as:

$$K_{eq} = \frac{1}{a_C^{0.9} \, a_{UC}}$$

Therefore, as the carbon activity increases towards unity, the UC activity must decrease. As the data in our paper indicate, deviations from ideality in the UC–UN solution are probably small. We appreciate that, at higher temperatures, the non-metal-to-metal ratios in the UC–UN solution can increase to a value greater than unity. Modelling these data will be more complex.

A. NAOUMIDIS: Katsura has tried to explain the large deviation in the carbon activity of different UC–UN solid solutions by the different configurations of carbon as the second solid phase in the phase region U(C,N)–C–N. There are probably difficulties in attaining equilibrium, because of the low diffusion coefficient of carbon in UC_xN_{1-x}.

P.E. POTTER: The major problem in the studies discussed here was that of attaining equilibrium rather than small differences in the activities of carbon due to its morphology.

ТЕРМОДИНАМИЧЕСКИЙ РАСЧЕТ
ФАЗОВОЙ ДИАГРАММЫ СИСТЕМЫ UC–UN

А.Л. УДОВСКИЙ, О.С. ИВАНОВ
Институт металлургии им. А.А. Байкова АН СССР,
Москва,
Союз Советских Социалистических Республик

Представлен А.С. Пановым

Abstract–Аннотация

THERMODYNAMIC CALCULATION OF THE PHASE DIAGRAM OF THE UC–UN SYSTEM.
In this paper the authors put forward a new version of the phase diagram for the UC–UN
section that they obtain by thermodynamic calculation using the regular-solution approximation
and taking into consideration the maximum melting point and the enthalpy of mixing of
experimentally obtained uranium carbonitride solid solutions. The results of the calculation
agree satisfactorily with experimental data on the melting points of uranium carbonitrides.
More particularly, the calculated alloy composition corresponding to the maximum melting
point is 0.6905, which is in good agreement with the experimental value of x_m = 0.7. The
analysis of the experimental concentration relationships between the adiabatic modulus of
elasticity, shear modulus and Debye temperature of the uranium carbonitride solid solutions
indicates, in accordance with the Lindemann relationship, the existence of a maximum melting
point in the region of compositions with x = 0.75–0.80 molar fractions of UN. The authors
also describe a method of calculating the thermodynamic parameters using a regular-solution
model from experimental phase diagrams for two-component systems. This method is
applicable to the bcc solid–liquid system phase equilibrium in the U–Pu system.

ТЕРМОДИНАМИЧЕСКИЙ РАСЧЕТ ФАЗОВОЙ ДИАГРАММЫ СИСТЕМЫ UC–UN.
В настоящей работе предложен новый вариант диаграммы состояния разреза UC–UN, полученный термодинамическим расчетом в приближении регулярных растворов и с привязкой к максимальной температуре плавления и энтальпии смешения твердых растворов карбонитридов урана,
полученных экспериментально. Результаты расчетов находятся в удовлетворительном согласии
с экспериментальными данными по температурам плавления карбонитридов урана. В частности,
вычисленная величина состава сплава, соответствующего максимальной температуре плавления,
равна 0,6905 и хорошо согласуется с экспериментальной величиной x_m = 0,7. Проведенный анализ
экспериментальных концентрационных зависимостей адиабатических модулей Юнга, сдвига и температуры Дебая твердых растворов карбонитридов урана также свидетельствует, согласно соотношению Линдемана, о наличии максимума температуры плавления в области составов
x = 0,75-0,80 мольных долей UN. Изложен также метод расчета термодинамических параметров в
модели регулярных растворов из экспериментальных диаграмм состояния двухкомпонентных
систем. Метод применен к фазовому равновесию ОЦК–жидкость (L) в системе U–Pu.

229

I. ВВЕДЕНИЕ

Вид фазовой диаграммы состояния системы $U-C-N$ существенным образом зависит от фазовой диаграммы разреза $UC-UN$. Непосредственных экспериментальных данных по кривым ликвидуса в системе $UC-UN$ в настоящее время нет. Известны лишь фазовые границы ликвидуса и солидуса для карбонитридов урана, полученные термодинамическим расчетом в приближении идеальных растворов [1]. Анализ термодинамических данных [2-4], проведенный в работе [5] свидетельствует о том, что карбонитриды урана в твердом состоянии не подчиняются модели идеальных растворов. Экспериментальные данные по температурам плавления сплавов $UC_{1-x}N_x$ [6] указывают на существование максимума температуры плавления $T_m = 2910 \pm 35°C$ при $UC_{0,3}N_{0,7 \pm 0,1}$, что свидетельствует об отклонении растворов карбонитридов урана от идеальных. Кроме того, из определения активности азота определена избыточная свободная энергия смешения карбонитридов урана в твердом состоянии [6], равная -0,7$\pm$1,5 ккал/моль для $UC-UN$. Экспериментальные данные по давлению газообразного азота в твердых растворах $UC-UN$ в интервале 1600- -2000°C [7] также показали, что избыточная свободная энергия имеет отрицательный знак, причем абсолютные значения избыточной свободной энергии для сплава $UC_{0,52}N_{0,48}$ при Т=2273, 2173 и 2073 К равны 870, 770 и 665 кал/моль, соответственно, что хорошо согласуется с данными Бенца [6].

В настоящей работе предложен новый вариант диаграммы состояния системы $UC-UN$, полученный термодинамическим расчетом в приближении регулярных растворов и с привязкой к максимальной температуре плавления и энтальпии смешения твердых растворов, полученных экспериментально [6], [7].

2. К термодинамике диаграммы состояния двух-компонентных систем с точкой максимума (минимума)

Термодинамический потенциал Гиббса i - фазы удобно представлять в виде:

$$G^i(x,T,P) = (1-x)G^i(0,T,P) + xG^i(1,T,P) +$$

$$+RT[x\ln x + (1-x)\ln(1-x)] + G^i_{изб.}(x,T,P), \qquad (2.1)$$

где х - концентрация второго компонента в фазе i
в атомных (мольных) долях, Т - температура, Р-давление, G^i (0,Т,Р) и G^i (I,Т,Р) - термодинамические
потенциалы Гиббса чистых компонентов (х=0 и х=I,
соответственно); $G^i_{изб}$ (х,Т,Р) - избыточный термодинамический потенциал Гиббса раствора в i -ой фазе,
имеющей вид:

$$G^i_{изб}(x,T,P) = x(1-x)\cdot f^i(x,T,P)$$
$$(2.2)$$

где $f^i(x,T,P)$ - аналитическая функция от своих аргументов.
Условия фазового равновесия i -ой и j -ой фаз при
Т= $const$ и Р= $const$

$$\left.\frac{\partial G^i(x,T,P)}{\partial x}\right|_{x^i} = \left.\frac{\partial G^j(y,T,P)}{\partial y}\right|_{x^j} = K(T,P) \left.\vphantom{\frac{\partial}{\partial}}\right\}$$

$$K(T,P)\left[x^i-x^j\right] = G^i(x^i,T,P) - G^j(x^j,T,P) \left.\vphantom{\frac{\partial}{\partial}}\right\}, (2.3)$$

где у - концентрация второго компонента в фазе j в
атомных долях, G^j (У,Т,Р) - имеет так же, как и
$G^i(x,T,P)$, вид уравнений (2.I) и (2.2); $x^i = x^{i/i+j}$,
$x^j = x^{j/i+j}$ - начало и конец ($i + j$) -фазной конноды,
в случае равенства нулю длины конноды, что реализуется на диаграммах состояния в точке максимума
или в точке равных концентраций (ТРК) с координатами (X_m, T_m), имеют вид

$$G^i\left(x^i_m,T_m,P\right) = G^j\left(x^j_m,T_m,P\right) \left.\vphantom{\frac{\partial}{\partial}}\right\}$$

$$\left.\frac{\partial G^i(x,T,P)}{\partial x}\right|_{x^i_m} = \left.\frac{\partial G^j(y,T,P)}{\partial y}\right|_{x^j_m} \left.\vphantom{\frac{\partial}{\partial}}\right\} \qquad (2.4)$$

$$x^i_m = x^j_m = Z \left.\vphantom{\frac{\partial}{\partial}}\right\}$$

Подставляя $G^i(x,T,P)$ и $G^j(y,T,P)$ в виде
(2.I) - (2.2) в соотношение (2.4), имеем

$$Z(1-Z)\left[f^i(Z,T_m,P) - f^j(Z,T_m,P)\right] = (1-Z)\Delta G^{i\to j}(0,T_m,P) + Z\Delta G^{i\to j}(1,T_m,P) \left.\vphantom{\frac{\partial}{\partial}}\right\}$$

$$Z(1-Z)\left(\left.\frac{\partial f^i}{\partial x}\right|_Z - \left.\frac{\partial f^j}{\partial y}\right|_Z\right) + (1-2Z)\left[f^i(Z,T_m,P) - f^j(Z,T_m,P)\right] = \left.\vphantom{\frac{\partial}{\partial}}\right\}$$

$$= \Delta G^{i\to j}(1,T_m,P) - \Delta G^{i\to j}(0,T_m,P) \qquad (2.5)$$

где для q -го компонента (q = 0 при x=0 и q = I при x=I)

$$\Delta G^{i-j}(q,Tm,P)\equiv G^j(q,Tm,P)-G^i(q,Tm,P)=-\int_{T^{i-j}(q,P)}^{Tm}\Delta S^{i-j}(q,T,P)dT$$

$$(2.6)$$

разность термодинамических потенциалов Гиббса между i и j - фазами q -го компонента, $T^{i-j}(q,P)$ - температура фазового перехода q -го компонента, $\Delta S^{i-j}(q,T,P)=S^j(q,T,P)-S^i(q,T,P)$ - разность энтропий j -ой и i -ой фаз q -го компонента.

Соотношения (2.5) и (2.6) являются точными и применимыми для фазовых границ с нулевой длиной конноды в ТРК, в рамках любых моделей. Иначе говоря, любые модели, описывающие диаграммы состояния с ТРК должны удовлетворять системе (2.5).

Рассмотрим применение (2.5) к некоторым простым феноменологическим моделям.

2.I. Модель идеальных растворов.

В рамках этой модели из (2.2) при $f^i(x,T,P)$ = $f^j(y,T,P)$ = 0 и (2.5) заключаем, что модель идеальных растворов для обоих i и j конкурирующих фаз не в состоянии описать диаграммы с ТРК. Однако, если $f^i(x,T,P)$ = 0, а $f^i(y,T,P)\neq0$, то в принципе диаграмма с ТРК может быть описана моделью идеальных растворов, но только для i -ой фазы.

2.2. Модель регулярных растворов.

В рамках модели регулярных растворов (2.2) принимает вид

$$G_{uзб}^i(x,T,P)=\mathcal{A}^i\cdot x(1-x),\ G_{uзб}^j(y,T,P)=\mathcal{A}^j\cdot y(1-y),\ (2.7)$$

где параметры взаимодействия (ПВ) $\mathcal{A}^i;\ \mathcal{A}^j=const$. Подстановка (2.7) в (2.5) дает:

$$(\mathcal{A}^i-\mathcal{A}^j)\cdot z\cdot(1-z)=(1-z)\cdot\Delta G^{i-j}(0,Tm,P)+z\cdot\Delta G^{i-j}(1,Tm,P)$$

$$(\mathcal{A}^i-\mathcal{A}^j)(1-2z)=\Delta G^{i-j}(1,Tm,P)-\Delta G^{i-j}(0,Tm,P)$$

$$(2.8)$$

Если $Z \neq \frac{1}{2}$, система (2.8) совместна при условии, что

$$\mathcal{A}^i - \mathcal{A}^j = \frac{\Delta G^{i \to j}(0, T_m, P)}{Z} + \frac{\Delta G^{i \to j}(1, T_m, P)}{1 - Z} = \frac{\Delta G^{i \to j}(1, T_m, P) - \Delta G^{i \to j}(0, T_m, P)}{1 - 2z} \qquad (2.9)$$

Если $Z = \frac{1}{2}$, то система (2.8) дает

$$\mathcal{A}^i - \mathcal{A}^j = 4 \Delta G^{i \to j}(0, T_m, P) = 4 \Delta G^{i \to j}(1, T_m, P). \qquad (2.9a)$$

Соотношения (2.9а) дают связь разности $\Pi B \mathcal{A}^i$ и $\mathcal{A}^j$ с координатами ТРК и с разностями термодинамических потенциалов Гиббса между j -ой и i -ой фазами чистых компонентов. Критерием проверки применимости модели регулярных растворов для описания диаграммы с ТРК является выполнимость условия

$$\frac{\Delta G^{i \to j}(0, T_m, P)}{\Delta G^{i \to j}(1, T_m, P)} = \left(\frac{Z}{1 - Z}\right)^2. \qquad (2.10)$$

что следует из (2.9). Если условие (2.10) не выполняется, то система (2.8) не совместна, т.е. такая диаграмма с ТРК не может быть описана в рамках модели регулярных растворов.

В качестве отрицательного примера получения параметров взаимодействия в модели регулярных растворов из диаграммы состояния, которая имеет точку экстремума на фазовых границах, без согласия с координатами этой точки укажем на работу [8], в которой для ОЦК и жидкой фаз в системе уран-плутоний приведены ПВ, равные B=3000 и L = 0 кал/г-атом, соответственно. Разность этих параметров согласно соотношению (2.9) не соответствует экспериментальным значениям: I) точки минимума на кривых солидуса и ликвидуса: T_m = 883 К и X_U = 0,12 [9], T_m = 897±2 К и X_m = 0,09 [10]; 2) энтальпий и температур плавления урана и плутония, равных согласно [11] 2185 и 675 кал/г-атом, 1408 и 913 К соответственно. Согласно данным [9] и [11] в линейном приближении по температуре для $\Delta G^{o \mu \kappa \to L}(T)$ для урана и плутония из (2.9) находим

$$L - B \simeq \frac{\Delta S_U^{L \to o \mu \kappa}(T_m - 1408) - \Delta S_{Pu}^{L \to o \mu \kappa}(T_m - 913)}{1 - 2 \cdot 0,12} = 1039 \frac{\kappa a \Lambda}{\Gamma - a \text{том}}. \quad (2.II)$$

Тогда как согласно данным /10/ и /11/ получаем
$\angle - B$ = 953 кал/г-атом.

В общем случае, когда для $i \rightarrow j$ фазовых переходов чистых компонентов известны не только теплоты перехода $\Delta H^{i \rightarrow j}$ при температуре фазового перехода $T^{i \rightarrow j}$, но и разности: I) теплоемкостей $\Delta C_p^{i \rightarrow j}$, 2) температурных наклонов теплоемкостей $\Delta \left(\frac{\partial C_p}{\partial T}\right)^{i \rightarrow j}$ и вообще, 3) $\Delta \left(\frac{\partial C_p}{\partial T^k}\right)^{i \rightarrow j}$ при температуре $T^{i \rightarrow j}$, температурную зависимость разности термодинамических потенциалов Гиббса между j-ой и i-ой фазами чистого компонента можно представить в виде /12/

$$\Delta G_{n+1}^{i \rightarrow j}(T) = -\Delta H^{i \rightarrow j}(T^{i \rightarrow j}) \cdot \tau \left\{ 1 + \frac{1}{\Delta S^{i \rightarrow j}(T^{i \rightarrow j})} \cdot \left[\Delta C_p^{i \rightarrow j}(T^{i \rightarrow j}) \cdot \right. \right.$$

$$\sum_{k=1}^{n} (-1)^{k-1} \cdot \frac{\tau^k}{k(k+1)} + T^{i \rightarrow j} \cdot A_1 \sum_{k=2}^{n} (-1)^{k-2} \cdot \frac{\tau^k}{k(k+1)} + (T^{i \rightarrow j})^2 \cdot \frac{A_2}{2!} \sum_{k=3}^{n} (-1)^{k-3} \cdot \frac{\tau^k}{k(k+1)}$$

$$+ (T^{i \rightarrow j})^3 \cdot \frac{A_3}{3!} \cdot \sum_{k=4}^{n} (-1)^{k-4} \cdot \frac{\tau^k}{k(k+1)} + \left. \left. \right] \right\} \qquad (2.12)$$

где $|\tau| = |T/T^{i \rightarrow j} - 1| \ll 1$, $A_k = \Delta \left(\frac{\partial^k C_p}{\partial T^k}\right)^{i \rightarrow j} \cdot T^{i \rightarrow j}$, $k = 1, 2 \ldots$.

Скачки энтальпии и энтропии, а также все разности теплоемкостей и их производных по температуре берутся в точке фазового перехода чистого компонента $T^{i \rightarrow j}$. Нижний индекс для $\Delta G_{n+1}^{i \rightarrow j}$ (Т) указывает до какой степени включительно по температуре рассматривается температурная зависимость разности термодинамических потенциалов Гиббса. Соотношение (2.12) позволяет по известным экспериментальным скачкам энтальпии, энтропии, разностям теплоемкости и разностям k - ых производных от теплоемкости по температуре, вычисленным при температуре фазового перехода $T^{i \rightarrow j}$, вычислить с любой наперед заданной точностью разность $\Delta G^{i \rightarrow j}$ при температуре достаточно удаленной от температуры фазового перехода, но удовлетворяющей неравенству $|\tau| \ll 1$. С помощью соотношения (2.12) и, используя экспериментальные данные для фазового перехода ОЦК $\rightarrow \angle$ для урана и плутония /11/, получаем:

$$\Delta G_{U, n+1}^{ОЦК \rightarrow L}(T) = -2185 \cdot \tau_U \cdot \left[1 + 1{,}600 \cdot \sum_{k=1}^{n} \frac{(-1)^{k-1} \tau^k}{k \cdot (k+1)} \right] \frac{кал}{г-атом} \quad (2.13)$$

$$\Delta G_{Pu,n+1}^{\text{ОЦК}\to L}(T) = -675\cdot\tau_{Pu}\cdot\left[1+2,5616\cdot\sum_{K=1}^{n}\frac{(-1)^{K-1}\cdot\tau^K}{K\cdot(K+1)}\right]\frac{\text{кал}}{\text{г-атом}},$$

$$(2.14)$$

где $\tau_U = \frac{T}{1408}-1$, $\tau_{Pu}=\frac{T}{913}-1$. В квадратичном по τ приближении из (2.13) и (2.14) имеем:

$$\Delta G_{U,2}^{\text{ОЦК}\to L}(T) = -2185\cdot\tau_U\cdot(1+0,800\cdot\tau_U)\frac{\text{кал}}{\text{г-атом}},\quad (2.15)$$

$$\Delta G_{Pu,2}^{\text{ОЦК}\to L}(T) = -675\cdot\tau_{Pu}\cdot(1+1,2808\cdot\tau_{Pu})\frac{\text{кал}}{\text{г-атом}}.\quad (2.16)$$

В квадратичном приближении по температуре разность ПВ согласно (2.9) для системы $U-Pu$ между ОЦК и L фазами сплавов по /9/ равна

$$(L-B)_2 = \frac{-2185\cdot\tau_U(1+0,800\cdot\tau_U)+675\cdot\tau_{Pu}\cdot(1+1,2808\,\tau_{Pu})}{1-2\cdot X_m}\bigg|_{\substack{T_m=883\ K\\ X_m=0,12}} = 724\frac{\text{кал}}{\text{г-атом}}$$

$$(2.17)$$

Тогда как по данным /10/ $(L-B)_2 = 672$ кал/г-атом. Полученные результаты (2.11) и (2.17) как на основании экспериментальных данных по ТРК /9/, так и по данным работы /10/ свидетельствуют о некорректности полученных в работе /8/ ПВ для жидкой и ОЦК фаз сплавов уран-плутоний.

Для получения самих ПВ из диаграммы состояния с ТРК возможны два подхода. I) Если известны аналитические зависимости одно/двухфазных границ диаграммы состояния $x^i = x^{i/i+j}(T)$ и $x^j = x^{j/i+j}(T)$ то как показано в работе /13/ в ТРК справедливо общее соотношение

$$\left[\left(\frac{\partial^2 x^i}{\partial T^2}\right)\bigg/\left(\frac{\partial^2 x^j}{\partial T^2}\right)\right]_{X_m,T_m} = \left[\frac{\partial^2 G^j}{\partial x^2}\bigg/\frac{\partial^2 G^i}{\partial x^2}\right]_{X_m,T_m}\equiv q,\,(2.18)$$

откуда в случае справедливости критерия (2.10), из соотношений (2.10) и (2.18) находим ПВ

$$\Lambda^i = \frac{a+bq}{q-1};\quad \Lambda^j=\frac{b-q}{q-1},\quad (2.19)$$

где $a=\frac{RT_m(1+q)}{2X_m(1-X_m)}$, $b\equiv\Lambda^i-\Lambda^j=\left[\Delta G^{i-j}(1,T_m)-\Delta G^{i-j}(0,T_m)\right]\cdot(1-2X_m)^{-1}$.

ТАБЛИЦА I. РАСЧЕТ ПАРАМЕТРА ВЗАИМОДЕЙСТВИЯ ДЛЯ ЖИДКОЙ ФАЗЫ (L) В МОДЕЛИ РЕГУЛЯРНЫХ РАСТВОРОВ ПО ЭКСПЕРИМЕНТАЛЬНЫМ СОЛИДУСУ И ЛИКВИДУСУ СИСТЕМЫ U–Pu [9] ($\Delta G_{Pu, U}^{OЦK \to L}$ в кал/г-ат)

T, K	$X_U^{L/L+OЦK}$	$X_U^{OЦK/OЦK+L}$	$\Delta G_{U, лин.}^{OЦK \to L}$	$\Delta G_{Pu, лин.}^{OЦK \to L}$	X_K по (2,23)	Y_K по (2,24)	L по (2,26)
1273	0,76	0,89	-266	209	-0,0386	-171,3	
1173	0,61	0,76	-192	364	-0,0664	-138,7	2582
1073	0,46	0,625	-118	519	-0,0842	-145,1	кал/г-атом
973	0,31	0,46	-44	674	-0,0727	-353,2	

В (2.19) ПВ $\mathcal{A}^i$ и $\mathcal{A}^j$ вычислены через координаты ТРК, отношение углов наклона одно/двухфазных границ X^i и X^j в ТРК и через разности ТПГ чистых компонентов при температуре T_m между i-ой и j-ой фазами.

2) Если фазовые границы ДС в окрестности ТРК точно неизвестны, как это имеет место в случае ОЦК/L равновесия в системе U–Pu [9,10], рационально применять следующий способ вычисления ПВ. В рамках модели регулярных растворов уравнения фазового равновесия (2.3) приводят к соотношениям:

$$\left. \begin{array}{l} RT \ell n \dfrac{1-X^i}{1-X^j} = \Delta G^{i \to j}(0,T) + \mathcal{A}^j(X^j)^2 - \mathcal{A}^i(X^i)^2 \\[2mm] RT \ell n \dfrac{X^i}{X^j} = \Delta G^{i \to j}(1,T) + \mathcal{A}^j \cdot (1-X^j)^2 - \mathcal{A}^i (1-X^i)^2 \end{array} \right\} \quad (2.20)$$

При известных зависимостях фазовых границ $X^i = X^{i/i+j}$ (T) и $X^j = X^{j/i+j}$ (T) в значительной части ДС, исключая лишь окрестность ТРК из (2.11) требуется вычислить ПВ $\mathcal{A}^i$ и $\mathcal{A}^j$. Записывая (2.20) в виде

$$\left. \begin{array}{l} \mathcal{A}^j (X^j)^2 - \mathcal{A}^i(X^i)^2 = RT \ell n \dfrac{1-X^o}{1-X^j} - \Delta G^{i \to j}(0,T) \\[2mm] \mathcal{A}^j (1-X^j)^2 - \mathcal{A}^i(1-X^i)^2 = RT \ell n \dfrac{X^i}{X^j} - \Delta G^{i \to j}(1,T) \end{array} \right\}, \quad (2.21)$$

имеем систему линейных уравнений относительно двух неизвестных ПВ, которая может быть решена различ-

ными способами: I) симплекс-методом, минимизируя
сумму модулей невязок уравнений равновесия /14/
для совокупности коннод /14/; 2) из условия наи-
лучшего среднеквадратичного приближения по всей
в интегральном смысле экспериментальной двухкомпо-
нентной диаграмме состояния в целом /15/; 3) методом наименьших квадратов (МНК) для дискретного на-
бора двухфазных коннод, минимизирующего также сред-
неквадратичные невязки на дискретном множестве двухфазных коннод. Применим МНК к преобразованной системе (2.2I)

$$Y = \mathcal{A}^i \cdot X \atop Z = \mathcal{A}^d \cdot X \Big\}, \tag{2.22}$$

где $X \equiv [x^i(1-x^i)]^2 - [x^d \cdot (1-x^i)]^2$, $\qquad$ (2.23)

$$Y \equiv (1-x^i)^2_\Delta G^{i \to i}(0,T) - (x^i)^2_\Delta G^{i \to d}(1,T) + RT \left[(x^i)^2 \ell n \frac{x^i}{x^d} - (1-x^i)^2 \ell n \frac{1-x^i}{1-x^i} \right], \tag{2.24}$$

$$Z \equiv (1-x^i)^2_\Delta G^{i \to i}(0,T) - (x^i)^2_\Delta G^{i \to i}(1,T) + RT \left[(x^i)^2 \ell n \frac{x^i}{x^i} - (1-x^i)^2 \ell n \frac{1-x^i}{1-x^i} \right], \tag{2.25}$$

Тогда согласно /16/ из (2.22) получаем ПВ $\mathcal{A}^i$ и $\mathcal{A}^d$

$$\mathcal{A}^i = \left(\sum_{K=1}^{N} \cdot X_K \cdot Y_K \right) \cdot \left(\sum_{K=1}^{N} X_K^2 \right)^{-1}, \quad \mathcal{A}^d = \left(\sum_{K=1}^{N} X_K \cdot Z_K \right) \cdot \left(\sum_{K=1}^{N} X_K^2 \right)^{-1}, \tag{2.26}$$

где N -число $(i \to j)$ фазных коннод. Применительно к
кривым солидуса и ликвидуса в системе $U-Pu$ достаточно вычислить один ПВ, например, $\mathcal{A}^i = L$ по (2.23),
тогда как второй ПВ $\mathcal{A}^d = B$ можно найти из соотношения (2.9), например, в виде (2.II) или (2.I7).
Результаты расчета параметра L для жидкой фазы в
системе $U-Pu$ (см.таблицу I) дают $L = \left(\sum\limits_{K=1}^{4} X_K Y_K \right)$.
$\left(\sum\limits_{K=1}^{4} X_K^2 \right)^{-1}$ = 2580 кал/г-атом, тогда, используя
данные /I0/ по координатам ТРК, получаем B=3530
кал/г-атом. При I273 К согласно результатам расчетов /7/ в системе $U-Pu$ составы фаз, находящихся
в равновесии на кривых ликвидуса и солидуса рав-
ны $X_U^{\mathcal{M}}$ = 0,705 и $X_U^{\textit{ОЦК}}$ = 0,88. Как следует из сравнения с данными таблицы I состав жидкой фазы при
I273 К на кривой ликвидуса отличается от расчетной в /8/ соответствующей величины примерно на
5,5 ат % U. По вычисленным нами ПВ для жидкой и
ОЦК фаз в системе $U-Pu$ были рассчитаны солидус и
ликвидус в этой системе /15/ в хорошем согласии с
экспериментальными данными /9.I0/.

2.3. Модель субрегулярных растворов

В рамках модели субрегулярных растворов (2.2) примет вид

$$G^i_{u3\delta}(x,T,P) = x(1-x)\cdot(A^i + B^i\cdot x),$$

$$G^j_{u3\delta}(y,T,P) = y(1-y)\cdot(A^j + B^j\cdot y),\qquad (2.27)$$

где A^i, A^j, B^i, B^j — ПВ. Тогда (2.5) дает:

$$\left.\begin{array}{l}[A^i - A^j] + Z[B^i - B^j] = M(Z,T_m,P)\\[2mm](1-2z)\cdot[A^i - A^j] + Z(2-3z)\cdot[B^i - B^j] = N(T_m,P)\end{array}\right\},\ (2.28)$$

где

$$M(Z,T_m,P) = \frac{\Delta G^{i-j}(0,T_m,P)}{Z} + \frac{\Delta G^{i-j}(1,T_m,P)}{1-Z},$$

$$N(T_m,P) = \Delta G^{i-j}(1,T_m,P) - \Delta G^{i-j}(0,T_m,P).$$

Система (2.28) при известных z, T_m и P представляет собой систему двух линейных уравнений относительно двух неизвестных $(A^i - A^j)$ и $(B^i - B^j)$ и, следовательно, разрешима относительно последних, поскольку детерминат, составленный из коэффициентов при этих неизвестных, равен $z(1-z)$ и является невырожденным, когда $z \neq 0$ или $z \neq 1$.

В случае наличия на кривых фазового равновесия, как точки максимума, так и минимума (например, на кривых солидус-ликвидус в частной диаграмме $(UC_1 - UC_2)$ (z_1, T_{m_1}) и (z_2, T_{m_2}) для каждой точки экстремума справедлива система типа (2.28). Из этих двух систем, исключая неизвестные $(A^i - A^j)$ и $(B^i - B^j)$, получаем критерий проверки применимости модели субрегулярных растворов диаграмм состояния с точками максимума и минимума

$$[M(z_2,T_{m_2},P) - M(z_1,T_{m_1},P)]\cdot(z_2 \cdot z_1)^{-1} =$$

$$= \frac{[N(T_{m_1},P)/(1-2z_1)] - [N(T_{m_2},P)/(1-2z_2)]}{[z_1(2-3z_1)/(1-2z_1)] - [z_2(2-3z_2)/(1-2z_2)]}.\qquad (2.29)$$

Более усложненные методы, чем рассмотренные, могут быть проанализированы аналогичным образом.

Таким образом, перед применением какой-либо
модели к расчету диаграммы состояния системы, в которой экспериментально обнаружено наличие максиму-
ма (минимума) на $i/i \to j$ и $j/i \to j$ фазовых границах, следует убедиться в правомочности применения этой модели, т.е. в выполнении условий (2.5) в общем виде,
либо (2.10) и (2.28) в случае модели регулярных и
субрегулярных растворов соответственно.

3. Выбор приближений и значений параметров

3.1. Выбор приближений.

Расчеты будем вести при $P = const$, поэтому ниже
этот аргумент опускается. Разложим
подынтегральную функцию $\Delta S^{i \to j}(T)$ в (2.6) в ряд
по малому параметру $\tau = \dfrac{T}{T^{i \to j}} - 1$ в окрестности тем-

пературы $i \to j$ перехода чистого компонента: и,
ограничиваясь нулевым членом разложения, после ин-
тегрирования в (2.6) из (2.12) получаем

$$\Delta G^{i \to j}(q,T) \simeq -\Delta S^{i \to j}\left[q, T^{i \to j}(q)\right] \cdot \left[T - T^{i \to j}(q)\right] \qquad (3.1)$$

где $\Delta S^{i \to j}\left[q, T^{i \to j}(q)\right]$ — разность энтропий i и j
фаз для чистого q-го компонента при температуре
фазового перехода $T^{i \to j}(q)$. Исходя из эксперимен-
тального факта существования максимальной температуры плавления карбонитридов урана, примем модель
регулярных растворов для жидкой (L) и твердой (Т)
фаз. Тогда критерий применимости модели регулярных
растворов к описанию кривых солидуса и ликвидуса с
наличием точки максимума примет вид

$$\frac{\Delta S_{UC}^{T \to L} \cdot \left[T_m - T_{UC}^{T \to L}\right]}{\Delta S_{UN}^{T \to L} \cdot \left[T_m - T_{UN}^{T \to L}\right]} = \left(\frac{Z}{1-Z}\right)^2 \qquad (3.2)$$

Экспериментальные данные по энтропии плавления
карбида и нитрида урана $\Delta S_{UC}^{T \to L}$ и $\Delta S_{UN}^{T \to L}$ в настоящее
время нам не известны. В работе [1] принято $\Delta S_{UC}^{T \to L} =$
$= \Delta S_{UN}^{T \to L}$ = 6,3 кал/(моль·град) В работах [6],
следуя правилу Кубашевского $\Delta S^{T \to L} = 2/2,2 + R ln2) =$
$= 7,2$ кал/(моль·град), принято, что $\Delta S_{UC}^{T \to L} = \Delta S_{UN}^{T \to L} =$
$= 7,2$ кал/(моль·град) Согласно [6], T_m = 2910±35 °С=
$= 3183$ К и $Z = X_m = 0,7$ мольных доли UN. Таким об-
разом, следуя [1] и [6] примем, что $\Delta S_{UC}^{T \to L} = \Delta S_{UN}^{T \to L}$,

тогда соотношение (3.2) дает

$$(4,98 \pm \frac{2,91}{1,18}) \approx 5,44$$

Здесь в левой части стоит значение левой части соотношения (3.2), что соответствует среднему значению T_m = 3183 К, положительное отклонение - значению 3148 К и отрицательное - значению T_m + 35 К= = 3228 К; в правой части записан результат правой части соотношения (3.2). Таким образом, можно заключить, что в пределах ошибки опыта модель регулярных растворов применима для описания кривых солидуса и ликвидуса карбонитридов урана.

3.2. Выбор значений параметров.

Следуя Бенцу [6], примем

$$\Delta S_{UC}^{T \to L} = \Delta S_{UN}^{T \to L} = 7,2 \text{ кал/(моль·град)}$$
$$T_{UC}^{T \to L} = 2495 \pm 35\,^{0}C \approx 2700 \text{ К} \qquad (3.3)$$
$$T_{UN}^{T \to L} = 2830 \pm 35\,^{0}C \approx 3100 \text{ К}$$
$$\mathcal{A}^{i} = B = -2800 \text{ кал/моль}$$
$$\mathcal{A}^{d} = L = 3300 \text{ кал/моль}$$

Значение второго параметра A^d = энергии смешения в жидкой фазе между UC и UN проверим по уравнению (2.9)

$$B - L = \Delta S_{UC}^{T \to L} \cdot \frac{T_{UN}^{T \to L} - T_{UC}^{T \to L}}{1 - 2z}, \qquad (3.4)$$

где Z может быть рассчитано из (2.10) по формуле:

$$Z = \left\{ 1 + \left[\frac{\Delta G^{T \to L}(1, T_m)}{\Delta G^{T \to L}(0, T_m)} \right]^{1/2} \right\}^{-1} \qquad (3.5)$$

Из (3.1), (3.3) и (3.5) имеем $Z = X_m = 0,6905$, что хорошо согласуется с экспериментальными данными Бенца X_m =0,70 [6]. Оценим к какой вариации X_m приводит ошибка определения T_m, равная $\pm 35\,^{0}$С. Имеем из (3.5) $X_m = 0,6905 {\pm}^{0,046}_{0,041}$. В дальнейшем будем пользоваться координатами точки максимума кривых ликвидуса и солидуса T_m =3183 К и X_m= 0,6905. Подстановка этих значений и величин (3.3) в соотношение (3.4) дает В − L = -6240 кал/моль. Сопоставление с данными Бенца (3.3) указывает на хорошее согласование результатов (3.3) с термодинамическими условиями.

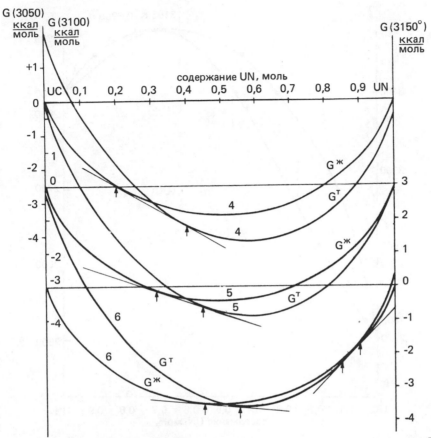

*Рис. 1. Расчетные зависимости термодинамического потенциала от концентрации для жидкой (G^L)
и твердой (G^T) фаз системы UC– UN при различных температурах (1–2800 К, 2–2900 К, 3–3000 К,
4–3050 К, 5–3100 К, 6–3150 К).*

4. Результаты расчетов.

Так как условия фазового равновесия (2.3) инвариант-
ны относительно прибавления к $G^L(x,T) = G^L(x,T)$
и $G^2(x,T) = G^T(x,T)$ любой линейной функции от состава, то будем производить расчет кривых солидуса и
ликвидуса, исходя из термодинамических потенциалов
Гиббса для жидкой и твердой фаз:

$$G^L(x,T) = RT\left[x\ln x + (1-x)\ln(1-x)\right] + 3300\,x\,(1-x), \quad (4.1)$$

$$G^T(x,T) = RT\left[x\ln x + (1-x)\ln(1-x)\right] - 2800\cdot x\,(1-x) +$$
$$+\,7,2\cdot\left[(1-x)\cdot(T-2770) + x\,(T-3100)\right]. \qquad (4.2)$$

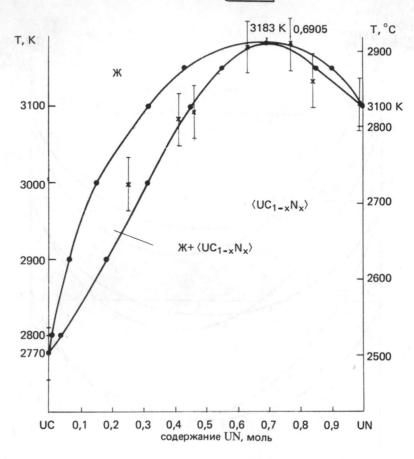

Рис. 2. Кривые солидуса и ликвидуса в системе UC– UN, рассчитанные в приближении модели регулярных растворов (●- расчетные точки, × – экспериментальные данные (ошибка ±35К) [6]). 1 – кри-
вая $T_0(x)$, вдоль которой $G^L(x, T_0) = G^T(x, T_0)$.

На рис.1 приведены полученные по формулам (4,1) –
(4.2), изотермические сечения (Т=2800, 2900, 3000,
3050, 3100 и 3150 К) поверхностей термодинамических
потенциалов для жидкой (G^L) и твердой (G^T) фаз
в зависимости от концентрации. Общие касательные,
проведенные к $G^L(x)$ и $G^T(x)$, указывают на протя-
женность двухфазной области. На рис.2 сопоставлены
результаты расчетов кривых ликвидуса и солидуса с
экспериментальными данными по температурам плавления сплавов карбонитридов урана [6]. Видно хорошее
согласие результатов расчета с экспериментальными
данными.

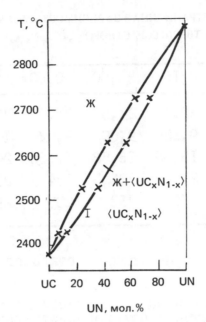

Рис. 3. Кривые ликвидуса и солидуса для системы UC– UN, рассчитанные в приближении идеальных растворов, 1–результат эксперимента [1].

5. Обсуждение

5.I. Вариант диаграммы системы $UC-UN$,предложенный Ондрачеком и Петцовым [I].

На рис.3 представлены кривые ликвидуса и солидуса для карбонитридов урана, рассчитанные в модели идеальных растворов [I]. На этом же рисунке приведена одна экспериментальная точка состава 70 мол.% UC +30 мол.% UN, для которой определена температура начала плавления (2485°C) при давлении 0,I ат N_2. По мнению Ондрачека и Петцова она подтверждает расчетную кривую солидуса, хотя авторы приводят состав этого образца после опыта, содержащего 5 вес % вольфрама. (Опыт проводился в вольфрамовом тигле). По-видимому, в работе [I] измерена температура плавления эвтектики $UC_x N_{1-x}$ -W, что согласуется с результатами работы [I7], в которой показано, что состав эвтектики системы $UN-W$ соответствует составу сплава с 96 вес % UN.

Кроме того сопоставление экспериментальных данных Бенца [6] рис.2 с результатами расчетов кривых солидуса и ликвидуса (рис.3),проведенных в [I],

ТАБЛИЦА II. КОНЦЕНТРАЦИОННАЯ ЗАВИСИМОСТЬ МОДУЛЕЙ УПРУГОСТИ И ТЕМПЕРАТУРЫ ДЕБАЯ ТВЕРДЫХ РАСТВОРОВ $UC_{1-x} N_x$

X	0	0,15	0,50	0,60	0,80	1,00
$E \cdot 10^{11}, \frac{дин}{см^2}$	22,49	22,15	24,57	25,0	27,8	26,68
$\mu \cdot 10^{11}, -"-$	8,73	9,15	9,80	9,92	10,95	10,39
$K \cdot 10^{11}, -"-$	17,68	13,86	16,58	17,36	20,15	20,59
θ_D, K	269	274,3	283,5	284	298	291
θ_D, K	254	263	263	287	295	232

указывает на их существенное отклонение от расчетной диаграммы.

5.2. Концентрационные зависимости модулей упругости и температуры Дебая карбонитридов урана

Падель с сотрудниками /18/, измеряя скорость звука при комнатной температуре на спеченных образцах карбонитридов урана, определили модули упругости (модуль Юнга-E, модуль сдвига-μ, объемный модуль - K) и температуру Дебая (θ_D) из упругих констант при нулевой пористости (табл.II). Полученные в /18/ данные для изотропных модулей упругости для спеченных UC и UN несколько превышают значения соответствующих величин, рассчитанных из адиабатических упругих модулей при 25°C измеренных на монокристаллах монокарбида урана, $E=21,3$, $M=10,36$, K=15,81 (все в единицах 10^{11} дин/см^2) /19/ и хорошо согласуются с изотропными модулями, рассчитанными из адиабатических упругих модулей, измеренных при 25°C на монокристалле мононитрида урана, $E = 26,51$, μ = 10,43 и K=19,11 (в единицах 10^{11} дин/см^2 /20/). В последней строке таблицы II приведены температуры Дебая, полученные из измерений теплоемкости при T=1,3÷10 К в работе /21/. Исключая значения θ_D для мононитрида урана, температуры Дебая твердых растворов карбонитридов урана, полученные расчетом из упругих модулей и из измерений теплоемкости, хорошо согласуются между собой.

Из табл.II видно, что E,μ и θ_D при x=0,80 имеют максимальные значения. По формуле Линдемана A. $(VN)^{2/3}$

$\theta_{\mathcal{Д}}/T_{nn.} = const$ (А-масса атома, N - число Авогадро, V-объём, $T_{nn.}$ - средняя температура плавления сплава, которую можно отождествлять с температурой $T_c(x)$ - рис.2, вдоль которой $G^{L}(x,T_c)=G^{T}(x,T_c)$), в работе $\sqrt{18\sqrt{}}$ была вычислена максимальная температура плавления карбонитридов урана $T_{nn.}^{max}$ = 3005 К при x=0,80.

В работе Меньшиковой с сотрудниками $\sqrt{22\sqrt{}}$ резонансным методом определены модули Юнга твердых растворов карбонитридов урана в области 20-1300°С. Концентрационная зависимость Е^{20}°С для системы имеет максимум при x≃0,70÷0,75. Причем значения модулей Юнга для монокарбида урана равно 21.10^3 кг/мм^2, а для мононитрида урана – 26.10^3 кг/мм^2, что хорошо согласуется с соответствующими данными $\sqrt{18\sqrt{}}$. Таким образом, результаты измерений упругих модулей и температуры Дебая твердых растворов карбонитридов урана также свидетельствуют о максимуме температуры плавления в системе $UC-UN$ в области составов 0,70-0,80 мольных долей UN, что хорошо согласуется с экспериментальными данными Бенца $\sqrt{6\sqrt{}}$ и результатами наших расчетов.

ЛИТЕРАТУРА

[1] ONDRACEK, J., PETZOV, J., J. Nucl. Mater. 25 (1968) 132.

[2] NAOUMIDIS, A., STÖCKER, H.-J., Thermodynamics of Nuclear Materials (Proc. Symp. Vienna, 1967), IAEA, Vienna (1968) 287.

[3] SANO, T., KATSURA, M., KAI, H., Thermodynamics of Nuclear Materials (Proc. Symp. Vienna, 1967), IAEA, Vienna (1968) 301.

[4] LEITNAKER, J.M., Thermodynamics of Nuclear Materials (Proc. Symp. Vienna, 1967), IAEA, Vienna (1968) 317.

[5] KATSURA, M., NAOUMIDIS, A., NICKEL, M., J. Nucl. Mater. 36 (1970) 169.

[6] BENZ, R., J. Nucl. Mater. 31 (1969) 93.

[7] IKEDA, Y., TAMAKI, M., MATSUMOTO, G., J. Nucl. Mater. 59 2 (1976) 103.

[8] HOLLECK, H., Thermodynamics of Nuclear Materials (Proc. Symp. Vienna, 1974) Vol. 2, IAEA, Vienna (1975) 213.

[9] ЭЛЛИОТ, Р.П., Структуры Двойных Сплавов, 2, М., Металлургия, 1970, стр. 355.

[10] ШАНК, Ф.А., Структуры Двойных Сплавов, М., Металлургия, 1973, стр. 790.

[11] OETTING, F.L., RAND, M.H., ACKERMANN, R.J., The Chemical Thermodynamics of Actinide Elements and Compounds, Part 1. The Actinide Elements, IAEA, Vienna (1976).

[12] УДОВСКИЙ, А.Л., Сплавы для Атомной Энергетики, М., Наука, 1979.

[13] УДОВСКИЙ, А.Л., Доклад на Всесоюзном совещании по диаграммам состояния металлических систем, М., ИМЕТ, (30. X-I/XI, 1978).

[14] RAO, M.V., HICKES, R., TILLER, W.A., Acta Metall. 21 (1970) 733.

[15] ИВАНОВ, О.С., УДОВСКИЙ, А.Л., ГАЙДУКОВ, А.М., Доклад на Всесоюзном совещании по диаграммам состояния металлических систем, М., ИМЕТ (30.X-1.XI-1978).

[16]   ГУТЕР, Р.С., ОВЧИНСКИЙ, Б.В., Элементы Численного Анализа и Математической Обработки Результатов Опыта, М., Наука, 1970, гл. VIII.
[17]   POLITIS, C., THUMMLER, F., WEDEMEYER, H., J. Nucl. Mater. 38 (1971) 132.
[18]   PADEL, A., et al., J. Nucl. Mater. 36 (1970) 297.
[19]   ROUTBOURT, J.L., J. Nucl. Mater. 40 (1971) 17.
[20]   GUINAN, M., CLINE, C.F., J. Nucl. Mater. 43 (1972) 43.
[21]   COMBARIEN, A., CONTA, P., MICHEL, J.C., C.R. Hebd. Seances Acad. Sci. 256 (1963) 5518.
[22]   МЕНЬШИКОВА, Т.С., РЕШЕТНИКОВ, Ф.Г., МУХИН, В.С., РЫМАШЕВСКИЙ, Т.А., ЛЕБЕДЕВ, И.Г., ЭПШТЕЙН, А.А., Ат. Энерг. 31 (1971) 393.

ФАЗОВЫЕ РАВНОВЕСИЯ В СИСТЕМЕ
УРАН – ХРОМ – УГЛЕРОД

З.М.АЛЕКСЕЕВА, О.С.ИВАНОВ
Институт металлургии им.А.А.Байкова
Академии наук СССР,
Москва,
Союз Советских Социалистических Республик

Представлен А.С.Пановым

Abstract–Аннотация

PHASE EQUILIBRIA IN THE URANIUM–CHROMIUM–CARBON SYSTEM.

The alloys for this study (40 different compositions) were prepared by melting a charge comprising appropriate amounts of the various components and a 'master' alloy under an argon atmosphere on a water-cooled copper tray in an arc furnace equipped with a tungsten electrode. The starting components were uranium of 99.7% purity, graphite with an 0.004% ash content and electrolytically prepared chromium. Heat treatment of the alloys at $1300-1600°C$ was carried out in graphite crucibles in a vacuum furnace ($\sim 1 \times 10^{-5}$ torr). The alloys were analysed by X-ray spectroscopy, X-ray diffraction analysis and microstructural methods, and a selective chemical analysis was made. As a result of these studies, a new compound of composition 45 at.% Cr – 10 at.% U – 45 at.% C was found; it was subjected to X-ray diffraction analysis. The composition of the compound $x(UCr_2C_3)$ was refined. The crystallization characteristics of the alloys were established over the entire composition range, a diagram of the non- and monovariant equilibria was drawn and the section for the sub-solidus temperature was constructed over the entire range of concentrations. The isothermal section at $1500-1600°C$ was constructed for the $UC_2-UC-UCrC_2$ region; this is characterized by a broad region consisting of a homogeneous solid solution of chromium in the high-temperature $(UC-\beta-UC_2)$ solid solution, which the chromium stabilizes down to a lower temperature. The observed increase of the carbon content in the solid solution with increase in the content of iron can apparently explain the improvement of chromium-modified monocarbide nuclear fuel reported in the literature. The carbon content in uranium-based monocarbide fuel can be increased to ~ 55 at.% by the addition of ~ 7 at.% chromium without change in the crystal structure. For higher chromium and carbon contents there is a change to a tetragonal structure in the solid solution. The parameters of the tetragonal $(UC-\beta-UC_2)$ lattice were determined to be a = 5.041 Å and c = 5.128 Å. This study of the crystallization of the alloys has helped to explain certain earlier experimental data published by other workers.

ФАЗОВЫЕ РАВНОВЕСИЯ В СИСТЕМЕ УРАН – ХРОМ – УГЛЕРОД.

Сплавы для исследования (40 составов) готовили сплавлением в дуговой печи на медном водоохлаждаемом поддоне вольфрамовым нерасходуемым электродом в атмосфере аргона шихты, составленной из исходных компонентов и лигатуры. Исходными компонентами служили уран чистотой 99,7%, графит 0,004% зольности, электролитический хром. Термическая обработка сплавов при 1300-1600°C произведена в графитовых тиглях в вакуумной печи ($\sim 1 \cdot 10^{-5}$ мм рт. ст.).

Сплавы анализировали рентгеноструктурным, рентгеноспектральным и микроструктурным методами, выборочно произведен химический анализ. В результате проведенных исследований обнаружено новое соединение состава Cr 45 – U10 – C 45 (ат %), проведен рентгеноструктурный анализ. Уточнен состав соединения x(UCr$_2$C$_3$). Установлен характер кристаллизации сплавов во всей области составов, составлена схема нон- и моновариантных равновесий, во всей области концентраций построено сечение для подсолидусной температуры. Для области UC$_2$ – UC – UCrC$_2$ построено изотермическое сечение при 1500-1600°C, характеризующееся широкой областью гомогенности твердого раствора хрома в высокотемпературном твердом растворе (UC – β – UC$_2$), который хром стабилизирует до более низкой температуры. Обнаруженным увеличением содержания углерода в твердом растворе с увеличением содержания железа, по-видимому, можно объяснить отмеченный в литературе факт улучшения модифицированного хромом монокарбидного ядерного топлива. Содержание углерода в монокарбидном топливе на основе урана можно увеличить до ~55 ат % при добавлении хрома ~7 ат % без изменения его кристаллической структуры. При более высоком содержании хрома и углерода происходит тетрагонализация твердого раствора. Определены параметры тетрагональной решетки (UC – β – UC$_2$) a = 5,041 Å, c = 5,128 Å. Установление деталей в характере кристаллизации сплавов позволило объяснить некоторые опытные данные в опубликованных ранее работах других исследователей.

1. ВВЕДЕНИЕ

Карбидное топливо весьма перспективно для высокотемпературных ядерных реакторов. Однако при разработке этого вида горючего возникает ряд трудностей, как например, нежелательное образование высших карбидов урана. В литературе систематически появляются сведения об использовании присадок хрома для улучшения карбидного топлива. Так в работе [1] предложено модифицировать урано-плутониевое карбидное топливо хромом и железом, чтобы уменьшить распухание. В работе [2] обнаружено, что добавки хрома и ванадия в монокарбид урана подавляют образование высших карбидов урана, улучшают спекание и обеспечивают получение высокой плотности при низкой температуре спекания.

Однако сведения о фазовом строении сплавов тройной системы U – Cr – C скудны и противоречивы, некоторые экспериментальные наблюдения не получили удовлетворительного объяснения. Так Х. Новотный с сотр. [3,4], исследовавшие 7 сплавов по разрезу UC – Cr$_3$C$_2$ (с 5,10,20,40,60,90 и 95 мол.% Cr$_3$C$_2$)* обнаружили в образце с 5 мол.% Cr$_3$C$_2$ две фазы на основе UC: фаза, присутствовавшая в большем количестве, имеет больший параметр решетки (4,98Å), параметр другой изоморфной фазы меньше параметра стехиометрического UC. Авторы работ [3,4] считают,

---

* В переводе на ат % состав первых трех сплавов: Cr : C : U = 7 : 49 : 44, 13 · 48 : 39, 23 : 46 : 31.

что в этом сплаве, а также в сплаве с 10 мол.% Cr_3C_2
состояние равновесия недостигнуто (образцы для исследования готовили горячим прессованием при 1000°С
из соответствующих пропорций UC и Cr_3C_2 с добав-
лением 1% C_o . Затем прессовки отжигали 4 ч при
2000°С). В образце с 20 мол.% Cr_3C_2 преимущественной фазой является двойной карбид $UCrC_2$ [5].
Слабые дополнительные к UC и Cr_3C_2 рефлексы,
присутствующие на рентгенограмме с 10 мол.% Cr_3C_2 ,
не соответствуют рефлексам $UCrC_2$.

В работах [6,7] исследовали сплавы разреза
$UC - Cr$, приготовленные металлокерамическим методом и спеченные в ториевых тиглях в атмосфере очи-
щенного аргона при 1000-2000°С до достижения равновесия. Никаких других фаз кроме UC и Cr не обнаружено. Нет также растворимости в твердом состоянии. Определено, что при 1315±5°С в сплаве с 34мол.%
Cr жидкость находится в равновесии с UC и Cr .
В работе [8] точка плавления эвтектики в этом разрезе первоначально была определена равной 1425±5°С.
Эта температура была установлена металлографическим
исследованием отожженных при повышенных температурах смесей порошков UC и хрома по первому появлению эвтектической структуры. Впоследствии было
установлено, что на термических кривых смесей UC -
Cr эвтектическая остановка появляется при гораздо
более низкой температуре. Избранные смеси нагрева-
лись от 2 до 20 ч при 1300°С и исследовались
металлографически для обнаружения эвтектической
структуры. Установлено, что следы эвтектики образуются при $t <$ 1315°С, однако полное плавление
эвтектических композиций не происходит ниже 1425°С
даже при продолжительных отжигах. Авторы [8] пола-
гают, что составы смесей сдвинуты с двойного разреза, так что при 1315°С образуется небольшое количество тройной эвтектики.

В работе [8] утверждается существование двухфазных коннод: $UC - Cr_{23}C_6$ (построено поли-
термическое сечение этого разреза), $UC - Cr_3C_2$
(утверждается без доказательства), $UC - UCrC_2$
(по определению состава - UC , первичнокристаллизую-
щейся фазы одного сплава).

В системе обнаружено соединение $UCrC_2$, плавя-
щееся инконгруэнтно при 1625±15°С [8],а также сообщается о тройном соединении неопределенного состава $UCrC_x$ с параметрами тетрагональной решетки
$a = 3,686 Å$, $c = 15,739 Å$ [9], полученного путем сплавления $UCrC_2$ с углеродом. В [8], по-
видимому, то же соединение, обозначенное через x ,

ТАБЛИЦА I. СОСТАВЫ СПЛАВОВ И РЕЖИМЫ ТЕРМООБРАБОТОК

Атомные проценты			Вес слитка* (г)	Температура (^{0}C) и время (ч) отжига			
Cr	U	C		1300	1400	1500	1600
33	33	34	4,980	6	–	–	–
32	32	36	4,938	6	–	–	–
31	31	38	4,957	6	–	–	–
30	30	40	4,928	6	8	–	–
29	29	42	4,934	–	8	6	–
28	28	44	4,903	–	–	6	–
27	27	46	4,942	–	–	6	–
26	26	48	4,962	–	–	6	–
25	25	50	4,936	–	–	6	–
30	20	50	4,985	–	–	–	4 4
35	15	50	4,601	–	–	–	4 +24
40	10	50	3,746	–	–	–	4
45	5	50	3,376	–	–	–	4
45	10	45	4,934	7	–	–	–
50	5	45	3,992	–	–	–	4
50	10	40	4,837	7	–	–	–
55	5	40	4,841	–	–	–	4
60	5	35	4,915	–	–	–	4
63	5	32	4,840	7	–	–	–
5	43	52	4,866	–	–	–	4
5	45	50	4,684	–	–	–	4
5	47	48	4,989	–	–	–	4
10	38	52	4,927	–	–	–	4
10	40	50	4,778	6	6	6	–
10	42	48	4,990	–	–	–	4
15	33	52	4,925	–	–	–	4
15	35	50	4,771	6	6	6	4
15	37	48	4,880	–	–	–	4
5	37	58	3,260	–	–	6	–
10	34	56	5,788	–	–	6	–
15	31	54	4,850	–	–	6	–
20	28	52	4,898	–	–	6	–
5	31,5	63,5	4,810	–	–	6	–
10	30	60	4,754	–	–	6	–
15	28,4	56,6	4,826	–	–	6	–
20	30	50	4,875	–	6	6	–
15	25	60	4,780	–	–	6	–

* При весе шихты, равном 5 г.

нанесено на концентрационный треугольник составов приблизительно в точке U 25 - Cr 15 - C 60 (ат %).

Проанализировав вышеописанную информацию, авторы доклада задались целью исследовать ограниченное число сплавов этой системы, сконцентрировать свое внимание на область составов $UC - UC_2 - UCrC_2$, столь мало затронутую предыдущими исследователями, сравнить полученные результаты с литературными данными и по возможности выявить ошибки в толковании экспериментальных данных.

2.   МЕТОДИКА ЭКСПЕРИМЕНТА

2.1. Приготовление сплавов для исследования

Для исследования приготовили сплавы по разрезу U : Cr = 1:1 от 34 до 50 ат % C (9 составов), в области $UCrC_2 - Cr_7C_3 - C$ (12 составов) и в области $UC - UCrC_2 - UC_2$ (18 составов). Всего вместе с эталонами были приготовлены сплавы 40 составов (табл.I).

Сплавы готовили путем сплавления в дуговой печи на медном водоохлаждаемом поддоне вольфрамовым электродом в атмосфере аргона шихты, составленной из исходных компонентов и лигатуры. В качестве лигатуры использовали монокарбид урана, предварительно выплавленный в тех же условиях. Исходными компонентами служили уран чистотой 99,7%, графит 0,004% зольности и хром электролитический. Следует особо остановиться на приготовлении карбидов хрома, служивших в качестве эталонов. Так как все карбиды хрома образуются по перитектической реакции, то для их приготовления использовали металлокерамический метод. Для каждого из трех карбидов были зашихтованы и спрессованы порошковые стехиометрические смеси. Смесь для получения $Cr_{23}C_6$ спекалась при 1450°С в течение 24 ч, смесь для получения Cr_7C_3 спекалась при 1300°С 5 ч , смесь для получения Cr_3C_2 спекалась при 1800°С 15 ч. Cr_3C_2 может быть получен также дуговой плавкой.

2.2. Термическая обработка

Высокотемпературные отжиги сплавов производили в вакууме при разрежении порядка 1.10^{-5} мм рт.ст. в печи ТВВ в графитовых тиглях. Температуру в печи определяли по кривой зависимости температуры от мощ-

ТАБЛИЦА II. КРИСТАЛЛОГРАФИЧЕСКИЕ ХАРАКТЕРИСТИКИ ЭТАЛОННЫХ ЭЛЕМЕНТОВ И СОЕДИНЕНИЙ

Эталон	Параметры, Å			Сингония	Источ-ники
	α	β	c		
U	2,854	5,869	4,955	ромбическая	[10]
Cr	2,8829	–	–	кубическая	[10]
C	2,461	–	6,708	гексагональная	[11]
UC	4,960	–	–	кубическая	[12]
U_2C_3	8,088	–	–	кубическая	[13]
$\alpha - UC_2$	3,529	–	5,995	тетрагональная	[14]
$Cr_{23}C_6$	10,660	–	–	кубическая	[10]
Cr_7C_3	14,01	–	4,532	гексагональная	[10]
Cr_3C_2	5,545	2,830	11,470	ромбическая	[10]
$UCrC_2$	5,433	3,231	10,636	ромбическая	[5]

ности, подаваемой на нагреватель печи. При построении градуировочной кривой был использован оптический пирометр ОППИР-17, которым измеряли температуру абсолютно черного тела, помещенного в печь. В таблице I даны режимы отжигов сплавов.

2.3. Рентгенофазовый анализ

Рентгенофазовые исследования проводили по стандартной методике с использованием рентгеновских камер типа РКД диаметром 57,3 мм. Исследуемый сплав в виде порошка накатывали при помощи коллодия на тонкую стеклянную палочку диаметром 0,2 мм. Закалка пленки – асимметричная, для съемки дебаеграммы применяли нефильтрованное излучение. Рентгенофазовый анализ проводили путем сравнения рентгенограмм сплавов с рентгенограммами эталонов, т.е. всех известных соединений ограничивающих двойных и исследуемой тройной систем (таблица II). В таблицу II не включено соединение $UCrC_x$ или x, описанное во введении, так как выплавленный сплав с шихтовым составом U 25- Cr 15 - C 60 (ат %) не обладал соответствующей кристаллической решеткой. Рентгенограммы, которые

необходимо было рассчитать для определения параметров кристаллической решетки, пересиимали в камере
РКУ-86, которая обеспечивает более высокую точность
измерения и расчет за счет увеличения диаметра камеры и специального механизма, плотно прижимающего
пленку к цилиндрической поверхности камеры.

2.4. Количественный анализ

Контроль химического состава выплавленных сплавов в первую очередь проводили путем взвешивания
полученных слитков. Отклонение от шихтового состава
всегда было отрицательно и не превышало нескольких
процентов. Уменьшение веса слитков по сравнению с
шихтой происходит за счет разбрызгивания расплава.
Установлено, что не происходит разбрызгивания исходных компонентов, обычно слиток разбрызгивается
при повторном плавлении (таблица I).

Химический анализ.* В таблице Ш представлены результаты химического анализа сплавов на содержание
хрома. Пересчет остальных компонентов: урана и углерода произведен в предположении, что их количество изменились по сравнению с хромом незначительно.
Сплавы №№ I-I0 анализировали калориметрическим методом, остальные объёмным.

Спектральный анализ. Произведен полуколичествен-
ный анализ на спектроскопической приставке к раст-
ровому микроскопу $JSM - U3$ путем счета интенсив-
ности $M\alpha$ - излучения урана и $K\alpha$ - излучения хрома
за определенный промежуток времени (I0 сек). Затем
интенсивность счета пересчитывали на приведенный
ток зонда (0,I.I0^{-8}а) и эти числовые значения подвергали анализу. Произведены также съемки микрошли-
фов в лучах хрома.

3. РЕЗУЛЬТАТЫ ЭКСПЕРИМЕНТА

3.1. Область составов $UC - UCrC_2 - Cr_7C_3 - Cr$

На рис.I(а)дана микроструктура сплава Cr 32 -
- U 32 - C 36 (ат %), отожженного при I300°C. При
температуре отжига сплав находится частично в расплавленном состоянии. На микроструктуре видны большие зерна UC и $Cr_{23}C_6$, которые при температуре

---

* Химический анализ проведен химиком-аналитиком Г.И.Исаевой.

ТАБЛИЦА III. СОСТАВЫ СПЛАВОВ

№№ п/п	ат % по шихте Cr	U	C	вес % по шихте Cr	U	C	Вес % Cr по хим. анализу	Пересчет вес % U	C	Пересчет ат % U	C	Cr
1	5	47	48	2,16	93,04	4,79	1,19	93,96	4,84	48,1	49,1	2,8
2	5	45	50	2,25	92,56	5,19	1,16	93,59	5,25	46,4	51,6	2,0
3	5	43	52	2,34	92,05	5,62	1,59	92,76	5,66	43,7	52,5	3,4
4	5	37	58	2,66	90,24	7,10	2,01	90,84	7,15	37,6	58,6	3,8
5	5	31	64	3,05	88,00	8,95	1,96	88,99	9,05	32,09	64,68	3,23
6	10	42	48	4,69	90,12	5,20	3,05	91,67	5,29	43,5	49,8	6,6
7	10	40	50	4,89	89,47	5,64	2,81	91,43	5,76	41,84	52,26	5,88
8	10	38	52	5,10	88,77	6,13	3,72	88,90	7,38	39,11	53,5	7,4
9	10	34	56	5,60	87,16	7,24	4,23	90,06	6,22	35,25	58,00	6,75
10	10	30	60	6,20	85,20	8,60	5,05	86,24	8,71	30,58	61,22	8,20
11	15	37	48	7,68	86,65	5,67	5,63	88,57	5,79	38,6	50,1	11,2
12	15	35	50	8,03	85,79	6,18	—	—	—	—	—	—
13	15	33	52	8,42	84,83	6,74	7,56	85,63	6,80	33,6	52,9	13,5
14	15	31	54	8,83	83,84	7,33	6,95	85,57	7,48	32,21	55,81	11,97
15	15	28	57	9,49	82,24	8,27	7,36	84,18	8,46	29,48	58,72	11,80
16	15	25	60	10,47	79,86	9,67	6,82	83,12	10,06	26,51	63,54	9,94
17	20	30	50	11,84	81,32	6,84	9,57	83,41	7,02	31,31	52,24	16,44
18	20	28	52	12,48	80,02	7,50	11,28	81,12	7,60	28,62	53,16	18,22
19	25	25	50	16,56	75,79	7,65	14,49	77,67	7,84	25,94	51,91	22,15
20	26	26	48	16,66	76,24	7,10	15,19	77,58	7,23	26,71	49,35	23,93
21	27	27	46	16,75	76,66	6,59	15,25	78,04	6,71	27,79	47,34	24,85
22	28	28	44	16,83	77,06	6,11	13,83	79,83	6,33	29,72	46,71	23,56
23	29	29	42	16,91	77,43	5,65	14,29	79,87	5,83	30,62	44,30	25,07

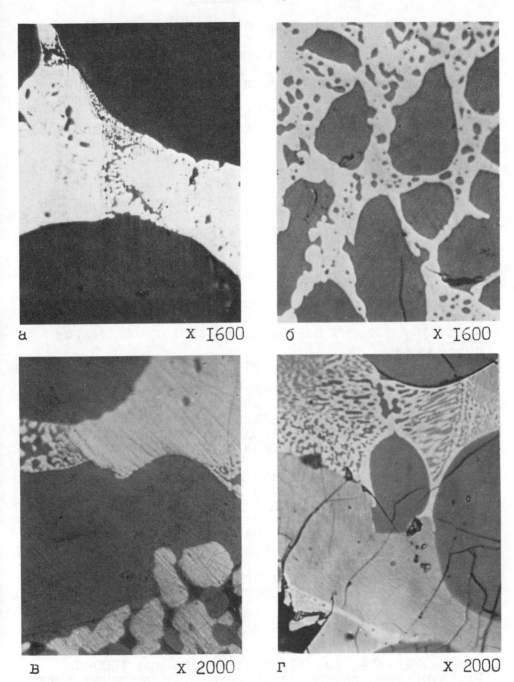

Рис.1. Микроструктура сплавов разреза U : Cr = 1 : 1. Cr 35 – U 32 – C 36 (ат %), отжиг при
1300°C (а); Cr 30 – U 30 – C 40 (ат %), отжиг при 1300°C (б) и при 1400°C (в);
Cr 29 – U 29 – C 42 (ат %), отжиг при 1500°C (г).

отжига были в твердом состоянии. Зерна $Cr_{23}C_6$ окаймлены перитектическими ободками. Часть сплава, находившаяся при температуре отжига в жидком состоянии, закристаллизовалась в виде крупной эвтектики $UC + Cr$, в которую вкраплены участки третичных выделений $U + Cr + UC$. Наличие перитектических ободков у зерен $Cr_{23}C_6$ и присутствие металлического урана в третичных выделениях свидетельствует о существовании реакции

$$ \text{Ж} + Cr_{23}C_6 \rightleftharpoons UC + Cr \qquad (A) $$

температура этой реакции лежит чуть ниже 1300°С. В области существования реакции расположены сплавы Cr 33- U 33- C 34, Cr 32 - U 32 - C 36 и Cr 31- U 31- - C 38 (ат %). По характеру микроструктур перечисленных сплавов пришли к заключению о том, что точка " Ж " близко расположена от прямой Cr - UC и лежит значительно правее разреза $U : Cr$ = 1:1, что согласуется с [7,8].

На рис. I(б) и I(в) представлены микроструктуры сплава Cr 30- U 30- C 40 (ат %), отожженного при 1300 и 1400°С, которые подтверждают существование реакции [8].

$$ \text{Ж} + Cr_7C_3 \rightleftharpoons UC + Cr_{23}C_6 \qquad (Б) $$

Действительно, при температуре отжига 1400°С сплав находится частично в расплавленном состоянии. В твердом состоянии остались зерна UC и Cr_7C_3 , последний при охлаждении подвергся перитектическому распаду, что привело к появлению перитектических ободков из UC + $Cr_{23}C_6$. Остаток жидкости кристаллизуется в виде вторичных выделений UC + Cr . При 1300°С сплав находится в твердом состоянии. На микроструктуре видны зерна UC и не до конца скоагулированные вторичные выделения UC + $Cr_{23}C_6$. Реакция Б имеет место при температуре между 1300 и 1400°С. В области распространения реакции Б лежит только один этот сплав.

На примере оставшихся сплавов нами показано существование четырехфазной перитектической реакции П класса

$$ \text{Ж} + UCrC_2 \rightleftharpoons UC + Cr_7C_3 \qquad (В) $$

На рис. I(г) представлена микроструктура сплава Cr 29- U 29- C 42 (ат %), отожженного при 1500°С. При этой температуре сплав находится частично в жидком состоянии. Твердыми при этой температуре остались фазы UC (темно-серые зерна) и $UCrC_2$ (се-

ТАБЛИЦА IV. РЕЗУЛЬТАТЫ РЕНТГЕНОФАЗОВОГО АНАЛИЗА СПЛАВОВ ИЗ ОБЛАСТИ C – UCrC$_2$ – Cr$_7$C$_3$

Состав образца, ат %	Состояние образца	Фазы						
		x	C-45	UCrC$_2$	C	C$_3$C$_2$	C$_7$C$_3$	UC
Cr 25- U25- C50	литое	+		+				
	отожженное	+		+				
Cr 30- U20- C50	литое	+	+	+				
	отожженное	+	+	+				
Cr 35- U15- C50	литое	+	+					
	отожженное	+	+			+		
Cr 40- U10- C50	литое	+						
	отожженное				+			
Cr 45- U 5- C50	литое	+			+	+		
	отожженное	+			+			
Cr 45- U10- C 45	литое		+					
	отожженное		+					
Cr 50- U 5- C 45	литое			+		+		
	отожженное			+		+		
Cr 50- U10- C 40	литое		+	+		+	+	
	отожженное		+	+			+	
Cr 55- U 5- C 40	литое		+			+		
	отожженное		+					
Cr 60- U 5- C 35	литое		+				+	
Cr 63- U 5- C 32	литое						+	+
	отожженное						+	+

рые большие зерна). При охлаждении жидкость начинает взаимодействовать с $UCrC_2$ по реакции В , образуя продукты реакции $UC + Cr_7C_3$.

3.2. Область составов C – UCrC$_2$ – Cr$_7$C$_3$

В таблице IУ даны результаты рентгенофазового анализа сплавов из области C – $UCrC_2$ – Cr_7C_3 . Эталонами служили рентгенограммы элементов и соединений,

ТАБЛИЦА V. РЕЗУЛЬТАТЫ ИССЛЕДОВАНИЯ СПЛАВОВ НА РАСТРОВОМ МИКРОСКОПЕ (ИСХОДНЫЕ ДАННЫЕ)

Состав образцов, ат % (по шихте)	Ток зонда, 10^{-8} а	Интенсивность счета в импульсах за 10 сек									Ток зонда 10^{-8} а
		M_α излучения урана					K_α излучения хрома				
		UC	$UCrC_2$	x	$Cr45$	UC	$UCrC_2$	x	$Cr45$	Cr_3C_2	
Cr 26–U26–C48	0,24	1660	1300	–	–	–	5450	–	–	–	0,2
Cr 30–U20–C50	0,18	1400	1400	930	–	–	4000	10500	–	–	0,3
Cr 35–U15–C50	0,3	–	–	1300	–	–	–	4000	–	28500	0,19
Cr 45–U10–C45	0,3	–	–	–	650	–	–	–	12500	30000	0,25

Пересчет исходных данных на приведенный ток зонда 0,1.10^{-8} а

Состав образцов, ат % (по шихте)	Интенсивность счета в импульсах за 10 сек при приведенном токе зонда								
	M_α – излучение урана				K_α– излучение хрома				
	UC	$UCrC_2$	x	$Cr45$	$UCrC_2$	x	$Cr45$	Cr_3C_2	
Cr 26–U26–C48	700	550	–	–	2720	–	–	–	
Cr 30–U20–C50	–	800	500	–	1330	3500	–	–	
Cr 35–U15–C50	–	–	430	–	–	2100	–	15000	
Cr 45–U10–C45	–	–	–	220	–	–	5000	1200	

перечисленных в таблице П. Остальные кристаллические структуры, выявленные на рентгенограммах, назывались неизвестными, таких структур обнаружено две, названные x и $Cr45$. При анализе табл.I и IУ обнаружили, что наибольшее отклонение от шихтового состава наблюдается в трех сплавах: Cr 45 - U5 - C 50, Cr 40- - U10 - C 50 и Cr 50 - U5 - C 45 (ат %), в которых обнаружены либо графит, либо высший карбид хрома Cr_3C_2, либо обе эти фазы вместе, форма слитков этих сплавов также отличается от обычной.

В таблице У приведены результаты исследования некоторых сплавов с помощью спектроскопической приставки на растровом микроскопе, из которых следует, что содержание урана убывает, а содержание хрома возрастает в ряду соединений $UCrC_2$, x, Cr 45. Более того, составы $UCrC_2$ и x близки друг другу. Состав Cr 45 более отличается от первых двух.

3.3. Область составов $UC_2 - UCrC_2 - UC$

В этой области исследовали сплавы 18 составов (таблица I).

Согласно химическому анализу (табл. IУ) сплавы, зашихтованные по разрезу $UC-UCrC_2$, на самом деле содержат больше 50 ат % углерода.

Сплав Cr 5 - U 45 - C 50 (ат %). На микроструктуре литого сплава выявлены крупные зерна UC, интенсивно окрашиваемые азотной кислотой, по границе зерна - небольшие утолщения, не окрашиваемые азотной кислотой. После отжига утолщения по границам зерна исчезли. На рентгенограмме литого сплава выявлена одна решетка типа UC с чуть размытыми линиями, после отжига рефлексы стали острыми, расчет дал величину параметра, равную 4,961 Å.

Сплав Cr 10 - U 40 - C 50 (ат %). На микроструктуре литого сплава выявлены зерна, обогащенные UC и интенсивно окрашиваемые HNO_3, выделения по границам зерна азотной кислотой не окрашиваются. После отжига утолщения по границам зерна исчезли. На рентгенограмме литого сплава есть довольно четкие рефлексы UC с чуть увеличенным параметром решетки. Наблюдается увеличение параметра решетки с увеличением температуры отжига:

t, °C	1500	1400	1320
a, Å	4,985	4,970	4,834

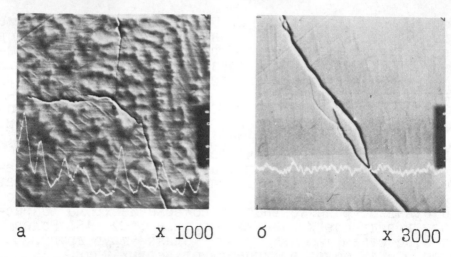

а   x 1000   б   x 3000

Рис.2. Изображение сплава Cr 15 – U 35 – C 50 (ат %) в литом (а) и отожженном при 1500° С (б) состояниях в упруго отраженных электронах с линейным профилем распределения в лучах $K_{a_1} Cr$ при сканировании вдоль горизонтальной прямой, проходящей через светлую точку в центре снимка.

При 1320° замечен распад, сопровождающийся мартенситными выделениями внутри зерна и размытием дифракционных колец.

Сплав Cr 15 – U 35 – C 50 (ат %). микроструктурно картина литого и отожженного сплавов аналогична предыдущей, только увеличились утолщения по границам зерна. На рис.2 демонстрируются линейные профили распределения в лучах K_a, Cr в литом и отожженном состояниях, которые подтверждают увеличение концентрации хрома по границам зерна в литом состоянии и равномерное распределение хрома в отожженном образце. Также наблюдается увеличение параметра решетки с увеличением температуры отжига:

t, °C	1500	1400	1320
a, Å	5,022	4,975	4,966

Сплав Cr 20 – U 30 – C 50 (ат %). Согласно рентгеноструктурному анализу образец в литом и отожженном состояниях демонстрирует две фазы $UCrC_2$ и (UC), первая является преимущественной.

Были специально выплавлены сплавы, которые по шихте имели 2 ат % избытка углерода относительно разреза UC – $UCrC_2$.

Сплав Cr 5 - U 43 - C 52 (ат %). На микроструктуре литого сплава выявлены зерна, окрашиваемые азотной кислотой; по границам зерна - утолщения, которые азотной кислотой не окрашиваются; внутри зерна - видманштеттова структура. На рентгенограмме литого сплава обнаружены отражения UC с параметром решетки 4,963Å, а также наиболее сильные рефлексы (2II и II4) решетки $\alpha - UC_2$. На микроструктуре отожженного сплава зерно укрупнилось, утолщения по границам зерна исчезли, укрупнились также иглы видманштеттовой структуры. На рентгенограмме отожженного сплава кроме рефлексов UC (α = 4,964Å) обнаружены рефлексы U_2C_3.

Сплав Cr I0 - U 38 - C 52 (ат %). На микроструктуре литого сплава выявлена дендритная структура, внутри дендритов видна видманштеттова структура; дендриты интенсивно окрашиваются азотной кислотой, прожилки между дендритами азотной кислотой не травятся (рис.3). На рентгенограмме литого сплава обнаружены очень размытые рефлексы UC с увеличенным параметром решетки и наиболее сильные рефлексы соединения x. После отжига микроструктура резко меняется: все поле шлифа - однофазное, азотной кислотой не окрашивается, границы полиэдров - тонкие, изломанные. На рентгенограмме отожженного сплава обнаружены две изоморфные решетки на основе UC с параметрами 4,961 и 5,052Å.

Сплав Cr I5 - U 33 - C 52 (ат %). Согласно микроструктурному анализу в литом состоянии сплав двухфазен, ни одна из фаз не окрашивается азотной кислотой, что свидетельствует о присутствии хрома в обеих фазах. Вид микроструктуры отожженного сплава аналогичен литому (по-видимому, время отжига было недостаточно). Рентгеноструктурно в обоих состояниях обнаружено три фазы: две - изоструктурные UC с параметрами 4,961 и 5,058Å, третья фаза - $UCrC_2$.
    Перейдем к описанию сплава разреза U_2C_3 - $UCrC_2$.

Сплав Cr 5 - U 37,I - C 57,9 (ат %). Согласно рентгенофазовому анализу в литом состоянии присутствуют две фазы: $\alpha - UC_2$ и UC с увеличенным параметром, рефлексы обеих фаз размытые. После отжига рефлексы двух вышеназванных фаз стали четкими, значения вычисленных параметров для $\alpha - UC_2$ a = 3,527Å, c = 6,00IÅ, для (UC) - 5,0I9Å, обнаружены также рефлексы U_2C_3.

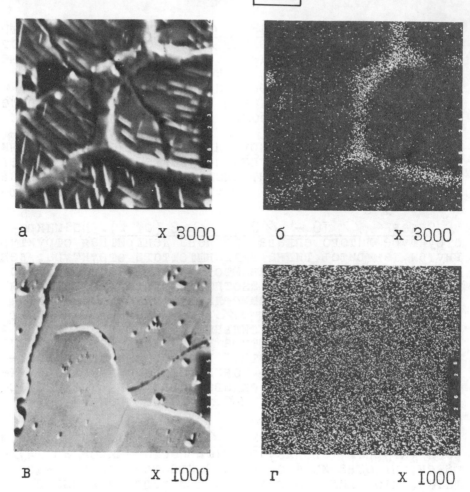

а   x 3000   б   x 3000

в   x 1000   г   x 1000

Рис. 3. Изображение сплава Cr 10 – U 38 – C 52 (ат %) в упруго отраженных электронах (а, в) и в лучах K_{a_1} Cr (б, г) в литом (а, б) и отожженном при 1500°C (в, г) состояниях.

Сплав Cr 10 – U 34 – C 56 (ат %). В литом состоянии фазовый состав сплава аналогичен предыдущему сплаву. После отжига фаза на основе UC подвергалась тетрагонализации с параметрами $a = 5,041\text{Å}$, $c = 5,124\text{Å}$.

Сплав Cr 15 – U 31,1 – C 53,9 (ат %). Согласно рентгеноструктурному анализу сплав в литом состоянии имеет две фазы: соединение x и твердый раствор на основе UC с очень размытыми рефлексами. После отжига фаза на основе UC подвергалась тетрагонализации с параметрами $a = 5,044$, $c = 5,128Å$.

Согласно рентгенофазовому анализу фазовый состав сплавов с шихтовым составом, соответствующим разрезу UC_2 - $UCrC_2$, Cr 5 - U 31,5 - C 63,5; Cr 10 - U 30 - C 60; Cr 15 - C 28,4 - C 56,6 (ат %) в литом и отожженном состояниях одинаков: $\alpha - UC_2$ и x, только в последнем сплаве после отжига появились довольно интенсивные линии (UC) (тетр.), кроме того в этом сплаве микроструктурно отмечено равенство объемных количеств $\alpha - UC_2$ и x. Во всех сплавах после отжига отмечено также превращение, "разъедающее" зерна $\alpha - UC_2$, что можно расценивать как реакцию перитектического образования x:

$$\alpha - UC_2 + c + ж \rightleftharpoons x$$

Сплав Cr 15 - U 25 - C 60 (ат %), который по данным работы $[8]$ является составом соединения x, имеет фазовый состав, близкий к фазовому составу сплава Cr 15 - U 28,4 - C 56,6 (ат %). Используя данные по фазовому составу сплавов Cr 15 - U 28,4 - C 56,6 и Cr 35 - U 15 - C 50 (ат %), по правилу рычага оценили состав соединения x. Действительно, в сплаве Cr 15 - U 28,4 - C 56,6 (ат %) количества $\alpha - UC_2$ и x приблизительно равны, а в сплаве Cr 35 - U 15 - C 50 (ат %) обнаружены некоторые количества Cr 45 и Cr_3C_2, поэтому провели окружность с центром в точке $o = Cr$ 15 - U 28,4 - C 56,6 (ат %) радиусом, равным отрезку $UC_2 - O$, которая пересекает прямую, проведенную через точки Cr 35 - U 15 - C 50 (ат %) и Cr_3C_2 (так как Cr 45 лежит почти на этой прямой) в точке Cr 30 - U 17,5 - C 52,5 (ат %), которая, по нашему мнению, близка к составу соединения x. Не исключено, что это соединение имеет формульный состав UCr_2C_3.

Изложенное в данном разделе дало возможность сконструировать изотермическое сечение при 1500- 1600^{0}С диаграммы состояния фрагмента UC - $UCrC_2$ - - x - UC_2 (рис.4), составы исследованных сплавов нанесены согласно данным хим. анализа. Наиболее интересным следует считать образование широкой области твердых растворов хрома на основе высокотемпературного твердого раствора ($UC - \beta - UC_2$), который хром стабилизирует до более низкой температуры. Обогащенный хромом твердый раствор ($UC - \beta - UC_2$) имеет более низкую температуру плавления, о чем свидетельствует его расположение по границам дендритов в литых сплавах. Согласно строению изотермического

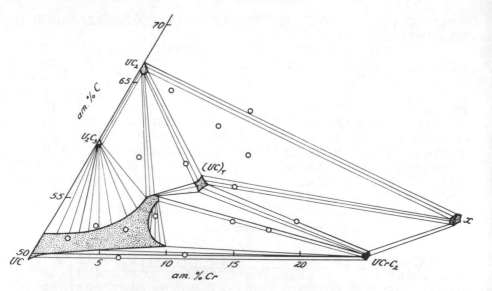

Рис. 4. Изотермическое сечение при 1500-1600°C диаграммы состояния фрагмента $UC - UCrC_2 - x - UC$

сечения при 1500-1600°C UC может обогащаться углеродом до ~55 ат % при добавлении хрома ~7 ат % не изменяя кристаллической структуры, по-видимому, это справедливо вплоть до температуры плавления этого раствора. Вне области существования твердого раствора ($UC - \beta - UC_2$) добавки хрома не подавляют образование U_2C_3 .

3.4. Схема кристаллизации сплавов системы U – Cr – C

Процесс кристаллизации начинается с образования соединения x по перитектической реакции (рис. 5).

$$UC_2 + C + \mathcal{K} \rightleftharpoons x \qquad (I)$$

Состав соединения оценен как Cr 30 – U 17 – C 53 (ат %). От нонвариантного равновесия I отходят два моновариантных равновесия $\mathcal{K} = UC_2 + x$ и $\mathcal{K} = C + x$ с жидкостью и одно трехфазное равновесие в твердом состоянии, выделяющее трехфазную область $UC_2 + C + x$ (а) (рис. 6). Моновариантное равновесие $\mathcal{K} = C + x$, пересекаясь с моновариантным равновесием $\mathcal{K} + C = Cr_3C_2$, образует нонвариантное перитектическое равновесие II рода.

$$C + \mathcal{K} \rightleftharpoons x + Cr_3C_2 \qquad (II)$$

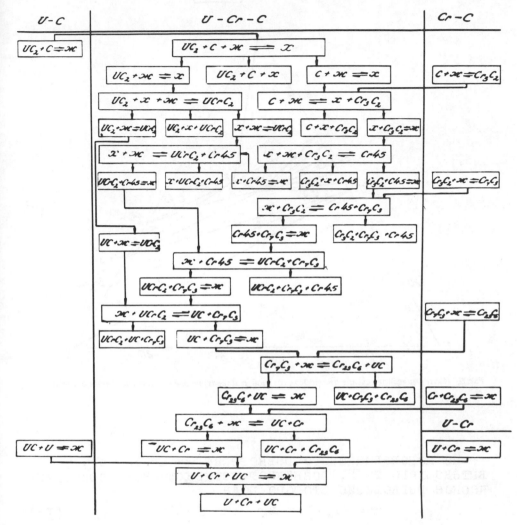

Рис.5. Схема моно- и нонвариантных равновесий в системе U – Cr – C.

а из моновариантного равновесия $\mathcal{Ж} = UC_2 + x$ возникает нонвариантное перитектическое равновесие

$$UC_2 + x + \mathcal{Ж} \rightleftharpoons UCrC_2 \qquad\qquad (\text{Ⅲ})$$

Ниже равновесия Ⅱ возникает трехфазная область $C + Cr_3C_2 + x$ (б) – сплав Cr 45– U 5 – C 50 (ат %), а ниже равновесия Ⅲ – трехфазная область $x + UC_2 + UCrC_2$ (в). Перитектический характер образования соединения x затрудняет достижение равновесия в сплавах, поэтому x – соединение мы находим в сплавах Cr 10 – U 34 – C 56 и Cr 15–U 31– C 54 (ат %) даже после длительного отжига.

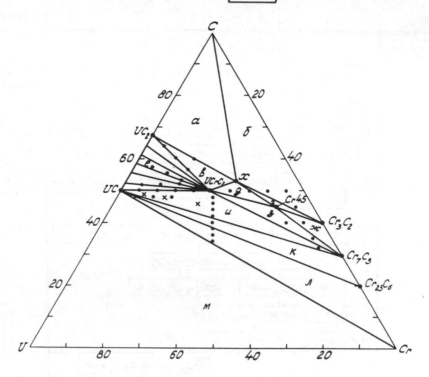

Рис. 6. Изотермическое сечение при подсолидусной температуре диаграммы состояния системы $U - Cr - C$.

Из моновариантного равновесия $\mathcal{Ж} \rightleftarrows x + Cr_3 C_2$ вытекающего из П, возникает четырехфазное перитектическое равновесие второго рода

$$x + Cr_3 C_2 + \mathcal{Ж} \rightleftarrows Cr 45 \tag{IУ}$$

в результате которого образуется тройное химическое соединение состава Cr 45 – U 10 – C 45 (ат %). Ниже равновесия IУ выделяется трехфазная область в твердом состоянии $x + Cr$ 45 + $Cr_3 C_2$ (г) (сплав Cr 35-– U 15 – C 50, ат %).

Моновариантное равновесие $\mathcal{Ж} = x + UC_2$, вытекающее из Ш, и моновариантное равновесие, вытекающее из IУ, пересекаясь, образуют четырехфазное перитектическое равновесие П рода

$$\mathcal{Ж} + x \rightleftarrows UCrC_2 + Cr 45 \tag{У},$$

ниже которого расположена трехфазная область x + + C 45 + $UCrC_2$ (д) – сплав Cr 30- U 20- C 50 (ат %).

Другое моновариантное равновесие $UC_2 + ж = UCrC_2$, вытекающее из Ш, пересекает двухфазную область (е) и в конце ее превращается в равновесие $UC + ж = UCrC_2$, ниже температуры солидуса в двухфазной области (е) идут сложные превращения в твердом состоянии.

Третье моновариантное равновесие $ж = Cr_3C_2 + Cr\,45$, отходящее от реакции IУ, пересекаясь с моновариантным равновесием $ж + Cr_3C_2 = Cr_7C_3$ образует четырехфазное перитектическое равновесие II рода.

$$ж + Cr_3C_2 \rightleftharpoons Cr_7C_3 + Cr\,45, \qquad (УI)$$

ниже которого выделяется трехфазная область $Cr_3C_2 + Cr_7C_3 + Cr\,45$ (Ж), в которой находится сплав $Cr\,55 - U\,5 - C\,40$ (ат %). Моновариантное равновесие, вытекающее из равновесия УI, пересекаясь с моновариантным равновесием $ж = UCrC_2 + Cr\,45$, вытекающим из равновесия У, образует четырехфазное перитектическое равновесие II рода.

$$ж + Cr\,45 \rightleftharpoons UCrC_2 + Cr_7C_3, \qquad (УП)$$

ниже которого расположена трехфазная область $Cr\,45 + Cr_7C_3 + UCrC_2$ (З), в которой находится сплав $Cr\,50 - U\,10 - C\,40$ (ат %), а второе моновариантное равновесие $ж = UCrC_2 + Cr_7C_3$, пересекаясь с моновариантным равновесием $ж + UC = UCrC_2$, образует четырехфазное перитектическое равновесие II рода

$$ж + UCrC_2 \rightleftharpoons UC + Cr_7C_3, \qquad (УШ)$$

ниже которого расположена трехфазная область $UCrC_2 + UC + Cr_7C_3$ (И), в которой находятся сплавы $Cr\,29 - U\,29 - C\,42$, $Cr\,28 - U\,28 - C\,44$, $Cr\,27 - U\,27 - C\,46$ и $Cr\,26 - U\,26 - C\,48$ (ат %). Второе моновариантное равновесие $ж = UC + Cr_7C_3$, пересекаясь с моновариантным равновесием $ж + Cr_7C_3 = Cr_{23}C_6$, образует четырехфазное перитектическое равновесие II рода.

$$ж + Cr_7C_3 \rightleftharpoons Cr_{23}C_6 + UC, \qquad (IX)$$

ниже которого расположена трехфазная область $Cr_7C_3 + Cr_{23}C_6 + UC$ (К), характеризуемая сплавом $Cr\,30 - U\,30 - C\,40$ (ат %). Второе моновариантное равновесие $ж = Cr_{23}C_6 + UC$, пересекаясь с моновариантным равновесием $ж = Cr + Cr_{23}C_6$, образует четырехфазное перитектическое равновесие II рода

$$ж + Cr_{23}C_6 \rightleftharpoons UC + Cr, \qquad (X)$$

ниже которого расположена трехфазная область $UC + Cr + Cr_{23}C_6$ (Л), характеризуемая сплавами $Cr\,33 - U\,33 - C\,34$, $Cr\,32 - U\,32 - C\,36$, $Cr\,31 - U\,31 - C\,38$ (ат %). Второе моновариантное равновесие $ж \rightleftharpoons UC + Cr$,

пересекаясь с моновариантными равновесиями $ж \rightleftharpoons U +$
$+ Cr$ и $ж \rightleftharpoons U + UC$ образует нонвариантное четырехфазное эвтектическое равновесие

$$ж \rightleftharpoons U + Cr + UC, \qquad (XI)$$

завершающее процесс кристаллизации в тройной системе
$U - Cr - C$, ниже этой реакции расположена трехфазная область $U + Cr + UC$ (М).

3.5. Кристаллическое строение тройных фаз

Порошковая рентгенограмма тройной фазы x ,
состав которой близок к составу сплава Cr 35 - U 15-
-C 50 (ат %), была проиндицирована путем сравнения
экспериментальных межплоскостных расстояний с расчет-
ными, полученными при подстановке в квадратичную
форму тетрагональной сингонии параметров a = 3,636Å,
C = 15,739Å $[9]$. Получено отличное совпадение.
На порошкограмме сплава Cr 35 - U 15 - C 50 (ат %),
отожженного при 1600°С 4 ч., кроме рефлексов соединения x обнаружены наиболее сильные рефлексы Cr 45
и Cr_3C_2 . Состав соединения оценен как UCr_2C_3 .
Для монокристалла соединения, состав которого
близок составу сплава Cr 45 - U 10- C 45 (ат %),
проведен первый этап рентгеноструктурного анализа.
Соединение относится к тетрагональной сингонии с параметрами элементарной ячейки a = 7,933Å, C =
3,073Å. По законам погасания установлена рентгенов-
ская пространственная группа $4/mmm$.

4. ЗАКЛЮЧЕНИЕ

В результате настоящего исследования обнаружено
новое соединение вблизи состава Cr 45 - U 10 -C 45
(ат %), определена его кристаллическая структура.
Уточнен состав соединения x (UCr_2C_3). Установлен
характер кристаллизации сплавов во всей области составов, составлена схема нон- и моновариантных равновесий. Доказано отсутствие конноды $UC - Cr_3C_2$,
поэтому стало возможным отнести замеченные, но не-
опознанные рефлексы в сплаве с 20 мол.% Cr_3C_2 разреза $UC - Cr_3C_2$ $[3,4]$ к Cr_7C_3 , поскольку
они не принадлежат ни UC , ни UCr_2C_2 , ни Cr_3C_2 .
Показано существование четырехфазной перитектической реакции

$$ж + Cr_{23}C_6 \rightleftharpoons UC + Cr$$

что объясняет экспериментальные факты изложенные в
$[6-8]$, а именно, что в сплавах разреза $UC - Cr$
следы эвтектики образуются при $1315^{\circ}C$, а полное
плавление эвтектических композиций не происходит ниже $1425^{\circ}C$ даже при продолжительных отжигах. Это происходит вследствие того, что разрез $UC - Cr$ не является бинарным, а пересекает область существования
четырехфазной перитектической реакции, поэтому поверхность солидуса и ликвидуса на разрезе $UC - Cr$
не имеет точек соприкосновения, что и наблюдается
экспериментально. Этим же объясняется эксперименталь-
ный факт $[8]$ несовместимости твердых растворов
$Cr - Fe$ с UC и прохождения реакции $UC + Cr + Fe$
$\longrightarrow UFe_2 + Cr_{23}C_6$, так как в результате реакции
$UC + Cr \rightleftharpoons \text{ж} + Cr_{23}C_6$ появляется жидкий уран,
связывающий железо в UFe_2.

Наиболее интересные результаты получены в области
составов $UC - UC_2 - UCrC_2$. Построенное изотермическое сечение при $1500-1600^{\circ}C$ характеризуется широкой областью гомогенности твердого раствора хрома в
высокотемпературном твердом растворе ($UC - \beta - UC_2$),
который хром стабилизирует до более низкой температуры. По-видимому, этим явлением можно объяснить отмеченный во введении факт улучшения модифицированного
хромом монокарбидного топлива $[2]$, так как содержание углерода в монокарбидном топливе можно увеличить
до ~ 55 ат % при добавлении хрома ~ 7 ат % без
изменения кристаллической структуры топлива. При более высоком содержании хрома и углерода происходит
тетрагонализация твердого раствора хрома на основе
($UC - \beta - UC_2$). Форма однофазной области твердого
раствора хрома в ($UC - \beta - UC_2$) позволяет объяснить обнаружение в $[3,4]$ в сплаве с 5 мол.% Cr_3C_2
двух решеток на основе UC в предположении выгора-
ния хрома при приготовлении сплава и перемещения его
в область составов $UC - UC_2 - UCrC_2$.

ЛИТЕРАТУРА

[1] HARBOURNE, B.L., et al., Proc. of the Conference on Fast Reactors, Fuel Elements, Technology,
     New Orleans, Louisana (1971) 869.
[2] GORLE, F., COHEUR, L., TIMMERMANS, W., J. Nucl. Mater. 51 3 (1974) 343.
[3] NOWOTNY, H., KIEFFER, R., LAUBE, E., Monatsh. Chem. 88 3 (1956) 336.
[4] NOWOTNY, H., KIEFFER, R., BENESOVSKY, F., Rev. Metall. (Paris) 55 5 (1958) 453.
[5] NOWOTNY, H., et al.,Monatsh. Chem. 89 6 (1958) 692.

[6]   BARTA, J., BRIGGS, G., WHITE, J., J. Nucl. Mater. 4 3 (1961) 322.

[7]   BRIGGS, G., et al., Trans. J. Br. Ceram. Soc. 62 (1963) 221.

[8]   BRIGGS, G., DUTTA, S.K., WHITE, J., Carbides in Nuclear Energy, v.I. Physical and Chemical Properties Phase Diagrams, (RUSSELL, L.E., Ed.), Macmillan, New York (1964) 7.

[9]   FARR, J.D., BOWMAN, M.G., ibid., 184.

[10] СТОРМ, Э., Тугоплавкие Карбиды, М., Атомиздат (1970) 304.

[11] КОТЕЛЬНИКОВ, Р.Б. и др., Особотугоплавкие элементы и Соединения, М. Металлургиздат (1969) 10.

[12] FROST, B.R.T., J. Nucl. Mater. 10 4 (1963) 265.

[13] MALLETT, M.W., GERDS, A.F., VAUGHAN, D.A., J. Electrochem. Soc. 98 (1951) 505.

[14] IMOTO, S., et al., Carbides in Nuclear Energy, v.I. Physical and Chemical Properties Phase Diagrams (RUSSELL, L.E., Ed.), Macmillan, New York (1964) 7.

DISCUSSION

A. NAOUMIDIS: In the U–Cr–C phase diagram for $T \approx 1350°C$ you indicate a high hyperstoichiometry in the UC binary region in comparison with the well known U–C phase diagram. Could you please comment?

A.S. PANOV: The paper clearly shows that addition of chromium to uranium carbide widens the region of homogeneity of UC, and the dicarbide phase disappears at higher values of carbon. This fact is apparently due to changes in the thermodynamic activity of carbon in uranium monocarbide consequent upon the addition of chromium.

M.H. RAND: Were any measurements made of the lattice parameters of the fcc phase containing 7 at.% Cr and 55 at.% C?

A.S. PANOV: Yes, the lattice parameters of alloys of this system were measured, but only at room temperature.

CALCULABILITY OF
MULTIPLE CHEMICAL INTERACTIONS

S. FENYI, H. SUNDERMANN*
Kernforschungszentrum Karlsruhe,
Karlsruhe,
Federal Republic of Germany

Abstract

CALCULABILITY OF MULTIPLE CHEMICAL INTERACTIONS.

Chemical interactions are taken as a basis for calculating the equilibria and kinetics in multiple-component systems. Integration methods, the use of statistical solid-state and chemical models, and model-free thermodynamic relationships are discussed.

Almost all questions of practical importance concerning the behaviour of substances derive from the problem of drawing from known initial conditions conclusions regarding future thermodynamic behaviour. For example, statements about the dynamic behaviour of nuclear fuels or about their interactions with the clad under changing concentrations of fission products are kinetic questions that can only be treated if, in addition to knowledge of the non-equilibrium of the initial state, the equilibrium of the final state is also known. In order to calculate this equilibrium, however, it is necessary to know all the interactions in the total, usually multi-component system. Hence the determination of interactions and of quantities to be derived from these interactions (e.g. solubility, distributions, excess functions and others) become the basis for the start of each thermodynamic treatment. And it is only in this way that the kinetics of exchange reactions, which transform an initial state into a final state [1], can be treated meaningfully.

There are, in principle, two ways of tackling the problem of determining interactions (multi-component activity coefficients): the interactions can be determined experimentally or with the help of theoretical relationships. The experimental procedure already becomes impracticable for ternary systems because of the large number of analyses to be performed. On the theoretical side, various authors have different angles of attack, for example:

    (a) Integration of the Gibbs-Duhem equation [2, 3];

    (b) Derivation of thermodynamic relations between binary and multi-component quantities with the help of solid-state statistical models [4, 5]; or

    (c) Generation of simple relations via simplified chemical models [6, 7].

---

* At present at the International Atomic Energy Agency, Vienna.

KEY TO SYMBOLS

---

x_i concentration of the solvent components; $\sum_i x_i = 1 - x_{0,m}$

f_i binary activity coefficients of the solvent components

$j_{0,i}$ activity coefficient of the dissolved component 0 in **dilute solution** in component i alone ($j_{0,i} x_{0,i} = \sqrt{p_0}$, where p is the partial pressure of component 0)

$j_{0,m}$ corresponding multi-component activity coefficient

ϵ interaction coefficient

All variants of the method involving integration of the Gibbs-Duhem equation are limited to ternary systems and require a knowledge of the dependence of the partial molal Gibbs function, $\bar{G}_i$, upon the ternary composition.

We have attempted to expand these methods to the case of n substances ($n \geqslant 3$). The condition

$$\sum_{i=1}^{n} N_i = 1 \tag{1}$$

for the mole fractions is a simplex [8] in the composition space. We kept one arbitrarily chosen N_i free and the following ratios of mole fractions constant:

$$\frac{N_1}{N_n}, \ldots \ldots \frac{N_{i-1}}{N_n}, \frac{N_{i+1}}{N_n}, \ldots \ldots \frac{N_{n-1}}{N_n}$$

Together, these relations determine a straight line on the simplex [8]. Use of the differentiation rules for implicit functions yielded, after the expenditure of some effort, a set of separable ordinary differential equations:

$$\frac{\bar{G}_i - G}{1 - N_i} = \left(\frac{\partial G}{\partial N_i}\right)_{\frac{N_1}{N_n}, \ldots, \frac{N_{i-1}}{N_n}, \frac{N_{i+1}}{N_n}, \ldots, \frac{N_{n-1}}{N_n}} \qquad i = 1, \ldots n-1 \tag{2}$$

which for n = 3 is identical to the set of Darken equations; in fact, we have used Darken's terminology [2]. This set of equations can be integrated and G can be determined as a function of N_i.

This method calls for measurements of $\overline{G}_i$, which essentially require the reversible mixing of the i-th component into the multi-component solution. Such measurements are commonly undertaken with galvanic cells.

It should be noted that this procedure does not reduce the problems mentioned initially. It merely enables one to check whether the results of the measurements are, in fact, consistent with the Gibbs-Duhem equation.

Wagner attempted to derive thermodynamic relations with the help of solid-state statistical models [4, 5]. Simply summarized, Wagner started from the number of nearest neighbours of a dissolved atom and the configurational enthalpies resulting from the permutation of the neighbours. The relations derived were fitted to experimental results from various systems, but for each system considered the values determined for the unknown parameters were different, and did not form a basis for a general theory.

The attempts made using simplified chemical models [6, 7] had to assume special configurations of the dissolved and solvent components, and were even more restricted than Wagner's approach.

All these model-dependent relationships must be thermodynamically inconsistent, as the concentration of a dissolved component in a multi-component system results from a large number of clusters of its atoms with the atoms of the solvent — the clusters representing many different microstates; one clearly cannot simply select specific clusters and permute only these.

For these reasons we have elected not to use a model and determined, through consideration of partial equilibria and with the help of an ansatz [9] that suits certain regular solutions, that the partial solubility of a dissolved component changes linearly with its activity, according to the following relation:

$$\frac{1}{j_{0,m}} = \sum_{i=1}^{n} \frac{x_i f_i}{j_{0,i}} \tag{3}$$

For a ternary system this equation reduces to:

$$\frac{1}{j_{0,m}} = \frac{x_1 f_1}{j_{0,1}} + \frac{x_2 f_2}{j_{0,2}} \tag{4}$$

For $x_{0,m} \ll 1$ the activity coefficients f_i are purely binary quantities. It should be noted that Eq. (4) is similar to approximations which have been reported by several authors [7, 10]: those approximations have, however, resulted from assumptions and relationships for parameters they have introduced and which are in conflict with their definitions deriving from the original model.

From Eq. (4) one obtains for the interaction coefficient according to:

$$\epsilon_{0,1}^{(n)} = \left. \frac{\partial \ln j_{0,m}}{\partial x_n} \right|_{x_1 \to 1} \tag{5}$$

the following relation:

$$\epsilon_{0,1}^{(n)} = 1 - \frac{j_{0,1}\, f_{n,1}\,(x_n = 0)}{j_{0,n}\, f_{n,1}\,(x_n = 1)} = F(0, n) \tag{6}$$

If one compares Eq. (4) and Eq. (6) with data reported in the literature for ternary systems, the relations given here are only approximately valid. One has to take into account, however, that for identical systems the data of different authors differ (by orders of magnitude) and the activity coefficients which have, for comparison purposes, to be inserted into Eqs (4) and (6) are equally unreliable.

The obvious symmetries of the periodic system, which are obtained if one plots the interaction coefficients versus their atomic numbers, can be understood easily after the explicit representation of Eq. (6), as the expression $\epsilon_{0,1}^{(n)} = F(0, n)$ defines matrix elements involving atomic numbers when the subscript and superscript are treated as new elemental variables; the converse can also be postulated. We assume that beyond the relations presented here there are, in addition, further ones.

REFERENCES

[1] FENYI, S., Kernforschungszentrum, Karlsruhe, Rep. KFK-2484 (1978).
[2] DARKEN, L.S., J. Am. Chem. Soc. 72 (1950) 2909.
[3] SCHUMANN, R., Acta Metall. 3 (1955) 219.
[4] WAGNER, C., Thermodynamics of Alloys, Cambridge, Mass. (1952) 41.
[5] WAGNER, C., Acta Metall. 21 (1973) 1297.
[6] RICHARDSON, F.D., Scripta Metall. 3 (1969) 161.
[7] JACOBS, K.T., ALCOCK, C.B., Acta Metall. 20 (1972) 221.
[8] STROUD, A.H., Approximate Calculation of Multiple Integrals, Prentice Hall, Englewood Cliffs, N.J. (1971) 28.
[9] SUNDERMANN, H., Kernforschungszentrum, Karlsruhe, Rep. KFK-2302 (1976).
[10] OHTANI, M., GOKEN, N.A., Trans. Metall. Soc. AIME 218 (1960) 533.

Section K

STUDIES RELATED TO CLAD PERFORMANCE

THERMODYNAMIC STUDIES OF THORIUM CARBIDE FUEL PREPARATION AND FUEL/CLAD COMPATIBILITY*

T.M. BESMANN, E.C. BEAHM
Chemical Technology Division,
Oak Ridge National Laboratory,
Oak Ridge, Tennessee,
United States of America

Abstract

THERMODYNAMIC STUDIES OF THORIUM CARBIDE FUEL PREPARATION AND FUEL/CLAD COMPATIBILITY.

The carbothermic reduction of thorium and uranium-thorium dioxide to monocarbide has been assessed. Equilibrium calculations have yielded Th-C-O and U-Th-C-O phase equilibria and CO pressures generated during reduction. The CO pressures were found to be at least five orders of magnitude greater than any of the other 15 gaseous species considered. This confirms that the monocarbide can successfully be prepared by carbothermic reduction. The chemical compatibility of thorium carbides with the Cr-Fe-Ni content of clad alloys has been thermodynamically evaluated. Solid solutions of $\langle ThNi_5 \rangle$ and $\langle ThFe_5 \rangle$ and of $\langle Cr_7C_3 \rangle$ and $\langle Fe_7C_3 \rangle$ were the principal reaction products. The Cr-Fe-Ni content of 316 stainless steel showed much less reaction product than that of any of the other six alloys considered.

1. INTRODUCTION

Carbides of thorium are presently being considered as alternative fertile materials in breeder reactor systems. The monocarbide is especially useful due to its high metal-to-carbon ratio and its wide homogeneity range, extending from $\langle ThC_{0.67} \rangle$ to $\langle ThC_{0.97} \rangle$ even below 1300 K [1]. Thus, carbon from single-phase $\langle ThC_{0.97} \rangle$ can form significant quantities of carbide fission products without generating metallic thorium. Similarly, fissile material may be solid solutions of uranium and thorium monocarbides.

This study considers two aspects of the chemical thermodynamics of carbide fuel. The first is the carbothermic conversion of thorium or thorium-uranium dioxide to monocarbide. This involves the computation of phase equilibria and gaseous species

* Research sponsored by the Division of Materials Sciences, United States Department of Energy, under contract W-7405-eng-26 with the Union Carbide Corporation.

equilibrium pressures. These are useful to know for optimizing the process and for identifying conditions under which actinide vaporization is a problem, as in plutonium-containing systems where plutonium pressures exceed (CO)* pressures in monoxycarbide plus sesqui- or dicarbide phase fields [2]. The second aspect of the work includes a method for assessing the chemical compatibility of thorium carbides with the Cr-Fe-Ni content of potential clad alloys. Basic chemical compatibility studies, at equilibrium, involve a determination of whether a chemical reaction can occur, as well as an evaluation of the products formed and the extent of reaction.

2. CALCULATIONAL METHOD

The calculation of equilibrium compositions in multicomponent systems, such as those considered here, can be complex and prohibitively tedious to perform manually. Thus a computer program for calculating equilibria in chemical systems (SOLGASMIX-PV) developed by Eriksson [3] and modified by Besmann [4] was employed. Equilibrium compositions are determined by the direct minimization of system free energy. The program can consider a gas phase with multiple condensed phases and contains a provision for the inclusion of condensed solutions, provided activity coefficient expressions are available for any nonideal systems. SOLGASMIX-PV was used for all the described calculations. Note that the thermodynamic data used in these calculations are at the temperature being considered rather than at 298 K.

3. CARBOTHERMIC REDUCTION

3.1. Th-C-O system

Potter [5] has presented a tentative Th-C-O phase diagram at 1773 K containing the dioxide and mono- and dicarbides, with the monocarbide dissolving significant amounts of oxygen. General features of this diagram have since been confirmed by Kanno et al. [6] and Pialoux and Zaug [7]. The mono- and dicarbides display wide homogeneity ranges which increase with temperature such that at $T \gtrsim 2150$ K the carbides form a single phase [8]. Assessed thermodynamic values, however, are available

---

*The following notation is used throughout: < > pure element or compound in the solid state; [ ] solid solution (subscript denotes solvent or other solution component); { } liquid state; ( ) gaseous state.

only for the 298 K carbon-saturated stoichiometries $<ThC_{0.97}>$
and $<ThC_{1.94}>$ (respectively mono- and dicarbides) [1], which are
appropriate for the envisioned single-phase monocarbide fuel.
These are the compositions which are assumed in the present equilibrium calculations.

Thorium monocarbide has successfully been carbothermically
reduced from the dioxide by Imai, Hosaka, and Naito [9], Heiss
and Djémal [10], Kanno et al. [6], and Pialoux and Zaug [7].
The overall reaction can be written

$$<ThO_2> + 2.97 \ <C> = <ThC_{0.97}> + 2 \ (CO)$$

Thus the system must pass through the $<ThO_2> + <ThC_{1.94}> + <C>$,
$<ThO_2> + <ThC_{1.94}> + <ThC_{0.97(1-y)}O_y>$, and $<ThC_{1.94}> +$
$<ThC_{0.97(1-y)}O_y>$ phase regions. At elevated temperatures, the
dioxide has a finite homogeneity range, but for the present calculations the sufficiently accurate approximation of O/Th = 2
was used throughout. Note that the monocarbide containing dissolved oxygen (monoxycarbide) is assumed to have oxygen substituting for carbon and carbon site vacancies. Potter [5] suggests
a maximum value for y of 0.2, although he assumes a thorium-to-
nonmetal ratio of unity.

Thermodynamic equilibrium calculations were performed for
Th-C-O system compositions spanning these phase regions. The
partial pressures of a total of 11 gaseous species — (O_2), (C_1),
(C_2), (C_3), (CO), (CO_2), (Th), (ThC_2), (ThC_4), (ThO), (ThO_2) —
were computed over the temperature range 1000 to 2500 K. The condensed phases $<ThC_{0.97(1-y)}O_y>$, $<ThC_{1.94}>$, $<ThO_2>$, $<Th>$, $\{Th\}$, and
$<C>$ were input to the calculations, the results of which indicate
those phases that are stable. The Gibbs standard free energy of
formation (ΔG_f^o) data for the thorium-containing species were
obtained from Rand et al. [1]. Data for all other species were
gathered from the JANAF Thermochemical Tables [11].

3.1.1.   *$<ThO_2> + <ThC_{1.94}> + <C>$ phase region*

The initial calculations using an overall composition expected
to lie in the dioxide-dicarbide-carbon phase region [5] indicated
that $<ThC_{0.97(1-y)}O_y>$ forms rather than $<ThC_{1.94}>$. To reconcile
the calculational results with the phase equilibria, it was necessary to modify the thermodynamic data of at least one condensed
phase. Since data for the monocarbide are the least certain of
the appropriate phases [1], its values were varied. It was found
that a 5 kJ/mol more positive standard free energy yielded the
expected phases and the modified values were used throughout the
calculations.

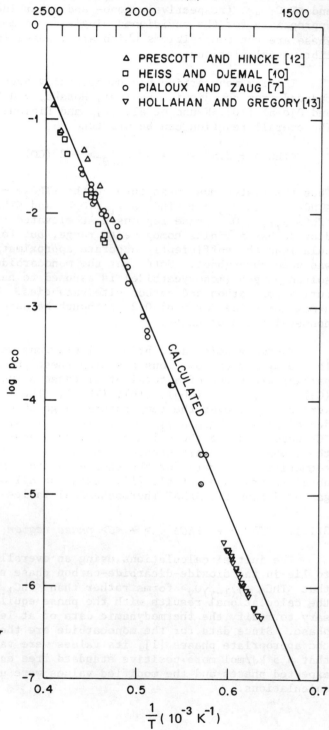

FIG.1. Experimentally
measured and calculated
equilibrium log p_{CO} versus
reciprocal temperature over
the phase region
$\langle ThO_2 \rangle + \langle ThC_{1.94} \rangle + \langle C \rangle$
(p in megapascals:
1 MPa = 9.87 atm).

Equilibrium calculations performed for the monovariant dioxide-dicarbide-carbon phase region yielded the (CO) pressure (p_{CO})-temperature relationship

$$\log p_{CO} \text{ (MPa)} = 8.483 - \frac{22\ 750}{T}$$

where T is temperature in Kelvin. All other gaseous species had partial pressures at least five orders of magnitude less than that of the (CO) and were considered to be unimportant.

This three-phase region is the only one for which experimental p_{CO} measurements have been reported [7, 10, 12, 13]. Figure 1 shows good agreement between the calculated and experimentally determined values.

3.1.2. *$<ThO_2>$ + $<ThC_{1.94}>$ + $<ThC_{0.78}O_{0.2}>$ phase region*

The phase $<ThC_{0.97(1-y)}O_y>$ has been modeled as an ideal solid solution of $<ThC_{0.97}>$ and $<ThO>$ after Potter [5]. The phase $<ThO>$, however, is only stable in solution with the monocarbide and thus $\Delta G^o_{f<ThO>}$ must be derived from the phase equilibria.

In the three-phase region $<ThO_2>$ + $<ThC_{1.94}>$ + $<ThC_{0.97(1-y)}O_y>$, the equilibrium reaction ($\Delta G = 0$)

$$0.5 \ <ThC_{1.94}> + [ThO]_{ThC_{0.97}} = 0.5 \ <ThO_2> + [ThC_{0.97}]_{ThO}$$

can be written. Where $<ThC_{0.97(1-y)}O_y>$ is the ideal solution $[ThC_{0.97}] + [ThO]$,

$$\Delta \overline{G}_{f[ThO]} = 0.5 \ \Delta G^o_{f<ThO_2>} + \Delta \overline{G}_{f[ThC_{0.97}]} - 0.5 \ \Delta G^o_{f<ThC_{1.94}>}$$

From the ideal solution relations

$$\Delta \overline{G}_{f[ThO]} = \Delta G^o_{f<ThO>} + RT \ln y$$

and

$$\Delta \overline{G}_{f[ThC_{0.97}]} = \Delta G^o_{f<ThC_{0.97}>} + RT \ln (1-y)$$

one can write the expression

$$\Delta G^o_{f<ThO>} = 0.5 \ \Delta G^o_{f<ThO_2>} + \Delta G^o_{f<ThC_{0.97}>} - \Delta G^o_{f<ThC_{1.94}>}$$
$$+ RT \ln \frac{1-y}{y}$$

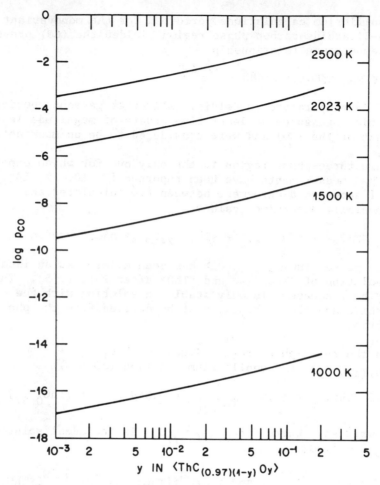

FIG.2.   *Calculated equilibrium* $\log p_{CO}$ *versus oxygen content of* $\langle ThC_{0.97(1-y)}O_y\rangle$ *in the presence of* $\langle ThC_{1.94}\rangle$ *(p in megapascals: 1 MPa = 9.87 atm).*

By using the known ΔG_f^o values for the dioxide, monocarbide, and dicarbide and assuming that y = 0.2 [5], $\Delta G_{f\langle ThO\rangle}^o$ values were found.

Equilibrium calculations indicate that the value of y for the monoxycarbide in equilibrium with the dioxide and dicarbide remains constant at 0.2 over the temperature range considered (1000 to 2500 K).

The equilibrium p_{CO} as a function of temperature was calculated to follow the relation

$$\log p_{CO} \text{ (MPa)} = 8.042 - \frac{22\ 340}{T}$$

3.1.3. $<ThC_{1.94}> + <ThC_{0.97(1-y)}O_y>$ *phase region*

The final phase region the system must pass through contains $<ThC_{1.94}> + <ThC_{0.97(1-y)}O_y>$. The dicarbide thus reacts with the monoxycarbide, evolving (CO) and decreasing y. The equilibrium p_{CO} as a function of y for 1000, 1500, 2023 (melting point of thorium), and 2500 K is shown in Fig. 2, where the pressure can be seen to decrease smoothly with decreasing y.

3.2. Th-U-C-O system

The carbothermic reduction of thorium-uranium dioxide to the monocarbide was studied in the same manner as the Th-C-O system. In addition to the gaseous species listed in Sect. 3.1, five additional gases — (U), (UO), (UO$_2$), (UO$_3$), (UC$_2$) — were included in the calculations. The thermodynamic data for (U) are from Oetting et al. [14], the data for the oxide species are from Ackermann and Chandrasekharaiah [15], and the data for (UC$_2$) are from Tetenbaum et al. [16]. The dioxide $<Th_{1-x_1}U_{x_1}O_2>$ was considered as an ideal solid solution of $<ThO_2>$ and $<UO_2>$ where the thermodynamic values for $<UO_2>$ are from Rand et al. [17]. As in the Th-C-O system, the dioxide was assumed to be stoichiometric [O/(U + Th) = 2]. The dicarbide was considered as an ideal solid solution of $<ThC_{1.94}>$ and $<UC_{1.9}>$, where the data for $<UC_{1.9}>$ are from Potter and Rand's [18] assessment of uranium carbides.

3.2.1. $<Th_{1-x_1}U_{x_1}O_2> + <C> + <Th_{1-x_2}U_{x_2}C_{1.94-0.04x_2}>$

$or <Th_{1-x_3}U_{x_3}C_{(0.97 + 0.03x_3)(1-y)}O_y>$

The quaternary $<Th_{1-x_3}U_{x_3}C_{(0.97 + 0.03x_3)(1-y)}O_y>$ was assumed to be an ideal solid solution of $<ThC_{0.97}>$, $<ThO>$, $<UC>$, and $<UO>$, where the thermodynamic data for $<UC>$ were also taken from the Potter and Rand [18] assessment. As in the case of $<ThO>$, $<UO>$ is stable only in solution with the monocarbide. In the manner similar to that described in Sect. 3.1.2 for the Th-C-O system, thermodynamic values for $<UO>$ were derived from the equilibria $<U> + <UO_2> + <UC_{0.65}O_{0.35}>$ [19], where the monoxycarbide was considered an ideal solid solution of $<UC>$ and $<UO>$.

The first calculational runs with compositions in the dioxide-dicarbide-carbon phase region indicated that $<UC_{1.5}>$ (data from Ref. 18) was more stable than the dicarbide at

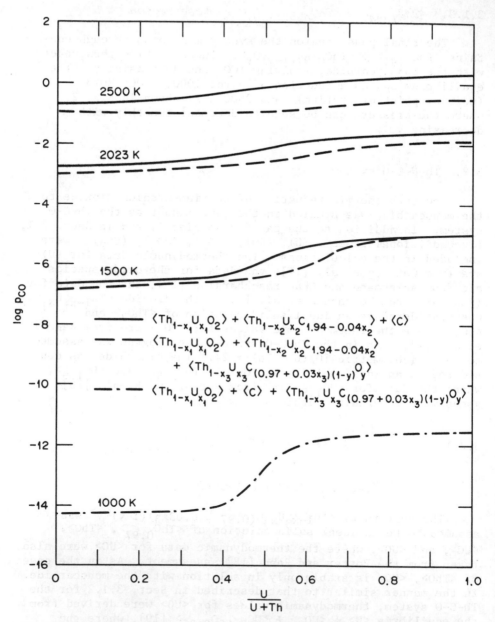

FIG.3. Calculated equilibrium log p_{CO} versus overall uranium fraction over the dioxide-dicarbide-carbon (overall O/C = 0.5 and (O + C)/(Th + U) = 3) or -monoxycarbide (overall O/C = 0.859 and (O + C)/(Th + U) = 1.642) phase regions at 1500, 2023 and 2500 K. In these compositional regions at 1000 K, the dicarbide is unstable with respect to carbon at U/(U + Th) > 0.95 (p in megapascals: 1 MPa = 9.87 atm).

overall $U/(U + Th) \geq 0.7$. In phase equilibria studies of the
U-C-O systems, however, the reverse has been demonstrated
due to some oxygen dissolving in the dicarbide phase and lower-
ing $\Delta G^0_{f<UC_{1.9}>}$ [19, 20-26]. Thus $\Delta G^0_{f<UC_{1.9}>}$ was modified in
the present calculations, and the addition of only -3 kJ/mol
was necessary to yield the formation of the appropriate phases.
These modified values were used throughout.

The results of the equilibrium calculations are plotted
in Fig. 3 as overall $U/(U + Th)$ versus log p_{CO}. Increased
uranium content increases p_{CO}, with the greatest influence
at lower temperatures. Note that at 1000 K, the dicarbide
was found no longer stable with respect to the monocarbide
and carbon at $U/(U + Th) >0.05$, although it is stable at
1500 K. This is in contrast to the U-C phase diagram of Storms
[8], which shows the dicarbide decomposing below 1787 K.
Apparently the addition of -3 kJ/mol, associated with dissolved
oxygen, to $\Delta G^0_{f<UC_{1.9}>}$ decreases the decomposition temperature.

As in the Th-C-O system, all gaseous species other than
(CO) exhibited negligible partial pressures.

3.2.2. $<Th_{1-x_1}U_{x_1}O_2> + <Th_{1-x_2}U_{x_2}C_{1.94-0.04x_2}>$

$+ <Th_{1-x_3}U_{x_3}C_{(0.97 + 0.03x_3)}(1-y)^O{}_y>$

The calculational results for the dioxide-dicarbide-
monoxycarbide phase region are also shown in Fig. 3, and the
p_{CO} values generally lie only slightly below those for the
dioxide-dicarbide-carbon region. It was again found that at
1000 K the dicarbide does not form at $U/(U + Th) >0.05$.

3.2.3. $<Th_{1-x_3}U_{x_3}C_{(0.97 + 0.03x_3)}(1-y)^O{}_y>$

$+ <Th_{1-x_2}U_{x_2}C_{1.94-0.04x_2}>$

During carbothermic reduction the system eventually enters
the dicarbide-monoxycarbide phase region in which progressive
reduction removes oxygen from the monoxycarbide. Figure 4
shows calculated p_{CO} versus y at $x_3 = 0.1$ and 0.5. Increased
uranium content increases the pressure somewhat and also lowers
the maximum value of y in the presence of the dicarbide. At

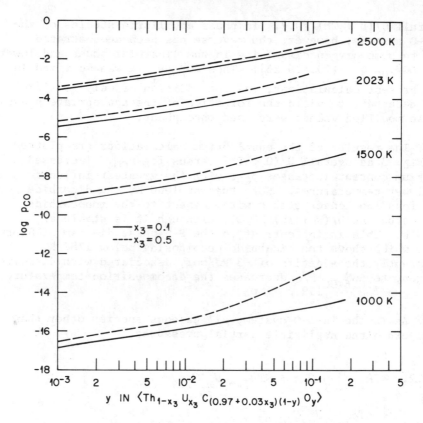

FIG.4.   Calculated equilibrium log p_{CO} *versus oxygen content of the monoxycarbide at 1000, 1500, 2023 and 2500 K for U/(U + Th) = 0.1 and 0.5.   The dicarbide is present at equilibrium in all cases except for U/(U + Th) = 0.5 at 1000 K, where carbon is more stable (p in megapascals: 1 MPa = 9.87 atm).*

1000 K and x_3 = 0.5, the dicarbide is calculated to be unstable with respect to carbon. At x_3 = 0.1, however, the dicarbide is predicted to form at these low oxygen-content compositions.

4.   CHEMICAL COMPATIBILITY

Thermodynamic modeling of the fuel-clad interface is a method developed for assessing chemical compatibility. In the thermodynamic modeling the phases present in the interface, or reaction area between bulk fuel and bulk clad alloy, are determined. The results are a profile of phases expected at equilibrium and the maximum limit of reaction products. This allows a screening of a large number of potential cladding alloys and permits an evaluation of maximum clad carburization.

Thermodynamic calculations were performed for the fuel-clad interface of the thorium carbides $<ThC_{0.97}>$, $<ThC_{0.90}>$, and 90% $<ThC_{0.97}>$ + 10% $<ThC_{1.94}>$ with the Cr-Fe-Ni compositions of seven alloys. The alloys considered are: 316 stainless steel, 330 stainless steel, A286, M813, PE-16, Inconel 706, and Inconel 718. The following species were considered as possible reaction products in the interface: $<Cr_{23(1-z_1)}Fe_{23z_1}C_6>$; $<Cr_{7(1-z_2)}Fe_{7z_2}C_3>$; $<Cr_3C_2>$; $<ThC_{z_3}>$ (z_3 = 0.67-0.97); $<ThC_{1.94}>$; $<C>$; $<Fe_3C>$; $<Cr_{z_4}Fe_{z_5}Ni_{1-(z_4 + z_5)}>$; $<Th_2Ni_{17(1-z_6)}Fe_{17z_6}>$; $<ThNi_{5(1-z_7)}Fe_{z_7}>$; $<Th_7Ni_{3(1-z_8)}Fe_{3z_8}>$; $<ThNi_2>$; $<Th_2Fe_7>$; $<ThNi>$; $<ThFe_3>$; $<Th>$.

The list of possible species includes all known binary compounds and solid solutions in the Th-C-Cr-Fe-Ni system. Compounds involving thorium with nickel and thorium with iron are assumed to be in solid solution if they have the same stoichiometric coefficients and crystal structure. The Th-Cr-C and Th-Fe-C systems were examined to determine if any ternary compounds (complex carbides) exist which would need to be considered in the interface calculations. Six arc-melted samples were prepared containing 25 at. % Th, 42-60 at. % C, and 15-34 at. % Cr or Fe. The samples were characterized by x-ray diffraction and metallography. There was no evidence of ternary compounds in any of the specimens. X-ray diffraction lines indicated the presence of only thorium carbides, chromium carbides, and iron.

Thermodynamic values used in the calculations were obtained from references [1, 27, 28, 29]. The Cr-Fe-Ni solid solutions were calculated from binary data [30] using the method of shortest distance composition path [31]. All other solutions were considered to be ideal.

4.1. Fuel-clad interface

The principal reaction products in all the calculated interfaces were found to be $<ThNi_{5(1-z_7)}Fe_{5z_7}>$ and $<Cr_{7(1-z_2)}Fe_{7z_2}C_3>$. Figure 5 presents the result for the PE-16-$<ThC_{0.97}>$ interface. This plot shows the gram-atom fraction of material in the interface versus position with the clad alloy on the left side and the fuel on the right side. The composition of the Cr-Fe-Ni alloy changes across the interface, becoming enriched in Fe and depleted in Ni and Cr from the clad side to the fuel side. At 0.2 on the interface, the clad alloy composition is 7.9 at. % Cr, 81.2 at. % Fe, and 10.9 at. % Ni. The interface plots for most of the

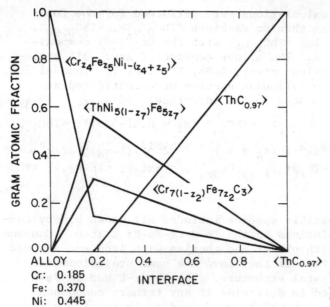

FIG.5.   Fuel-clad interface for
PE-16 with thorium monocarbide
⟨ThC₀.₉₇⟩ at 1173 K.

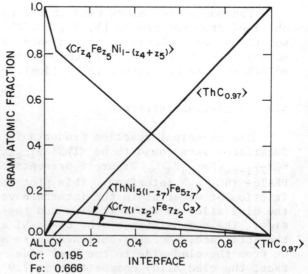

FIG.6.   Fuel-clad interface for
316 stainless steel with thorium
monocarbide ⟨ThC₀.₉₇⟩ at 1173 K.

alloys resemble this plot in extent of reaction product. The
only alloy which produced an interface differing greatly
from this type of diagram was 316 stainless steel, which showed
much less reaction than the others (Fig. 6). This is due to
the relatively higher iron content in 316 stainless steel compared to the other alloys considered. Nickel forms intermetallic compounds with large negative free energies of formation,
and chromium forms carbides with large negative free energies
of formation. Iron can take part in the formation of both
types of compounds, but its contribution to the free energy
of formation is much less than nickel in the intermetallic compounds and much less than chromium in the carbides.

Table I lists the maximum amounts of the reaction products
$<\text{ThNi}_{5(1-z_7)}\text{Fe}_{5z_7}>$ and $<\text{Cr}_{7(1-z_2)}\text{Fe}_{7z_2}\text{C}_3>$ in the interface for
all of the alloys.

To appraise the reliability of the results, calculations
were also performed using variations of $\pm 10\%$ of the listed
free energies of formation of the thorium carbides. In all
cases the principal reaction products were again
$<\text{ThNi}_{5(1-z_7)}\text{Fe}_{5z_7}>$ and $<\text{Cr}_{7(1-z_2)}\text{Fe}_{7z_2}\text{C}_3>$. The amounts of
these products were within $\pm 15\%$ of the amounts obtained using
the accepted values. Most of the change in amounts of reaction products was due to an increase or decrease in products
containing Fe.

4.2. Clad carburization

Austenitic steels with carbon contents over 0.6 wt % are
considered to be unacceptably embrittled [32]. Worst-case
carburization for the Cr-Fe-Ni contents of the seven alloys
was calculated by assuming that the point of maximum transition metal carbide in the interface extended throughout the
clad alloy. The carbon contents of all of the alloys under
the worst-case assumption are shown in Table II. Type 316
stainless steel was the only alloy giving a worst-case
carburization of less than 0.6 wt % C. With $<\text{ThC}_{0.97}>$ the worst-
case carburization for 316 stainless steel was calculated to
be 0.4 wt % C. However, with 90% $<\text{ThC}_{0.97}>$ + 10% $<\text{ThC}_{1.94}>$
the worst-case carburization for 316 stainless steel becomes
0.9 wt % C.

TABLE I. MAXIMUM AMOUNTS OF REACTION PRODUCTS (g.at. fraction) IN INTERFACE

	316 stainless steel	330 stainless steel	A286	M813	PE-16	Inconel 706	Inconel 718
Maximum							
$ThC_{0.97}$	0.11	0.42	0.28	0.44	0.54	0.51	0.61
$ThNi_5$–$ThFe_5$ $ThC_{0.90}$	0.13	0.42	0.29	0.44	0.55	0.51	0.63
Gram atom fraction in interface 90% $ThC_{0.97}$ + 10% $ThC_{1.94}$	0.11	0.41	0.28	0.43	0.54	0.50	0.58
Maximum							
$ThC_{0.97}$	0.06	0.22	0.15	0.23	0.29	0.27	0.33
Cr_7C_3–Fe_7C_3 $ThC_{0.90}$	0.05	0.20	0.14	0.21	0.27	0.25	0.32
Gram atom fraction in interface 90% $ThC_{0.97}$ + 10% $ThC_{1.94}$	0.14	0.25	0.17	0.25	0.32	0.30	0.34

TABLE II. WEIGHT PERCENT CARBON IN ALLOY FOR 'WORST CASE' CARBURIZATION[a] AT 1173K

	316 stainless steel	330 stainless steel	A286	M813	PE-16	Inconel 706	Inconel 718
$ThC_{0.90}$	0.3	1.3	0.9	1.4	1.7	1.6	2.0
$Th\dot{C}_{0.97}$	0.4	1.4	1.0	1.5	1.9	1.7	2.1
90% $ThC_{0.97}$ + 10% $ThC_{1.94}$	0.9	1.6	1.1	1.7	2.1	1.9	2.2

[a] In the "worst case" it is assumed that the point of maximum transition metal carbide in the interface extends throughout the clad alloy.

5. CONCLUSIONS

5.1. Carbothermic reduction

The present equilibrium thermodynamic calculations indicate that carbothermic reduction of thorium or thorium–uranium
dioxide to the monocarbide is feasible. No actinide species
exhibit significant partial pressures which would result in
large actinide losses or congruent vaporization, as is observed
in the monoxycarbide plus sesqui- or dicarbide phase fields in
plutonium-containing systems [1].

Although the assumption of ideality for the solid solutions
provides a good basis for these calculations, determination of
the actual relationships, if different from ideality, will allow
for further refinement. More accurate measurements of $\Delta G^o_{f<ThC_{0.97}>}$
and the stabilizing effect of oxygen dissolution in $<UC_{1.9}>$
will also improve the results.

5.2. Interface modeling

The relatively high iron content of 316 stainless steel
enables this alloy to react less with thorium carbide than any
of the other alloys we considered.

In thermodynamic modeling of fuel–clad interfaces, the
maxima in the reaction product intermetallic compounds and
transition metal carbides tend to occur in the same region.
This clearly shows that the formation of more than one product
greatly influences the chemical compatibility of a fuel–clad
couple. In an actual in-pile fuel–clad couple, the maxima
of the two types of reaction products will probably not occur
in the same area because of differing rates of migration of
carbon and metals. In-pile compatibility studies of He-bonded
uranium–plutonium fuel elements indicate that the intermetallic
compound $<U_{1-z_8}Pu_{z_8}Ni_5>$ is on the fuel side of the interface
and the transition metal carbide extends to the clad side [33].

REFERENCES

[1] RAND, M. H., et al., Thorium: Physico-Chemical Properties
     of its Compounds and Alloys, At. Energy Rev. Special Issue
     No. 5 (KUBASCHEWSKI, O., ed.), IAEA, Vienna (1975).

[2] BESMANN, T. M., LINDEMER, T. B., J. Nucl. Mater. 67
     (1977) 77.

[3]   ERIKSSON, G., Chem. Scr. 8 (1975) 100.

[4]   BESMANN, T. M., SOLGASMIX-PV, A Computer Program to Calcu-
      late Equilibrium Relationships in Complex Chemical Systems,
      Oak Ridge National Laboratory, ORNL/TM-5775 (1977).

[5]   POTTER, P. E., J. Inorg. Nucl. Chem., 31 (1969) 1821.

[6]   KANNO, M., et al., J. Nucl. Sci. Tech. 9 2 (1972) 97.

[7]   PIALOUX, A., ZAUG, J., J. Nucl. Mater. 61 (1967) 131.

[8]   STORMS, E. K., "2. Phase Relationships and Electrical
      Properties of Refractory Carbides and Nitrides," Solid
      State Chemistry (ROBERTS, L. E. J., ed.), University
      Park Press, Baltimore (1972) 37.

[9]   IMAI, H., HOSAKA, S., NAITO, K., J. Am. Ceram. Soc. 50 6
      (1967) 308.

[10] HEISS, A., DJÉMAL, M., Rev. Int. Hautes. Temp. Refract. 8
      (1971) 287.

[11] STULL, D. R., PROPHET, H. (Project Directors), JANAF Thermo-
      chemical Tables, 2nd ed., National Bureau of Standards,
      Washington (1971).

[12] PRESCOTT, C. H., HINCKE, W. B., J. Am. Chem. Soc. 49
      (1927) 2744.

[13] HOLLAHAN, J. R., GREGORY, N. W., J. Phys. Chem. 68 8
      (1964) 2346.

[14] OETTING, F. L., RAND, M. H., ACKERMANN, R. J., The Chemical
      Thermodynamics of Actinide Elements and Compounds. Part 1.
      The Actinide Elements, IAEA, Vienna (1976).

[15] ACKERMANN, R. J., CHANDRASEKHARAIAH, M. S., "Systematic
      Thermodynamic Properties of Actinide Metal-Oxygen Systems
      at High Temperatures," Thermodynamics of Nuclear Materials
      1974 (Proc. Symposium Vienna, 1974) Vol. II, IAEA, Vienna
      (1975) 3.

[16] TETENBAUM, M., SHETH, A., OLSON, W., A Review of the
      Thermodynamics of the U-C, Pu-C, and U-Pu-C Systems,
      Argonne National Laboratory, ANL-AFP-8 (1975).

[17] RAND, M. H., ACKERMANN, R. J., GRONVOLD, F., OETTING, F. L.,
      PATTORET, A., Thermodynamic Properties of the Urania Phase
      (part of an ongoing IAEA assessment) (1977).

[18] POTTER, P. E., RAND, M. H., Phase Diagrams and Thermo-
     dynamic Data for Carbide Systems (part of an ongoing IAEA
     assessment) (1977).

[19] POTTER, P. E., J. Nucl. Mater. 42 (1972) 1.

[20] LEITNAKER, J. M., GODFREY, T. G., J. Chem. Eng. Data 11 2
     (1966) 392.

[21] HENNEY, J., The Effects of Oxygen on Phase Equilibria in
     the Uranium Carbon System, Atomic Energy Research Establish-
     ment, Harwell, AERE-R-4661 (1966).

[22] BAZIN, J., ACCARY, A., Thermodynamics of ceramic systems,
     Proc. Brit. Ceram. Soc. 8 (1967) 175.

[23] HENRY, J. L., et al., Phase Relations in the Uranium
     Monocarbide Region of the System Uranium-Carbon-Oxygen
     at 1700°C, Bureau of Mines, BMR-6968 (1967).

[24] TAGAWA, H., FUJII, J., J. Nucl. Mater. 39 (1971) 109.

[25] HEISS, A., J. Nucl. Mater. 55 (1975) 207 and Thermo-
     dynamic Study of the U-O-C System by Measurement of
     the Carbon Monoxide Pressure at Equilibrium, Oak
     Ridge National Laboratory Translation, ORNL-tr-4326
     (1976).

[26] PIALOUX, A., DODE, M., J. Nucl. Mater. 56 (1975) 221
     and Contribution to the Establishment of the Phase
     Diagram for the U-O-C System by Means of X-ray Dif-
     fractometry at High Temperatures and Under Controlled
     Pressure, Oak Ridge National Laboratory Translation,
     ORNL-tr-4327 (1976).

[27] SMITH, J. F., J. Nucl. Mater. 51 (1974) 136.

[28] KULKARNI, A. D., WORREL, W. L., Met. Trans. 3 (1972)
     2363.

[29] LUNDBERG, R., WALDENSTRÖM, M., UHRENIUS, B., Calphad 1
     (1977) 159.

[30] MAZANDARANY, F. N., PEHLKE, R. D., Met. Trans. 4 (1973)
     2067.

[31] JACOB, K. T., FITZNER, K., Thermochim. Acta 18 (1977)
     197.

[32] ELKINS, P. E., Compatibility of Uranium Carbide Fuels
     with Cladding Materials, NAA-SR-7502 (1964).

[33] LATIMER, T. W., BARNER, J. O., KERRISK, J. F., GREEN, J. L.,
     Postirradiation Results and Evaluation of Helium-Bonded
     Uranium-Plutonium Carbide Fuel Elements Irradiated in
     EBR-II, Interim Report LA-6249-MS (1976).

DISCUSSION

D.D. SOOD: It is very nice to see such an interesting set of calculations.
I should just like to enquire how sensitive your program is. For example, we
know that addition of small amounts of oxygen to UC or $(U, Pu)C$ considerably
reduces the carburization attack on the clad. Do you think such a reduction of
attack could be predicted by your program?

T.M. BESMANN: The program SOLGASMIX-PV predicts equilibrium
compositions on the basis of thermodynamic data, overall composition, temperature
and total pressure (or volume). If changes in thermodynamic value due to the
addition of oxygen can be quantified, the program should indicate a decreased clad
attack.

P.E. POTTER: What values did you use for the maximum solubility of ThO
in ThC in order to estimate ΔG_f^0 (ThO, solid)?

T.M. BESMANN: We assumed the 20% solubility which you have published
as a tentative value.

P.R. BUSSEY: In the cases of $Th_{1-x_1} U_{x_1} O_2 + C + Th_{1-x_2} U_{x_2} C_{1.94-0.04x_2}$ and
$ThO_2 + ThC_{1.94} + C$, it was found necessary to adjust the free energy of formation
of one condensed phase in order to predict the experimentally observed phases.
Have you tried altering the free energy by more than the minimum amount to
determine how sensitive the model is to free energy values chosen?

T.M. BESMANN: For the compatibility studies we varied the thermodynamic
values by $\pm 10\%$. This resulted in a $\leqslant 15\%$ variation in the amounts of phases
formed.

A. NAOUMIDIS: Did you assume complete solubility between ThC_2 and
UC_2, or did you perform your calculations accepting a miscibility gap?

T.M. BESMANN: Although we recognize there is some evidence for a
miscibility gap, we simply assumed complete solubility for the purpose of the
calculations.

D.A. POWERS: Did you treat your steels as ideal solutions or did you
consider deviations from ideality? Does this have any significant impact on
your results?

T.M. BESMANN: We used available binary data and the method of the
shortest composition path to derive the steel alloy thermodynamic values.

THE EFFECT OF SMALL FOURTH-ELEMENT ALLOYING ADDITIONS ON THE CALCULATED PHASE STABILITY IN THE Fe—Cr—Ni SYSTEM

J.S. WATKIN
United Kingdom Atomic Energy Authority,
Springfields Nuclear Power Development
      Laboratories,
Springfields, Salwick, Preston, Lancashire,
United Kingdom

Abstract

THE EFFECT OF SMALL FOURTH-ELEMENT ALLOYING ADDITIONS ON THE CALCULATED PHASE STABILITY IN THE Fe—Cr—Ni SYSTEM.

Recent studies into the void swelling of Fe—Cr—Ni alloys have revealed that the magnitude of swelling depends upon alloy constitution and this, together with the fact that minor element additions also play a major role in swelling, necessitates a detailed knowledge of the influence of small fourth-element additions on phase stability. In this paper the effects of additions of niobium, titanium, aluminium, molybdenum, cobalt and carbon on the Fe—Cr—Ni ternary system are assessed by calculation. They confirm the ferritizing tendencies of niobium, titanium and aluminium and the strong austenitizing effect of carbon. Confirmation is also found for the scaling factors in the equivalent nickel and chromium equations in common usage and the paper presents Fe—Cr—Ni ternary sections at 400, 550 and 700°C modified for 1 at.% addition of each of the above elements.

1. INTRODUCTION

In the study of neutron-induced void swelling, attention is being given to alloy constitution [1], particularly in an attempt to explain differences in the swelling behaviour between alloys. Watkin [2] has shown, for the temperature range 400—600°C, that swelling is greatest in alloys predicted to be three phase, least in those predicted to be single phase, and of intermediate magnitude in two-phase alloys. During this investigation the importance of minor elements in solution was also recognized and it was suggested that they were responsible for the second-order variation in swelling, a fact which has been confirmed by other studies [3].

Additions of certain elements are known from experiment to alter phase boundaries in the Fe—Cr—Ni system. Attempts typified by the Schaeffler diagram and the use of equivalent chromium and nickel contents have been made to quantify the effect of certain additions. They are used, for the want of anything better,

297

over wide compositional and temperature ranges, although there is considerable doubt in the validity of such extrapolations. To obtain unequivocally data covering all compositions and temperatures of relevance by experimental means would be extremely expensive. However over the last few years it has become possible to calculate phase stability in ternary systems, and basic thermodynamic data are available for ten elements which include niobium, aluminium, molybdenum, cobalt and carbon, as well as iron, chromium and nickel. It has been possible, using this thermodynamic data and by making certain assumptions, to investigate the effect of these small fourth-element alloying additions on the stability of the Fe—Cr—Ni system.

The general method which has been used to calculate the phase stability in ternary alloys is that devised by Kaufman [4]. Briefly the technique involves the explicit description of the free energy of all competing phases in terms of lattice stability, mixing and compound parameters. The resulting equations permit the calculation of solution/solution and solution/compound interactions on a pair-wise basis. After each calculation has been performed the results are combined to yield the most stable configuration. A modification to these calculational procedures has been used here to determine the effect of fourth-element additions on the stability of Fe—Cr—Ni alloys containing up to 30% Cr, over the temperature range 400—700°C.

2.   CALCULATIONAL PROCEDURES

The method used for calculating the effect of fourth-element additions is a modification of that developed by Kaufman [4] for ternary alloy systems. The phase boundaries studied are those delineating the extent of the fcc phase field which run roughly from the Fe-corner to the Ni—Cr edge of the ternary. It has been possible to describe the total effect of a given addition from the summation of three correction terms, assuming that the fourth element is always in solution. In the following each correction will be considered in turn.

The first two relate to the fcc-bcc transus, where corrections on the Ni—Cr edge and in the Fe-corner will be determined, whilst the third is concerned with the fcc and bcc interactions with sigma phase.

2.1. Correction to the fcc-bcc transus along the Ni—Cr edge

The method used to correct the Ni—Cr edge of the diagram is the simplest and is illustrated for the case of aluminium at 600°C in Fig.1. Here the fcc-bcc interaction for the Ni—Cr—Al system has been jointed to the equivalent Fe—Cr—Ni interaction, both of these being calculated on the same basis and using the parameters listed by Kaufman [5, 6]. This diagram contains two

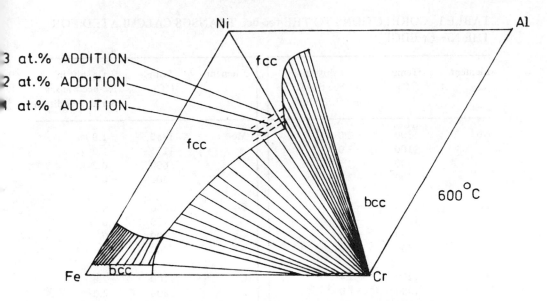

FIG.1.   *The fcc-bcc interactions for Fe—Cr—Ni and Ni—Cr—Al at 600°C.*

of the four outer surfaces of the Fe—Cr—Ni—Al quaternary. Using this diagram
an estimate of the changes in the fcc-bcc transus has been determined for the
addition of 1 at.% Al. The movement of the boundary has been measured in
terms of a change in chromium (equally well this could have been measured in
terms of nickel) and, from Fig.1, leads to $\Delta Cr/\Delta Al = -2.2$ at.%. The negative
sign indicates a destabilizing effect on fcc which, for the present situation, is
the same as a stabilizing effect on bcc. Clearly from this figure the correction is
not linear with increasing aluminium content; however, it can be assumed to be
so up to 3 at.% Al. In this way corrections have been determined for niobium,
titanium, aluminium, molybdenum, cobalt and carbon over a range of temperatures (Table I). In general the corrections are valid for additions up to 3 at.%
of an element except for carbon, which is restricted to concentrations less than 1 at.%.

2.2. Correction to the fcc-bcc transus in the Fe-corner

In principle the method used to correct the Ni—Cr edge of the ternary for
fourth-element additions could be used for the remaining two edges. However
the differences in free energy between the fcc and bcc phases in the Fe-corner
is small (~ 150 cal/g-at. at 650°C) and necessitate a more rigorous treatment.

TABLE I. CORRECTIONS TO THE fcc-bcc TRANSUS CALCULATED FOR THE Ni—Cr EDGE

Element	Temp. (°C)	% Cr shift in fcc-bcc transus per at.% addition	Element	Temp. (°C)	% Cr shift in fcc-bcc transus per at.% addition
Nb	1300	− 2.0	Mo	1350	− 1.0
	1100	− 2.0		1275	− 1.0
	600	− 2.5		600	− 0.2
	500	− 2.4		500	− 0.2
Ti	1350	− 2.0	Co	1300	0.2
	1275	− 2.0		750	0
	600	− 2.5		600	0
	500	− 2.6		500	0
Al	1200	− 1.4	C	1100	2.0
	1000	− 1.0		600	2.0
	600	− 2.2		500	2.0
	500	− 2.2			

This was achieved by deducing a corrective term for the lattice stability for pure iron as follows. Consider the Fe—X binary where X is the element under consideration. The free energy of fcc is given by:

$$G^{fcc} = (1-x) G_{Fe}^{fcc} + x G_X^{fcc} + RT\,[(1-x)\ln(1-x) + x\ln x] + G_E^{fcc} \qquad (1)$$

where G_{Fe}^{fcc} and G_X^{fcc} are the free energies of the pure constituents in the fcc phase, F_E^{fcc} is the excess free energy of mixing, R is the gas constant and T is the temperature in kelvin.

Similarly the free energy of the bcc phase is given by:

$$G^{bcc} = (1-x) G_{Fe}^{bcc} + x G_X^{bcc} + RT\,[(1-x)\ln(1-x) + x\ln x] + G_E^{bcc} \qquad (2)$$

where the symbols have similar meanings to those in Eq.(1), but for the bcc phase.

Subtracting and rearranging gives:

$$G^{fcc} - G^{bcc} - (G_{Fe}^{fcc} - G_{Fe}^{bcc}) = -x\,(G_{Fe}^{fcc} - G_{Fe}^{bcc}) + x\,(G_X^{fcc} - G_X^{bcc})$$

$$+ (G_E^{fcc} - G_E^{bcc}) = \Delta G_{Fe}^{bcc \rightarrow fcc} \qquad (3)$$

TABLE II. $\Delta G_{Fe}^{bcc \rightarrow fcc}$ FOR 1 at.% ADDITION OF ELEMENT X (cal/g-at.)

Element X	Temperature (°C)						
	400	450	500	550	600	650	700
Nb	1.54	2.73	3.87	4.97	6.00	6.96	7.84
Ti	28.98	30.19	31.36	32.48	33.54	34.53	35.43
Al	−3.64	−2.29	−1.00	0.24	1.42	2.53	3.56
Mo	3.58	4.42	5.21	5.96	6.64	7.25	7.78
Co	30.03	27.01	23.92	20.80	17.67	14.56	11.50
C	−110.55	−105.61	−100.72	−95.10	−91.10	−86.40	−81.78

The term underlined is the lattice stability for pure iron and the left-hand side of the equation can be interpreted as the change in the lattice stability of iron for the addition of x atomic fraction of X. The terms on the right-hand side include the lattice stability for iron and for element X and the excess free energy terms. These are all given in Ref. [6] for the elements being considered, and these values were used to compile Table II.

Instead of calculating fcc-bcc interactions for each $\Delta G_{Fe}^{bcc \rightarrow fcc}$, calculations were carried out at discrete energy intervals (± 20 and ± 60 cal/g-at) over the temperature range 400–700°C; a typical result is shown in Fig.2.

Thus for a given alloy the total effect on the fcc-bcc transus is obtained by interpolating to the appropriate energy change in the Fe-corner (taking $\Delta G_{Fe}^{bcc \rightarrow fcc}$ from Table II) and then smoothing out the resultant curve to meet the corresponding point on the Ni−Cr edge, deduced from Table I.

2.3. Corrections to the fcc-sigma and bcc-sigma transus

Quantification of the influence of fourth-element additions on the stability of the sigma phase has been achieved by determining the change in the free energy of σ caused by the addition. This change has then been simulated in subsequent interaction calculations by adjusting the 'CAB' parameter which appears in the equations below in the expression for the excess free energy of σ. Examination of this expression reveals that the magnitude of the change depends upon composition. The correction has been made at the most stable ternary composition (0.39 Fe, 0.51 Cr, 0.1 Ni). The object of the next few paragraphs is to determine the magnitude of the correction.

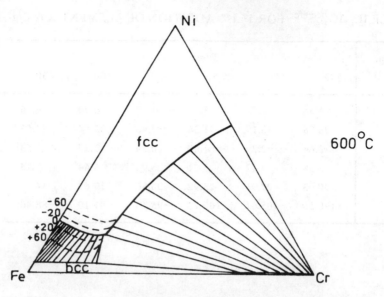

FIG.2.   The fcc-bcc interaction diagram at $600°C$ indicating changes in fcc-bcc transus for various $\Delta G_{Fe}^{bcc\rightarrow fcc}$.

The free energy of sigma phase in the Fe—Cr—Ni system is given by the following equations [5]:

$$G^\sigma = z_\sigma\, G_{Ni}^{bcc} + x_\sigma\, G_{Cr}^{bcc} + y_\sigma\, G_{Fe}^{bcc}$$

$$+ [1 - y_\sigma\,(1 - x_\sigma)^{-1}]\,\Delta G_A + [y_\sigma\,(1 - x_\sigma)^{-1}]\,\Delta G_B$$

$$+ RT\,[y_\sigma \ln y_\sigma + z_\sigma \ln z_\sigma - (1 - x'_*)\ln (1 - x'_*)] + \Delta G_E \qquad (4)$$

$$\Delta G_A = (1 - x'_*)\,x'_*\,[(1 - x'_*)\,LNICR + x'_*\,LCRNI - CNICR] \qquad (5)$$

$$\Delta G_B = (1 - x_*)\,x_*\,[(1 - x_*)\,LCRFE + x\,LFECR - CCRFE] \qquad (6)$$

$$\Delta G_E = (CAB)\,y_\sigma\,(1 - x'_* - (p+1)\,y_\sigma)\,(1 - x'_* - y_\sigma p)^{-1} \qquad (7)$$

$$p = (x_* - x'_*)\,(1 - x_*)^{-1} \qquad (8)$$

where G_{Ni}^{bcc}, G_{Cr}^{bcc}, G_{Fe}^{bcc} are the free energies of pure nickel, chromium and iron in the bcc phase, z_σ, x_σ, y_σ are the atomic concentrations of nickel, chromium and iron; x'_* is the fraction of chromium in σ on the Ni—Cr edge of the ternary, x_* is the fraction of chromium in σ on the Fe—Cr edge of the ternary, LNICR, LCRNI are the excess free energy parameters for liquid Cr—Ni, LCRFE, LFECR are the excess free energy parameters for liquid Cr—Fe and CNICR and CCRFE are the compound parameters for the sigma phase.

Restricting consideration to the composition at maximum sigma stability, i.e. 0.39 Fe, 0.51 Cr, 0.10 Ni, with CAB = − 1000 [5], substituting Eqs (5), (6), (7) and (8) into (4), and rearranging we obtain:

$$G^\sigma - z_\sigma G^{bcc}_{Ni} - x_\sigma G^{bcc}_{Cr} - y_\sigma G^{bcc}_{Fe} - RT \, [y_\sigma \ln y_\sigma + z_\sigma \ln z_\sigma$$

$$- (1-x'_*) \ln (1-x'_*)] = 0.0481 \, [0.38 \, LNICR + 0.62 \, LCRNI - CNICR]$$

$$+ \, 0.1983 \, [0.53 \, LCRFE + 0.47 \, LFECR - CCRFE] - 1000 \, [0.07982] = \Delta G_1 \quad (9)$$

Now a similar equation for the Fe—Cr—X system can be written with CAB = 0 (which is generally the case):

$$\Delta G_2 = 0.0481 \, [0.38 \, LXXCR + 0.62 \, LCRXX - CXXCR]$$

$$+ \, 0.1983 \, [0.53 \, LCRFE + 0.47 \, LFECR - CCRFE] \quad (10)$$

Hence ΔG_3 due to 1 at.% of a fourth-element addition will be:

$$\Delta G_3 = 0.99 \, \Delta G_1 + 0.01 \, \Delta G_2 \quad (11)$$

Thus the change in ΔG due to the addition is:

$$\delta G = \Delta G_3 - \Delta G_1 \quad (12)$$

which expressed as a $\delta(CAB)$ is:

$$\delta(CAB) = \frac{\delta G}{0.07982} \quad (13)$$

or explicitly:

$$\delta(CAB) = 0.01 \, \{1000 + 0.604 \, [0.38 \, (LXXCR - LNICR)$$

$$+ \, 0.62 \, (LCRXX - LCRNI) + CNICR - CXXCR]\} \quad (14)$$

TABLE III. FUNCTIONS WHICH DEFINE THE δ(CAB) CORRECTION FOR
THE SIGMA PHASE DUE TO THE ADDITION OF 1 at.% OF ELEMENT X

Element	δ(CAB) (footnote a)	Typical values of δ(CAB)	
		400°C	600°C
Nb	19.7 − 0.033 T	− 2.5	−91
Ti	33.3 + 0.015 T	43	46
Al	−44.9 + 0.015 T	−34	−32
Mo	43.9 − 0.013 T	35	32
Co	10.6 + 0.015 T	21	23
C	69.9 + 0.015 T	80	83

^a Temperature, T, in kelvin.

Because each parameter in this equation is a function of temperature, the
resultant δ(CAB) is also a temperature-dependent function. Using the values
recommended by Kaufman [5] these functions have been derived and are presented in Table III for 1 at.% niobium, titanium, aluminium, molybdenum,
cobalt and carbon, together with typical values for this function at 400 and
600°C.

The δ(CAB) values in Table III represent very small energy changes: for
example, a δ(CAB) of 100 corresponds to a free-energy change of only
7.98 cal/g-at. Negative values of δ(CAB) would stabilize sigma and positive
values destabilize sigma. In order to determine the effect of such changes on
the stability of sigma, Fig.3 was constructed. This figure is an fcc-sigma interaction diagram calculated at 600°C with CAB set at − 1000, − 950, − 1050 and
− 2000, corresponding to δ(CAB)s of 0, 50, − 50 and 1000 respectively. It can
be seen from this figure that changing CAB by ±50 has a negligible effect on the
stability of σ, and that values an order of magnitude higher are required to
produce a significant influence. This applies over the temperature range
400−700°C for both fcc/sigma and bcc/sigma interactions.

Thus from the above analysis it is concluded that fourth-element additions
of up to 3 at.% have their major influence only on the fcc/bcc interaction and
that the effect on fcc/sigma and bcc/sigma interactions can be neglected. However, this is not the same as saying that any phase field containing sigma will be
unaltered, as will be shown in the next section.

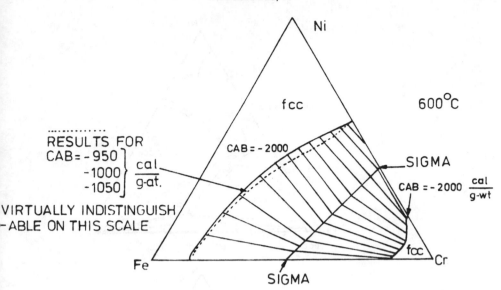

FIG.3. *Movement in the fcc-σ phase boundary as a result of changes in 'CAB'.*

3.   RESULTS AND DISCUSSION

Estimates of the effect of 1 at.% addition have been made and are presented
in Figs 4, 5 and 6 at 400, 550 and 700°C respectively. It is recognized that
1 at.% C solubility is unattainable in practice at these temperatures, but the
results are given in this manner to show the relative strengths of a given element.
They indicate carbon to possess a very strong influence on phase stability, while
cobalt and molybdenum have an almost negligible effect. The relative influence
of the three remaining elements has a tendency to increase with temperature,
with titanium and niobium having larger effects than aluminium.

The calculations also show the ferritizing tendencies of molybdenum,
niobium, titanium and aluminium, the virtual equivalence between cobalt
and nickel and the strong austenitizing effect of carbon. On the Ni—Cr edge
of the pseudo Fe—Cr—Ni ternary diagram, the present calculations can be
compared directly with the factors used in the equivalent nickel and chromium
approach (Table IV) and, with the exception of molybdenum, the agreement is
good — especially when one considers that most of the equivalent nickel and
chromium factors have been deduced from alloys with compositions close to
18 Cr 8 Ni. However, the present analysis allows the temperature sensitivity
to be studied and, over the temperature range 400–700°C, little influence is
seen; at temperatures > 1000°C there are indications of smaller effects for
niobium, titanium and aluminium, and an enhanced effect for molybdenum.

Text continues on page 312.

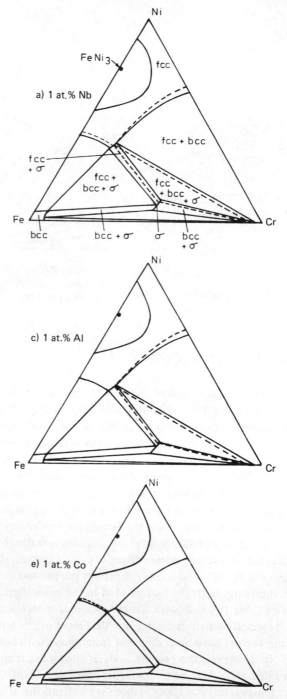

FIG.4. *The effect of 1 at.% additions of element X to the stabilit*

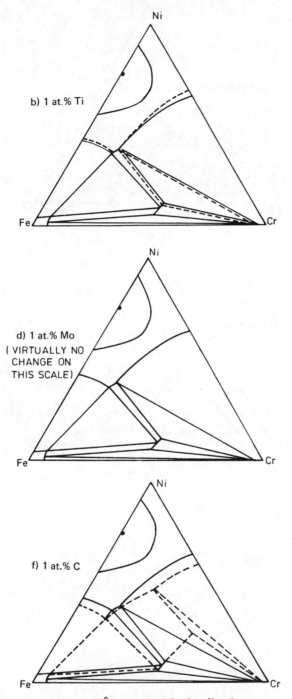

of the Fe—Cr—Ni system at 400°C (shown as broken lines).

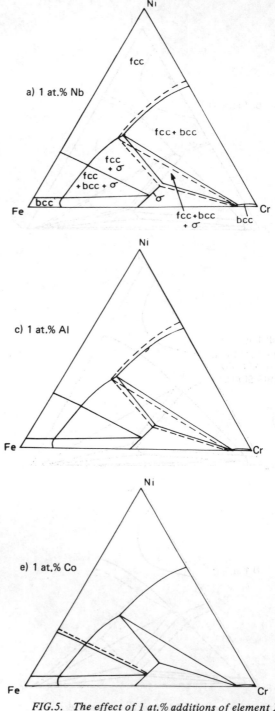

FIG.5. The effect of 1 at.% additions of element X to the stabilit

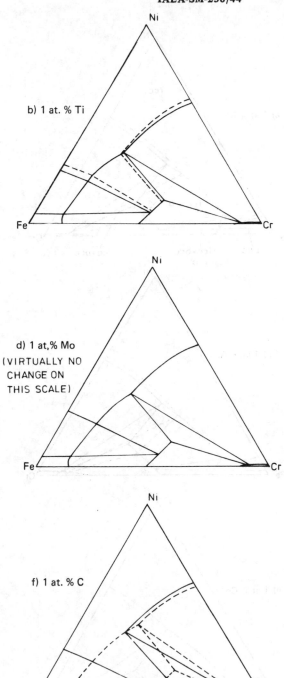

b) 1 at. % Ti

d) 1 at,% Mo
(VIRTUALLY NO
CHANGE ON
THIS SCALE)

f) 1 at. % C

f the Fe−Cr−Ni system at 550°C (shown as broken lines).

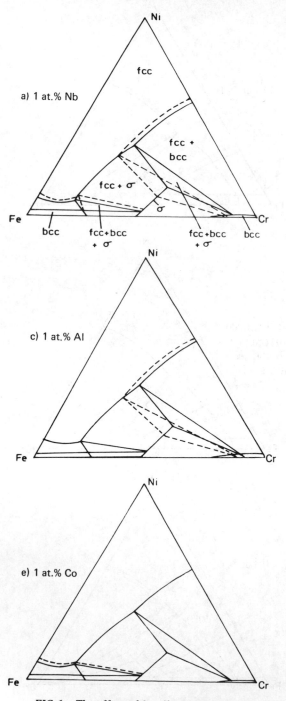

FIG.6.   *The effect of 1 at.% additions of element X to the stability*

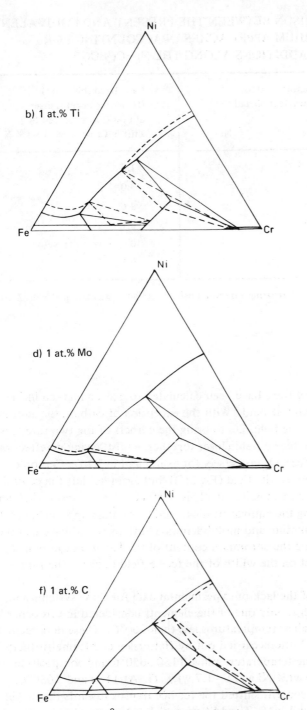

f the Fe—Cr—Ni system at 700°C (shown as broken lines).

TABLE IV. COMPARISON BETWEEN THE PRESENT AND EQUIVALENT
NICKEL AND CHROMIUM APPROACHES IN ACCOUNTING FOR
FOURTH-ELEMENT ADDITIONS ALONG THE Ni−Cr EDGE

Element X	Factors used for equivalent Ni and Cr		Calculated shift[a] in the (fcc)−(fcc+bcc) transus at 600°C as measured by a shift in Cr wt.% per 1 wt.% X
	Cr$'$	Ni$'$	
Nb	1.75		− 1.5
Ti	1.5		− 3.0
Al	5.5		− 4.6
Mo	1.5		− 0.1
Co		1	0
C		30	9.3

[a] A negative sign indicates a ferritizing effect, a positive sign an austenitizing effect. Zero indicates equivalence.

Fourth-element additions have been calculated to exert a marked influence on the size of the fcc+σ phase field. With the exception of carbon, the additions have tended to contract the field, i.e. to bring the corners of the two three-phase fields at each side of the fcc+σ field closer together, whilst having no effect on the central part of the fcc−σ transus (see, for example, Fig.5(b)). Carbon on the other hand has extended the field (Fig.5(f)), but again has left the central position of the fcc−σ transus unaltered. This conflicts with the result that would have been obtained using the equivalent nickel and chromium approach, because niobium, titanium, aluminium and molybdenum additions would have had the general effect of lowering the chromium content of the fcc−σ transus whilst having a negligible effect on the width of the fcc+σ field between the three-phase fields.

One is conscious of the lack of experimental data for these systems and, to the author's knowledge, only one of the elements considered in this paper has been studied in any detail at temperatures less than 700°C. These data relate to the effect of titanium additions carried out by Hattersley and Hume-Rothery [7]. These studies covered the temperature range 1150−650°C and compositions of 17−34 wt.% Cr, 24−30 wt.% Ni and 0−1.7 wt.% Ti. At 1150 and 1050°C, increasing the titanium content shifted the fcc-bcc transus towards the nickel corner by $\sim$ 1.2 wt.% per 1 wt.% Ti addition, which is in good agreement with the equivalent chromium approach (1.5 wt.%), but somewhat smaller than the present studies would suggest, i.e. 2.3 wt.% ($\equiv$ 2.0 at.%, Table I). At 800°C and below the effects of titanium are two-fold. Firstly it moves the fcc+σ+bcc boundary towards the Fe-corner until the level of titanium is such that Ni_3Ti is formed, that is to the limit of solubility of titanium in FeCrNi alloys. This is in good agreement with the trends presented in the present studies. Secondly the fcc—σ transus is shifted to lower chromium levels indicating that titanium possesses strong σ-stabilizing tendencies. This is contrary to the present calculations. The difference could be associated with the implicit assumptions made in the present calculations that the level of titanium is the same in all competing phases or, in other words, the tie-lines in the quaternary are not in the plane parallel to the Fe—Cr—Ni side. Because of the rather large effect shown by Hattersley and Hume-Rothery, the present assumptions may be too great a simplification and the titanium content of the σ-phase may be different from that in the bcc and fcc phases. Such a situation will have a larger effect on the relative free energies than that calculated above.

4.   CONCLUSIONS

A method has been described for estimating the effects of small fourth-element additions of niobium, titanium, aluminium, molybdenum, cobalt and carbon on the phase stability of Fe—Cr—Ni alloys from 400 to 700°C.

The calculations have confirmed and quantified the ferritizing tendencies of niobium, titanium and aluminium, the virtual equivalence between cobalt and nickel and the strong austenitizing tendencies of carbon.

With the exception of molybdenum, the calculations agree reasonably well with the scaling factors commonly used in the equivalent nickel and chromium approach and, further, show that the factors are only slightly temperature sensitive, a fact which has always been assumed until now.

Comparisons of calculated ternary sections corrected for fourth-element additions with experiment have only been possible for titanium, where the calculations have confirmed the movement in one of the 3-phase fields but failed to show any movement in the fcc—σ transus. Possible reasons are given for this last failing which, if correct, can only be overcome by a full quaternary treatment.

ACKNOWLEDGEMENTS

I am indebted to Dr. Larry Kaufman of Manlabs Inc. for many useful discussions.

REFERENCES

[1] HARRIES, D.R., "Void swelling in austenitic steels and nickel based alloys", Effects of Alloy Constitution and Structure (Proc. Consultants' Symp.), AERE Rep. 7934, Harwell (Sep.1974).

[2] WATKIN, J.S., GITTUS, J.H., STANDRING, J., "The influence of alloy constitution on the swelling of austenitic stainless steels and nickel based alloys", Int. Conf. Radiation Effects in Breeder Reactor Structural Materials (Scottsdale, Arizona, July 1977), Met. Soc. of American Institute of Mining, Metallurgical, and Petroleum Engineers, New York (1977).

[3] BATES, J.F., "Irradiation-induced swelling variations resulting from compositional modifications of type 316 stainless steel", Properties of Reactor Structural Alloys after Neutron Irradiation, American Society for Testing and Materials, ASTM STP 570 (1975) 369–386.

[4] KAUFMAN, L., Computer Calculation of Phase Diagrams, Academic Press, New York, London (1970).

[5] MANLABS-NPL, Materials Data Bank Manual (available from Manlabs Inc.).

[6] KAUFMAN, L., CALPHAD 1 1, Pergamon Press, Oxford (1977) 28.

[7] HATTERSLEY, B., HUME-ROTHERY, W., J. Iron Steel Inst. (Jul.1966) 683–701.

CALCULATION OF THE DRIVING FORCE FOR THE RADIATION INDUCED PRECIPITATION OF Ni₃Si IN NICKEL–SILICON ALLOYS

A.P. MIODOWNIK
University of Surrey,
Guildford, Surrey

J.S. WATKIN
United Kingdom Atomic Energy Authority,
Springfields Nuclear Power Development Laboratories,
Springfields, Salwick, Preston, Lancashire,
United Kingdom

Abstract

CALCULATION OF THE DRIVING FORCE FOR THE RADIATION INDUCED
PRECIPITATION OF Ni₃Si IN NICKEL–SILICON ALLOYS.

The appearance of precipitates which have been identified as Ni₃Si in irradiated stainless steels and nickel-rich alloys such as Inconel is of considerable interest in relation to the swelling behaviour of such materials. Work on binary nickel–silicon alloys has shown that Ni₃Si can be induced to precipitate in alloys whose silicon content is well below the accepted solubility limit, and it has also been shown that such precipitates redissolve when heat-treatment is continued at the same temperature in the absence of irradiation. Such effects imply an irradiation induced shift of chemical potential, and cannot be explained by merely involving accelerated diffusion. This paper represents an attempt to calculate the shift in chemical potential required to precipitate Ni₃Si in alloys containing 1–10 at.% Si over a range of temperatures (300–1000 K), and then proceeds to relate this calculated chemical potential to available information concerning the dose rates required to induce such precipitates at various temperatures. Presentation of the results is modelled on the well established methods for handling the time/temperature/transformation behaviour of ordinary alloy systems, with dose rate being substituted for the time axis. Analogous calculations are presented for nickel–germanium alloys, in order to check whether the numerical values deduced from the nickel–silicon system have more general applicability, and also to see whether there are any significant differences in a system where the size factor of the solute is of the opposite sign.

1. INTRODUCTION

The appearance of precipitates which have been identified as Ni₃Si in irradiated stainless steels [1] and nickel rich alloys [2-3] is of considerable interest in relation to the swelling behaviour of such alloys [4-5]. Studies of the binary nickel-silicon system have shown that Ni₃Si can be induced to precipitate in alloys whose silicon content is well below the accepted solubility limit [6-7], and it has been shown that such

315

TABLE I. SURVEY OF SYSTEMS IN WHICH PRECIPITATION FROM SUBSATURATED SOLID SOLUTIONS HAS BEEN OBSERVED

System	Precipitate	Radiation	Dose (displacements/atom)·s^{-1}	Ref.
Ni-Si	Ni$_3$Si	Ni$^+$	2.5×10^{-3}	[3]
Ni-Si	Ni$_3$Si	n	$\sim 10^{-6}$ (est)	[6]
Ni-Si	Ni$_3$Si	Ni$^+$	2×10^{-3}, 2×10^{-4}	[7]
Ni-Si	Ni$_3$Si	e	5×10^{-3}, 9×10^{-5}	[8]
Ni-Nb	Ni$_3$Nb	n	$\sim 10^{-6}$ (est)	[11]
Ni-Mo	NiMo	Ni$^+$, e	-	[12]
Ni-Be	NiBe	Ni$^+$	3×10^{-3}	[10]

Note: Although radiation effects have been reported in nickel-base systems containing aluminium and titanium [13,14], it is more likely that the precipitates in these systems are equilibrium phases formed through accelerated diffusion.

precipitates redissolve when heat treatment is continued at the same temperature in the absence of irradiation [8-9]. Reports of precipitation from other solid solutions are becoming more frequent [10-14] (Table I), but only for the nickel-silicon system is there substantial information on the temperature and dose rate dependence of the phenomenon [3,8,9] (Fig. 1).

The total accumulated evidence in this system makes it clear that the appearance of Ni$_3$Si in alloys containing as little as 1 at% silicon must imply an irradiation induced shift of chemical potential, and cannot be explained by merely invoking accelerated diffusion [15]. As indicated in Table I the appearance of Ni$_3$Si has been observed not only with neutron irradiation, but also after Ni$^+$ ion bombardment and 1 MeV electron irradiation in the HVEM. Although many different mechanisms have been proposed for the appearance of Ni$_3$Si [16-19], all theories tend to agree that such irradiation induced precipitation is the result of local supersaturation due to some solute-defect interaction. However free energy changes associated with the degree of supersaturation required to produce the precipitation of Ni$_3$Si at various solute contents have not been available to date. In this paper we describe an attempt to perform such a calculation, together with a presentation of the results which is modelled on the Time-Temperature-Transformation behaviour of ordinary alloys (with dose rate being substituted for the time axis). Analogous calculations have been performed for Nickel-Germanium alloys in order to check whether the numerical values deduced from the Nickel-Silicon system are of more general applicability, and also because it has been reported [20] that similar effects are produced in this system despite the fact that in this case the solute atom is larger than Nickel.

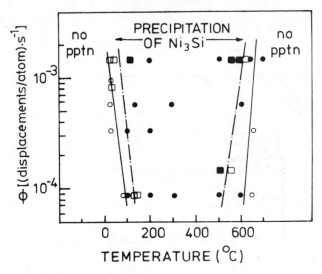

FIG.1. *Combinations of dose-rate and temperature which lead to precipitation of Ni₃Si (according to Barbu and Martin [8]).*

2.   THE THERMODYNAMIC CHARACTERISATION OF THE NICKEL-SILICON SYSTEM

The experimental data available to describe this system is surprisingly sparse. For instance, reference to the Hansen-Elliot-Schunk Bibliographies [21] does not even yield in formation on the FCC solidus of this relatively important system (Fig. 2). There does not appear to be any additional information available from later literature [22,23], although relatively detailed thermodynamic data exists for the liquid phase, and adequate though less accurate data for the compounds in this system, including Ni_3Si [24,25]. In general the magnitude of the Heats of Formation are consistent with empirical predictions [26] (which is useful in relation to assessing suitable operational data for the Ni-Ge system for which very much less information is available).

The position of the FCC/(FCC + Ni_3Si) transus, which controls any critical assessment of irradiation induced precipitation, has only been partially determined at high temperatures [21], although some results for the coherent phase boundary are available for the region 620-850°C [27]. The basic requirements in order to calculate the position of the FCC/(FCC + Ni_3Si) transus are as follows:

(a)   Suitable lattice stability parameters for nickel and silicon.

(b)   A suitable formulation for the excess free energy of the FCC solid solution.

(c)   A suitable formulation for the free energy of Ni_3Si, particularly any possible temperature variation due to ordering.

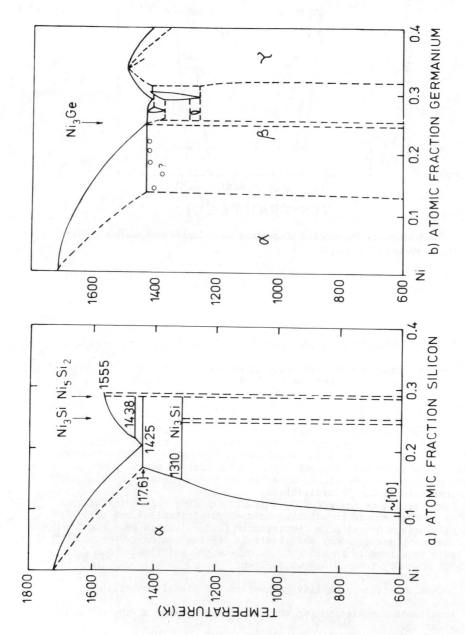

FIG.2. *Current phase diagrams of Ni–Si and Ni–Ge (according to Hansen et al. [21]).*

Lattice stability parameters were taken from the work of Kaufman [28,29]. The FCC solid solution parameters were calculated by finding numerical values which, when combined with the known liquidus data, give a reasonable fit for the observed liquid/solid transus (within the chosen formalism). The Heat of Formation of Ni_3Si (relative to Ni^{FCC} and Si^{DIA}) is quoted in the literature as 42 ± 4 $kJ \cdot mol^{-1}$ [24,25]. No experimental data is available for the temperature dependence of G_{Ni_3Si}, which will obviously hinge on the ordering temperature of this compound. However experimental ordering data on ternary $Ni_3(Fe,Si)$ alloys [30] yields an estimated value of 1812K for $T_o^{FCC/Ll2}$, which can then be used with a simplified ordering model to give a working approximation for the relevant changes in free energy. Such a model, based on the Inden-Bragg-Williams treatment [31], uses the following expressions:

$$H_{Ni-Si}^{FCC} = - N \, x(1 - x) \, [6W_1 + 3W_2] \tag{1}$$

$$H_{Ni_3Si}^{Ll2} = - N \, [x(6W_1 + 3W_2) - 12x^2W_2] \tag{2}$$

$$kT_o^{FCC/Ll2} = \chi \, x(1 - x) \, [4W_1 - 6W_2] \tag{3}$$

Here x is the atomic fraction of silicon, W_1 and W_2 are first and second-neighbour interaction coefficients, χ is a scaling factor (set equal to $\frac{1}{2}$ by Inden [31]), and N and k have their usual meanings.

The limited results available for ternary Fe-Ni-Si alloys are consistent with a high ratio for (W_1/W_2) so in the interests of simplicity W_2 has been set equal to zero, which then yields the following simple relationships:

$$H_{Ni_3Si}^{Ll2} \approx - 12.5 \, W_1 \tag{4}$$

$$T_o^{FCC/Ll2} \approx 3/8 \, W_1 \tag{5}$$

$$A_{NiSi} \approx - 50 \, W_1 \tag{6}$$

where A_{NiSi} is the parameter normally used to express the excess heat in the regular solution model at 0 K i.e.

$$E_H = A \, x(1 - x) \tag{7}$$

As must be evident, this ordering formalism assumes that the component elements of the system themselves crystallise in the FCC structure, so that some adjustment is necessary when comparing such prediction with experimental results referred to silicon in its normal diamond structure. Such adjustments are however easily performed through the availability of standard lattice stability parameters [28,29]. Thus a value of T_o = 1812K immediately yields W_1 = 4833K (Eqn. 5), which then corresponds to a value of **242 kJ·mol⁻¹** for A (Eqn. 6), and a value of $\sim$ 60 **kJ·mol⁻¹** for the Heat of Formation of Ni_3Si (referred to FCC Si). This may be compared to a value of 55 ± 4 **kJ·mol⁻¹** obtained by combining the experimental value of 42 ± 4 **kJ·mol⁻¹** with the appropriate lattice stability correction (Fig. 3). The agreement must be considered encouraging in the light of the simplifications made to the ordering formalism. The chosen description for the

TABLE II. CALCULATED AND OBSERVED EXCESS FREE ENERGY (kJ·mol⁻¹) FOR LIQUID Ni–Si ALLOYS [24, 25]

x_{Si}	${}^{E}G^{LIQ}_{observed}$; [1873K]	${}^{E}G^{LIQ}_{extrapolated}$; [1400K]	${}^{E}G_{calculated}$; [1400K]
0.10	– 13.36	– 18.71	– 19.13
0.15	– 19.63	– 26.66	– 27.83
0.20	– 25.32	– 34.07	– 34.93
0.25	– 30.01	– 40.18	– 41.03

ᵃ Calculated value derived from Eqn. 8 with pure liquid components as reference states.

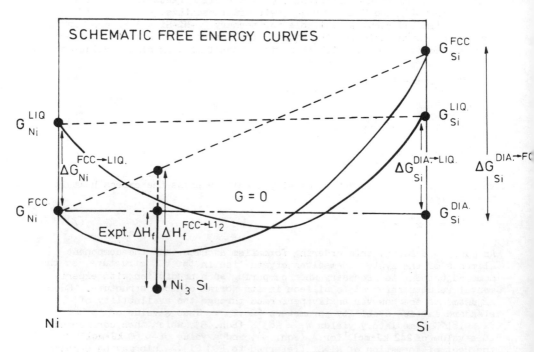

FIG.3.   Disposition of free-energy curves in relation to chosen reference states.

excess free energy of liquid Ni-Si alloys (Eqn. 8) yields the agreement shown in Table II, and combination of Eqn. 8 and the available information on the liquid-solid transus yields the corresponding expression for FCC Ni-Si solid solutions shown by Eqn. 9. A combination of the FCC data with an expression which approximates the rather complex temperature dependence of Ni_3Si (Eqn. 10) then yields the required FCC/Ni_3Si transus (the tangency points on the free energy curves being obtained by standard techniques). The resulting calculated values, together with the available experimental data are shown in Fig. 4.

$$^E_G G^{LIQ}_{Ni-Si} = - x(1 - x) [234.4 - 0.0272T] \tag{8}$$

$$^E_G G^{FCC}_{Ni-Si} = - x(1 - x) [242.7 - 0.0188T] \tag{9}$$

$$H^{L12}_{Ni_3Si} - {}^O H^{FCC}_{Ni} - {}^O H^{DIA}_{Si} = 41.85 \tag{10}$$

The supporting equations defining the necessary lattice stabilities of the elements are:

$$\Delta G^{FCC \to LIQ}_{Ni} = 17.37 - 0.0102T \tag{11}$$

$$\Delta G^{DIA \to FCC}_{Si} = 49.80 - 0.0179T \tag{12}$$

$$\Delta G^{DIA \to LIQ}_{Si} = 50.63 - 0.030T \tag{13}$$

The melting point of Nickel and Silicon were taken as 1728K and 1690K respectively. All data in Eqns (8-13) is expressed in $kJ \cdot mol^{-1}$.

The full equations representing the free energy of the various phases concerned relative to a reference state of ${}^O G^{FCC}_{Ni} = {}^O G^{DIA}_{Si} = 0$, are as follows:

$$G^{LIQ}_{Ni-Si} = (1 - x) \Delta G^{FCC \to LIQ}_{Ni} + x\Delta G^{DIA \to LIQ}_{Si} + x(1 - x) {}^E \Delta G^{LIQ}_{Ni-Si}$$

$$+ RT[x \ln x + (1 - x) \ln (1 - x)] \tag{14}$$

$$G^{FCC}_{Ni-Si} = x\Delta G^{DIA \to FCC}_{Si} + x(1 - x) {}^E \Delta G^{FCC}_{Ni-Si}$$

$$+ RT[x\ln x + (1 - x) \ln (1 - x)] \tag{15}$$

$$G^{L12}_{Ni_3Si} = \Delta H^{Ni_3Si}_{formation} + x\Delta G^{DIA \to FCC}_{Si} + B(T/T_o)^6 (T \not> T_o) \tag{16}$$

3. FREE ENERGY AND IRRADIATION INDUCED PRECIPITATION IN NICKEL SILICON ALLOYS

Radiation experiments indicate that Ni_3Si precipitates from solid solutions of much lower silicon contents than indicated by the FCC/Ni_3Si transus of Fig. 4. One can then specify a particular combination of

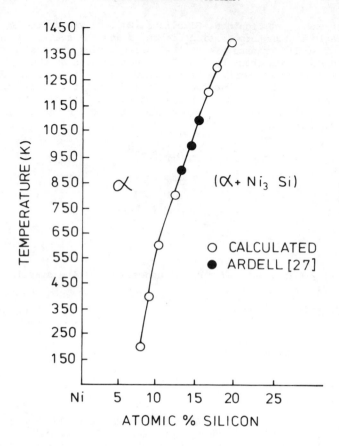

FIG.4. Comparison of calculated FCC/Ni$_3$Si transus with experimental results of Rastogi and Ardell [27].

matrix composition and temperature known to just precipitate Ni$_3$Si after a given radiation treatment, and hence it is possible to calculate the extra free energy (ΔG^*) necessary to displace the free energy curves in order to satisfy the new tangency conditions appropriate to the irradiated case (Fig. 5). Values of ΔG^* thus calculated for various solute contents are listed in Tables IIIa **and b.** It is possible in this way to plot the variation of ΔG^* with temperature as shown in Fig. **6.**

With increasing subsaturation (lower silicon contents) the presence of Ni$_3$Si is necessarily accompanied by an increasingly large driving force tending to redissolve the precipitate, and it is reasonable that (as shown by Fig. 6) a correspondingly larger dose rate is required to keep the precipitate in existence. On this line of reasoning, the correlation between ΔG^* and dose rate might be expected to be independent of composition, which is found to be more or less the case (Fig. 7).

TABLE III.   CALCULATED FREE ENERGY, ΔG^* ($J \cdot mol^{-1}$),
ASSOCIATED WITH THE PRECIPITATION OF Ni_3Si FROM
SUBSATURATED Ni–Si ALLOYS AS A FUNCTION OF
TEMPERATURE FOR VARIOUS SILICON CONTENTS

IIIa

Composition	Temperature					
(at%)	200K	300K	400K	500K	600K	700K
2% Si	377	695	845	1021	1197	1398
4% Si	244	314	435	557	703	858
6% Si	33	80	142	226	322	449

IIIb

Composition	800K	900K	1000K	1100K	1200K	1300K
2% Si	1615	1867	2151	2515	2967	3582
4% Si	1050	1256	1519	1846	2272	2863
6% Si	588	770	988	1276	1674	2243

Perhaps more interestingly, Fig. 7 also suggests that a threshold
level of radiation exists below which precipitation will not occur.   This
is also suggested by representing the data of Fig. 1 in the form of Fig. 8,
which can be seen to resemble a conventional Time-Temperature-Transformation
(TTT) curve with the time axis replaced by displacement dose rate.

It is worth pursuing this analogy further and consider whether it is
reasonable to formally relate   time and dose rate.   In conventional TTT
curves the start of transformation curves correspond to the appearance of a
precipitate of arbitrary dimensions (which depends on the sensitivity of
the detecting system).   The growth of such a precipitate to its critical
size can be expressed in the form:

$$r^n_{crit} \propto t \cdot \frac{\Delta G}{RT} \cdot D_o \cdot e^{-(Q_f+Q_m)/RT} \tag{17}$$

where r is the ruling dimension of the precipitate, t is the time, ΔG the
driving force, and D_o and Q are the usual diffusion parameters.   This leads
to the well known situation where ΔG is the rate controlling factor at
high temperatures whereas diffusion is the rate controlling step at low
temperatures.   For the purposes of this paper it has been
assumed that the critical dpa values[1] in the high temperature region of

---

[1]   dpa: displacements per atom.

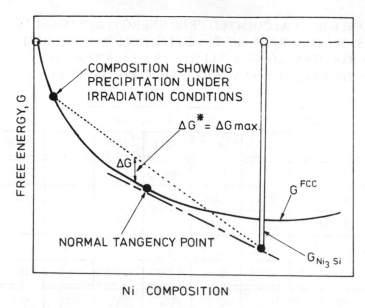

FIG.5.   Definition of the free energy (ΔG^*) required in order to precipitate Ni_3Si in subsaturated solid solutions.

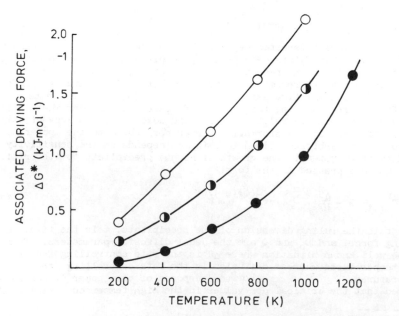

FIG.6.   Calculated variation of ΔG^* with temperature for 2 at% and 6 at% Si alloys.

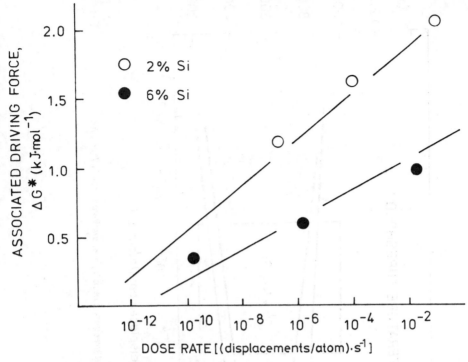

FIG.7. *Correlation of* Δ*G* with dose rate for 2 at% and 6 at% Si alloys.*

Fig. 8 depend only on ΔG*. Although this may not be entirely justified in the presence of irradiation induced defects, any errors will diminish as the temperature is raised. In the low temperature region it is generally accepted that the excess vacancy concentration will be proportional to the dose rate (or the square root of the dose rate) [32]. Since this extra vacancy concentration effectively substitutes for $(e^{-Q_f/RT})$ in Eqn. 17, we will have

$$r^n_{crit} \propto (t \times dpa) \qquad\qquad (18)$$

It is therefore reasonable to exchange a time axis by a dpa axis at least in qualitative terms. The temperature dependence of the dpa values in the low temperature region is in fact consistent with the activation energies generally found in other irradiated systems [15]. As in the normal TTT situation, one would expect that the driving force to precipitate Ni_3Si would increase as the temperature is decreased, but in the case under considera- tion, ΔG will also increase as the dose rate increases. At a given tem- perature the chemical potential offsetting the radiation induced driving force will clearly increase with lower solute contents, so that a higher dose rate is required to give the same degree of precipitation in alloys containing less silicon.

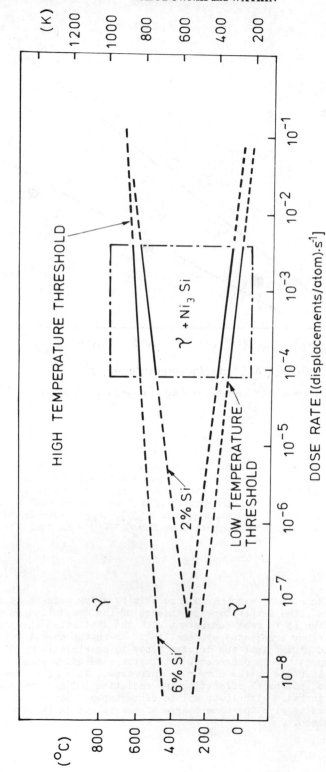

FIG.8. *Presentation of irradiation-induced precipitation in a form analogous to TTT graphs.*

TABLE IV.   COMPARISON OF THERMODYNAMIC INPUT PARAMETERS FOR Ni–Si AND Ni–Ge ALLOYS

Parameters	Nickel-Silicon	Ref.	Nickel-Germanium	Ref.
Free Energy change from Diamond to FCC $\Delta G_x^{DIA \rightarrow FCC}$	50.2 - 0.019T	[29]	36 - 0.016T	[36]
Heat of Formation relative to pure components $\Delta H_f(Ni_3X)$	- 42 ± 4	[24]	- 29.2 ± 2	[35]
Heat of Formation: First Order Prediction	- 29 ± 8	[26]	- 19 ± 8	[26]
Solid Solution Parameter $A[E_{G_{Ni-X}}^{FCC} = A\ x(1 - x)]$	- 242 + 0.019T		- 168 + 0.016T	
Liquid Solution Parameter $L[E_{G_{Ni-X}}^{UQ} = L\ x(1 - x)]$	- 234 + 0.027T		- 159 + 0.020T	
Ordering Temperature T_o	1812 K		1250 K/1600 K	
Nearest neighbour ordering parameter W_1 (in k units)[b]	4833 k		3300 k	
Values of B in Eqn. 10	5		2.5	

[a] Temperatures in kelvin, energies in kJ·mol^{-1}, entropies in kJ·mol^{-1}·K^{-1}.

[b] k = 13.8×10^{-24} J = 3.3×10^{-24} cal.

It should be noted that Fig. 8 was initially constructed entirely by heavily extrapolating the data of Fig. 1. However the additional data of Refs[6 and 3]fit well into the general picture, and further, as yet unpublished data [20] confirms that the original extrapolation is reasonable.

4.   ANALOGOUS CALCULATIONS FOR NICKEL GERMANIUM ALLOYS

Some of the theories which couple solute drift with defect flow are quite sensitive to the relative sizes of solute and solvent atoms [18,19], so it is of some interest to make similar calculations for the Nickel-Germanium system since the atomic diameters follow the order Si < Ni < Ge. There is very little consistent thermodynamic data available for Ni-Ge [33-35] particularly for the solid state. The internal consistency demanded by the Inden Formalism, and similar general principles [36], can be used to

TABLE V. CALCULATED FREE ENERGY, ΔG^* ($J \cdot mol^{-1}$), ASSOCIATED WITH THE PRECIPITATION OF Ni_3Ge FROM SUBSATURATED Ni–Ge ALLOYS AS A FUNCTION OF TEMPERATURE FOR VARIOUS GERMANIUM CONTENTS

Va

Composition (at%)	Temperature					
	200K	300K	400K	500K	600K	700K
2% Ge	314	448	598	770	950	1151
4% Ge	105	193	301	427	573	741
6% Ge	3	35	96	180	285	409

Vb

Composition	800K	900K	1000K	1100K	1200K	1300K
2% Ge	1368	1599	1858	2139	2465	2842
4% Ge	921	1126	1360	1615	1925	2289
6% Ge	565	741	942	1180	1469	1820

develop the data listed in Table IV; however this must be considered more tentative than the corresponding figures for Ni-Si, as there is little guidance from the equilibrium diagram (Fig. 2), which badly needs re-examination. It is in fact not at all clear how the critical transus between the FCC solid solution and Ni_3Ge behaves at high temperatures, and it is not possible to fit the diagram easily with one ordering parameter (W_1). In the light of the sketchy experimental evidence, it was, however, decided to make preliminary calculations setting $W_2 = 0$ as for Ni-Si, and using somewhat higher value of T_0 than is strictly determined by Eqn. 5. The resultant Free Energy composition relationships are listed in Tables Va and b, which may be compared with the figures for Ni-Si listed in Tables IIIa and b. It remains to be seen whether the actual behaviour of Ni-Ge alloys turn out to be closely similar to the published information on Ni-Si, but if the irradiation behaviour is indeed similar, this will reinforce the predictive power of the treatment outlined in this paper; it may, at the same time, necessitate a re-evaluation of the specific importance of solutes with negative size factors. In any event there should be an increased possibility of differentiating between the many theories of irradiation induced precipitation.

5. CONCLUSIONS

1. The magnitude of the driving forces involved in the precipitation of Ni_3Si have been found to be very similar to the values previously deduced in a quite different manner for irradiation effects in stainless steels

[37,38]. It would be very useful to find that precipitation effects in irradiated alloys were always associated with a fairly standard range of free energy changes (ΔG^*), which in turn bear a single relationship to the dose rate.

2. The calculation of the required free energies in more complex systems is not inherently difficult now that more powerful computing methods have become available, but the present treatment cannot be adequately tested because of a severe lack of information on the dose rate/temperature/composition relationships in other systems. This is therefore an area which deserves immediate attention.

3. A fruitful analogy may be drawn between irradiation induced precipitation and conventional TTT behaviour; plotting the experimental results as a function of temperature and dose rate indicates that there may be an absolute threshold value for the irradiation dose, below which no irradiation induced precipitates will form. Free energy calculations also indicate the existence of such a threshold, which corresponds to zero driving force for irradiation induced precipitation. In the case of nickel-silicon alloys this threshold dose appears to lie in the region $10^{-11} \rightarrow 10^{-12}$ (dpa)·s^{-1}.

REFERENCES

[1] CAWTHORNE, C., BROWN, C., J. Nucl. Mater. **66** 1/2 (1977) 201.

[2] BRAGER, H.R., GARNER, F.A., J. Nucl. Mater. **73** 1 (1978) 9.

[3] POTTER, D.I., RAHN, L.E., et al., Scr. Metall. **11** (1977) 1095.

[4] BATES, J.F., JOHNSTON, W.G., "The effect of alloy composition on void swelling", Proc. Int. Conf. Radiation Effects in Breeder Reactor Structural Materials (Scottsdale, Arizona, 1977), Metall. Soc. of American Institute of Mining, Metallurgical, and Petroleum Engineers, New York (1977).

[5] WATKIN, J., GITTUS, J.H., STANDRING, J., "Correlation of swelling with constitution", Proc. Int. Conf. Radiation Effects in Breeder Reactor Structural Materials (Scottsdale, Arizona, 1977), Metall. Soc. of American Institute of Mining, Metallurgical, and Petroleum Engineers, New York (1977).

[6] SILVESTRE, G., SILVENT, A., REGNARD, C., SAINFORT, G., J. Nucl. Mater. **57** (1975) 125.

[7] BARBU, A., ARDELL, A.J., Scr. Metall. **9** (1975) 1233.

[8] BARBU, A., MARTIN, G., Scr. Metall. **11** (1977) 771.

[9] MARTIN, G., BOCQUET, J.L., BARBU, A., ADDA, Y. (D. Tech/SRMP 4982), Proc. Int. Conf. Radiation Effects in Breeder Reactor Structural Materials (Scottsdale, Arizona, 1977), Metall. Soc. of American Institute of Mining, Metallurgical, and Petroleum Engineers, New York (1977).

[10] OKAMOTO, P.R., TAYLOR, A., WIEDERSICH, H., Proc. Int. Conf. Radiation Damage, USERDA Conf. 751006 (1975) 1188.

[11] APPLEBY, W.K., SANDUSKY, D.W., WOLFF, V.E., J. Nucl. Mater. **43** (1972) 43.

[12] CARPENTER, R.W., KENIK, E.A., BAYUZICK, R.J., American Institute of Mining, Metallurgical, and Petroleum Engineers Annual Mtg. (Atlanta, 1977), Oak Ridge National Lab. Rep. ORNL-5311 (1977).

[13] POTTER, D.I., HOFF, H.A., Acta Metall. **24** 12 (1976) 1155.

[14] HUDSON, J.A., J. Br. Nucl. Energy Soc. **14** (1975) 127.

[15] ADDA, Y., BEYELEV, M., BREBEC, G., Thin Solid Films **25** (1975) 107.

[16] MARTIN, G., Philos. Mag. **32** 3 (1975) 615.

[17] WILKES, P., LIOU, K.Y., LOTT, R.G., Radiat. Eff. **29** (1976) 249.

[18] MAYDET, S.I., RUSSELL, K.C., J. Nucl. Mater. **64** (1977) 101.

[19] OKAMOTO, P.R., WIEDERSICH, H., J. Nucl. Mater. **53** (1974) 336.

[20] MARTIN, G., private communication.

[21] HANSEN, M., ELLIOTT, R.P., SHUNK, F.A., Constitution of Binary Alloys, McGraw-Hill, New York (1958/69).

[22] BADTIEV, E.B., et al., Vestn. Mosk. Univ., Ser. 2, Khim. **29** 3 (1974) 367.

[23] CHART, T.G., private communication.

[24] CHART, T.G., National Physical Laboratory (UK) Publication No. 18 (1972).

[25] CHART, T.G., High Temp.-High Press. **5** (1973) 241.

[26] MIEDEMA, A.R., J. Less-Common Met. **46** (1976) 67.

[27] RASTOGI, P.K., ARDELL, A.J., Acta Metall. **17** (1969) 595.

[28] KAUFMAN, L., CALPHAD **1**, Pergamon Press, Oxford (1977) 28.

[29] KAUFMAN, L., private communication.

[30] INDEN, G., private communication.

[31] INDEN, G., J. Phys. (Paris) **38** 12, Suppl. C7 (1977) 373.

[32] HUDSON, J.A., NELSON, R.S., Vacancy '76 Symp. (SMALLMAN, R.E., HARRIES, J.E., Eds), Metals Society, London (1977) 126.

[33] ERDELYI, L., NECKEL, A., TOMISKA, I., NOWOTNY, H., Ber. Bunsenges. Phys. Chem. **81** 10 (1977) 1003.

[34] KOCHEROV, P.V., et al., Izv. Vyssh. Uchebn. Zaved., Chern. Metall. **4** (1976) 9.

[35] PRATT, J.N., MARTOSUDIRJO, S., 4th Int. Conf. Thermodynamique et Chimie (ROUQUEROL, J., SABBAH, R., Eds), Vol. 1, Centre national de la Recherche Scientifique, Marseilles (1975) 192.

[36] MIODOWNIK, A.P., NPL Symp. Metallurgical Chemistry (Brunel Univ., 1972), HMSO, London (1972) 233, 363.

[37] MIODOWNIK, A.P., GITTUS, J.H., WATKIN, J.S., KAUFMAN, L., CALPHAD **1**, Pergamon Press, Oxford (1977) 281.

[38] MIODOWNIK, A.P., GITTUS, J.H., WATKIN, J., KAUFMAN, L., National Bureau of Standards Special Publication No. 496, Vol. 2, US Dept. of Commerce, Washington, DC (1978) 1065.

PRECIPITATION REACTIONS IN AUSTENITIC STAINLESS STEELS

M. HOCH, YUNG-SHIH CHEN
Department of Materials Science
and Metallurgical Engineering,
University of Cincinnati,
Cincinnati, Ohio,
United States of America

Abstract

PRECIPITATION REACTIONS IN AUSTENITIC STAINLESS STEELS.

The precipitation reactions for commercial austenitic stainless steels (AISI type 347, 321, 316, 316L, 304 and 304L) and titanium-modified AISI type 316 SS were studied over the temperature range of 750–1350°C. Specimens were held at temperature for 15 to 25 hours to ensure that equilibrium conditions were reached. This was followed by a water quench to prevent further precipitation reactions during the cooling process. The precipitates were extracted from bulk specimens by anodic dissolution and identified by X-ray diffraction analysis. In titanium-stabilized 321 SS, large TiN and Ti_2S ($Ti_4C_2S_2$) precipitates were present in solution-treated and subsequently annealed specimens. Small TiC precipitates were present in specimens annealed below 1150°C. $M_{23}C_6$ precipitates were found to be present after annealing at 850°C for 25 hours. The amount of $M_{23}C_6$ was found to increase with decreasing titanium content, as shown in the titanium-modified 316 SS. In niobium-stabilized 347 SS, Nb(CN) precipitates were present in solution-treated as well as annealed specimens. The $M_{23}C_6$ precipitates were detected at an annealing temperature of 1050°C, which is higher than the precipitation temperature detected in 321 SS. Thermodynamic calculations were carried out to obtain the temperature at which precipitation starts, and the temperatures where 50, 90 and 99% of the precipitates should be formed. The experimental results are in very good agreement with the calculations. The $M_{23}C_6$ precipitates were found to be present in 316 SS specimens annealed below 1050°C, in 304 and 304L SS specimens annealed below 950°C, and in 316L SS specimens annealed at 850°C. In the 316L SS specimens, higher concentrations of titanium and niobium impurities were present. The formation of $M_{23}C_6$ in stabilized 321 and 347 SS can be explained in terms of slower diffusion rates of titanium and niobium than that of carbon atoms and in terms of the large concentration of chromium atoms, which need only diffuse a much shorter distance than titanium and niobium to form a precipitate.

1. INTRODUCTION

Austenitic stainless steels have wide applications in food, chemical and power industries because of their good mechanical properties and corrosion resistance at elevated temperatures. They are also used extensively as economical structural materials due to their excellent fabricability. They have also gained considerable importance in nuclear reactor applications as fuel cladding and structural materials.

331

For nuclear reactor applications, austenitic stainless steels suffer two major problems due to irradiation effects: one is the severe loss of high-temperature ductility, and the other is the swelling of the steels themselves. In general, the presence of finely dispersed precipitates, such as NbC and TiC, decrease swelling during irradiation; however, the association of void formation with $M_{23}C_6$ precipitates is not beneficial [1].

For the purpose of enhancing the properties of the steels, small amounts of certain elements, such as aluminium, titanium, niobium, vanadium, zirconium, etc. are sometimes added. For instance, the intergranular corrosion problem of type 304 SS is minimized by adding enough titanium (type 321 SS) or niobium (type 347 SS) as a stabilizer; they combine with the carbon to prevent $M_{23}C_6$ formation. Because of irradiation effects, the property enhancement is especially needed in nuclear reactor applications. The effect of adding small amounts of titanium to type 316 SS (the titanium-modified type 316 SS) has been studied, the results showing enhancement of post-irradiation ductility and creep properties in comparison with the standard type of 316 SS [2].

Since the precipitation reactions play such an important role in property changes, many studies have been carried out on the precipitation reactions of austenitic stainless steels under various thermal-mechanical treatments and with additions of minor elements. However, some of the phenomena of the reactions are still not well understood. For instance, $M_{23}C_6$ has been found to form in steels to which enough titanium or niobium had been added in the hope that it would impede such a formation [3].

In order to understand the precipitation reactions better and to resolve some of the questions, a study which focused on the thermodynamic considerations was carried out on commercial austenitic stainless steels and titanium-modified 316 SS. The objectives of this study were: (i) to characterize the precipitation reactions in different steels; (ii) to determine the effects of stabilization and the reasons for $M_{23}C_6$ formation; and (iii) to explain the precipitation reactions through thermodynamic considerations.

The detailed description of the work and especially the results of the scanning electron microscopy, transmission electron microscopy and optical microscopy studies are contained in the thesis of Chen [4].

2. THERMODYNAMIC CONSIDERATIONS

2.1. Thermodynamics of precipitation reactions

The precipitation reaction which takes place in steels can be expressed by the equation:

$$\underline{M} + \underline{X} \rightarrow MX \tag{I}$$

where M and X represent the metal and non-metal atoms, respectively. The bar underneath the element indicates the dissolved condition. The MX represents the pure compound formed. In order to obtain the free-energy change, ΔG_1^0, for this reaction, the data for three separate reactions are needed. One is the standard free energy of formation for the compound MX. The free-energy change is ΔG_2^0 for this reaction:

$$M + X \rightarrow MX \tag{II}$$

where both M and X are in their standard state, i.e. pure solid, liquid or gas. The other two are the free-energy changes, ΔG_3^0 and ΔG_4^0, for those two elements changing from their standard states into a hypothetical 1 wt% solution (for which, at low concentrations, the activity equals the weight per cent value in iron solution). The equations are represented by:

$$M \rightarrow \underline{M} \ (1 \ wt\%) \tag{III}$$

$$X \rightarrow \underline{X} \ (1 \ wt\%) \tag{IV}$$

The reason for chosing a 1 wt% solution as another standard state is that the solute concentrations in iron are usually less than 1 wt%, and concentrations are given in weight per cent.

The change in free energy is an additive function and is related to the heat and entropy change of the reaction by:

$$\Delta G^0 = \Delta H^0 - T \Delta S^0 \tag{1}$$

Information for reactions (II), (III) and (IV) is listed in various sources in the form of Eq. (1). The free-energy change for reaction (I), ΔG_1^0, is obtained after suitably combining ΔG_2^0, ΔG_3^0 and ΔG_4^0.

The relation between the change in free energy and the equilibrium constant, K, is:

$$\Delta G^0 = - RT \ln K \tag{2}$$

In reaction (I):

$$K = \frac{a_{MX}}{a_M \cdot a_X} \tag{3}$$

Since 1 wt% is the standard state and since Henry's law is applicable in dilute solution, the activities of M and X can be simply substituted by the weight per cent values of M and X in solution. Hence, at equilibrium, the relationship for reaction (I) can be expressed as:

$$\Delta G_1^0 = A + BT = - RT \ln \left(\frac{a_{MX}}{wt\% \ M \cdot wt\% \ X} \right) \tag{4}$$

The activity of MX is unity for pure compounds. Equation (4) can be simplified to read:

$$RT \ln (wt\% \ M) + RT \ln (wt\% \ X) = A + BT \tag{5}$$

From this relationship, the precipitation starting temperature can be calculated from the starting concentrations of the two components for reaction (I). Depending on the particular precipitation reactions, the starting concentrations are either the composition of the materials or the concentrations remaining in solution after earlier precipitation reactions have occurred.

The temperature ranges for precipitation reactions extend across liquid (ℓ), γ and α phases in iron and iron-alloy systems. Let T_ℓ, T_γ, and T_α represent the precipitation starting temperatures in the liquid, γ and α phases, while $T_{\ell\gamma}$ and $T_{\gamma\alpha}$ represent respectively the transition temperatures for $\ell \rightleftharpoons \gamma$ and $\gamma \rightleftharpoons \alpha$ phase transformations. Because the precipitation starting temperatures are different in the different phases, the precipitation reactions can take two forms:

Case 1. $T_{\ell\gamma} > T_\gamma$ or $T_{\gamma\alpha} > T_\alpha$
Precipitation reaction takes place in a single phase, which remains stable during cooling to low temperature.

Case 2. $T_\gamma > T_{\ell\gamma} > T_\ell$ or $T_\alpha > T_{\gamma\alpha} > T_\gamma$
Precipitation reaction takes place during phase transformation.

To exemplify the second case, the phase transformation from liquid to solid γ-phase is considered. The precipitation starting temperature is T_ℓ in the liquid phase and T_s in the solid phase; and the solidification temperature, $T_{\ell s}$, is above T_ℓ but below T_s, as shown graphically in Fig. 1. While the alloy is still in the liquid state, the precipitation reaction will not occur because the temperature is above the precipitation starting temperature in the liquid phase. When the solidification temperature is reached, solidification takes place. As soon as the solid phase starts to form during solidification, the precipitation reaction starts to occur in the solid phase or at the interface — because the temperature is below the precipitation starting temperature in the solid phase.

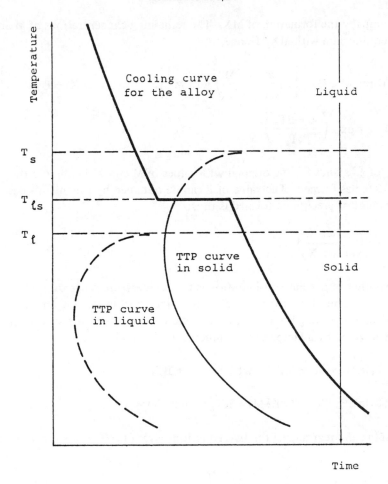

FIG.1. Schematic diagram for the case when precipitation reaction occurs during solidification.

To obtain the amount of precipitate formed at a particular temperature, T_a, e.g. the annealing temperature, Eq. (5) is first rewritten as:

$$\text{wt\% M} \cdot \text{wt\% X} = \exp\left(\frac{A + BT_a}{RT_a}\right) \tag{6}$$

Assuming P wt% of X forms precipitate MX; then M will be depleted by:

$$\left\{P\left(\frac{\text{at. wt M}}{\text{at. wt X}}\right)\right\} \text{wt\% M}$$

to accompany the formation of MX. The remaining concentrations of M and X are in equilibrium with MX. Hence:

$$\left[(\text{orig. wt\% M}) - \left\{P\left(\frac{\text{at. wt. M}}{\text{at. wt X}}\right)\text{wt\% M}\right\}\right] \times \left[(\text{orig. wt\% X}) - P\,\text{wt\% X}\right]$$

$$= \exp\left(\frac{A + BT_a}{RT_a}\right) \tag{7}$$

where 'orig.' stands for the original wt% values of M and X, i.e. before the precipitate has formed. The value of P can be obtained by solving this equation. The amount of precipitate formed is thus:

$$P + P\left(\frac{\text{at. wt. M}}{\text{at. wt. X}}\right)$$

Another important consideration is the temperature range over which the precipitates form. Rewriting Eq. (5) at the precipitation starting temperature, T_1, and then rewriting Eq. (5) for a case where the product of the two concentrations is lowered by a factor z at a temperature T_z yields:

$$RT_1\ln(\text{wt\% M}) + RT_1\ln(\text{wt\% X}) = A + BT_1 \tag{8}$$

$$RT_z\ln z + RT_z\ln(\text{wt\% M}) + RT_z\ln(\text{wt\% X}) = A + BT_z \tag{9}$$

Combining and rearranging the two equations gives $(T_1 - T_z)$ as a function of T_1 and z:

$$T_1 - T_z = \Delta T = T_1\left(1 + \frac{1}{RT_1\ln z}\right)^{-1} \tag{10}$$

It is important to see that there is a temperature range over which precipitation can occur. By considering the amount of precipitates formed as a percentage of the expected total, such as 50% or 90% of the total amount which can be expected to precipitate, a different precipitation temperature range (from T_1 to $T_{50\%}$ or T_1 to $T_{90\%}$) can also be obtained.

The overlapping of the temperature ranges over which two precipitates form can cause the coprecipitation of two compounds with the same structure. For example, if the carbide and nitride of metal M can form a solid solution with each other, then the precipitation reactions can be expressed separately as:

$$\underline{M} + \underline{C} \rightarrow MC \tag{V}$$

$$\underline{M} + \underline{N} \rightarrow MN \tag{VI}$$

TABLE I. STANDARD FREE ENERGIES OF SOLUTION AND FORMATION USED IN THERMODYNAMIC CALCULATIONS

Reactions[a]	ΔG^0 (cal/mol)	Ref.
Ti → Ti _in_ γ-Fe	−6742 − 8.84 T	[8]
Ti → Ti _in_ liq. Fe	−9700 − 8.84 T	[8]
Nb → Nb _in_ γ-Fe	−3420 − 5.58 T	[8]
Cr → Cr (20%) _in_ α-Fe	3584 − 4.718 T	[8]
Cr → Cr (20%) _in_ γ-Fe	3584 − 4.437 T	[8]
C → C _in_ α-Fe	19813 − 12.22 T	[9]
C → C _in_ γ-Fe	10300 − 9.6 T	[9]
C → C _in_ liq. Fe	5100 − 10.00 T	[9, 10]
N → N _in_ α-Fe	7200 + 4.62 T	[9]
N → N _in_ γ-Fe	−2060 + 8.94 T	[9]
N → N _in_ liq. Fe	860 + 5.71 T	[9, 10]
Ti + $\frac{1}{2}$N$_2$ → TiN	−80550 + 22.45 T	[9]
Ti + C → TiC	−45000 + 2.79 T	[9]
Nb + $\frac{1}{2}$N$_2$ → NbN	−57280 + 23.0 T	[11]
Nb + C → NbC	−34140 + 2.81 T	[11]
23Cr + 6C → Cr$_{23}$C$_6$	−98300 − 9.21 T	[10]

[a] Bar under element indicates the dissolved condition.

By assuming that an ideal solid solution is formed between the carbide and nitride, and that the mole fraction of MN is f, then the final coprecipitation reactions can be expressed as:

$$fMN + (1-f)MC \rightarrow M(fN + (1-f)C) \tag{VII}$$

Since the compounds are no longer pure compounds, the activities of the compounds in reactions (V) and (VI) are no longer unity. Instead, the mole fractions, f and (1−f), should be used in calculations. The coprecipitation temperature for impure nitride or carbide can be obtained by solving Eq. (4) with the value of a_{MX} being equal to the mole fraction. The coprecipitation temperature will then be higher than the precipitation starting temperature for the pure compound.

TABLE II.   THE CALCULATED FREE-ENERGY CHANGES FOR PRECIPITATE FORMATION

Precipitate	ΔG^0 in α-Fe	ΔG^0 in γ-Fe	ΔG^0 in liq. Fe
TiN[a]	$-79438 + 26.67$ T	$-70178 + 22.35$ T	$-70140 + 25.58$ T
TiN	$-81008 + 26.67$ T	$-71748 + 22.35$ T	$-71710 + 25.58$ T
TiC[a]	$-56501 + 23.85$ T	$-46988 + 21.23$ T	$-38830 + 21.63$ T
TiC	$-58071 + 23.85$ T	$-48558 + 21.23$ T	$-40400 + 21.63$ T
NbN	$-61060 + 23.96$ T	$-51800 + 19.64$ T	
NbC	$-50533 + 20.61$ T	$-41020 + 17.99$ T	
$\frac{1}{6}Cr_{23}C_6$	$-49932 + 28.77$ T	$-40419 + 25.07$ T	

[a] Changes in free energy due to Ti–Ni interaction are considered.

2.2. Thermodynamic calculations

The available thermodynamic data used for the precipitation reaction calculations are compiled from various sources and are listed in Table I.   The calculated free-energy changes for the precipitation reactions are presented in Table II.   The chromium contents of austenitic stainless steels are about 18 wt%; hence, 20 wt% Cr is used as the standard state for $Cr_{23}C_6$ precipitation reactions. The liquid Ti–Ni alloy data [5] indicated a very strong interaction between them. Since the value of free energy for the favoured interaction is -19.6 kcal/mol [6], the titanium activity will be lowered by -1.57 kcal/mol in stainless steels (they have about 9 at.% Ni and 0.7 at.% Ti).   The additional free-energy changes, corrected for interactions, are also listed in Table II.   The crystallographic data of the precipitates are given in Table III.

From the free-energy change and the composition of the alloy, the precipitation starting temperature can be calculated using Eq. (5).   The maximum amount of precipitate that can form is limited by the less concentrated component. Using Eq. (7), the temperatures at which 50%, 90% and 99% of the expected maximum amount of the precipitate are formed can also be calculated.   The calculated precipitation temperature ranges from starting to 99% formed are shown in Fig. 2 for different precipitates in different austenitic stainless steels. The depletion of the component after earlier precipitations is accounted for in the calculation of the temperatures for the subsequent precipitations.   The liquidus and solidus temperatures for 18–8 stainless steels were obtained from the Cr–Fe–Ni phase diagrams reported by Speich [7].

TABLE III. CRYSTALLOGRAPHIC DATA FOR PRECIPITATES

| Compound | Structure | Lattice parameters (Å) | | ASTM[a] Card No. or Reference |
		a	c	
$\alpha\text{-}Al_2O_3$	hex.	4.758	12.991	10−173
$\alpha\text{-}Fe$	bcc	2.8664		6−0696
$\alpha\text{-}MnS$	fcc	5.224		6−0518
$Cr_{23}C_6$	fcc	10.638		9−122, 14−407
$CuMn_2O_4$	fcc	8.33		11−480
NbC	fcc	4.4702		10−181
NbN	fcc	4.37		[12]
NbO	sc	4.21		12−607, 15−535
Nb(NO)	fcc	4.42		12−256
Nb(CNO)	fcc	4.40− 4.44		[13]
TiC	fcc	4.3285		6−0614
TiN	fcc	4.240		6−0642
$\tau\text{-}Ti_2S$	hex.	3.206	11.19	11−664
$Ti_4C_2S_2$	hex.	3.21	11.20	16−849
σ	tetr.	8.7995	4.5442	5−0708

[a] American Society for Testing and Materials, Philadelphia, Pennsylvania.

The theoretical precipitate weight per cent calculations at the annealing temperature were carried out using Eq. (7). By solving this equation and adding up all the different precipitates, the theoretical weight per cent is obtained. (It should be noted, however, that, since $M_{23}C_6$ formation is not under equilibrium conditions, the amount formed cannot be calculated.)

3. MATERIALS AND EXPERIMENTAL PROCEDURES

3.1. Austenitic stainless steels

Six different types of commercial austenitic stainless steels were used in this study. The materials, which were obtained from local suppliers, are: (i) niobium-stabilized AISI type 347 SS; (ii) titanium-stabilized AISI type 321 SS; and

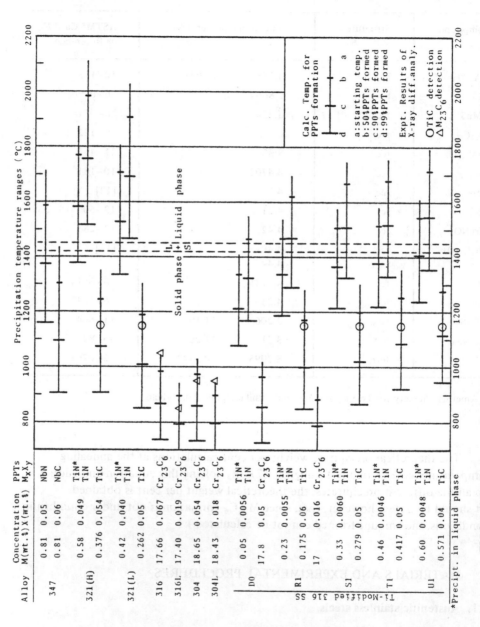

FIG.2. The temperatures for precipitate formation. The bars indicate the calculated temperature from starting to 99%-formed, while the symbol indicates the experimental result (PPT: precipitate).

TABLE IV.   CHEMICAL COMPOSITIONS (wt%) OF COMMERCIAL
AUSTENITIC STAINLESS STEELS USED IN THIS STUDY

Element	347	321(H)	321(L)	316	316L	304	304L
C	0.06	0.054	0.05	0.067	0.019	0.054	0.018
Mn	1.79	1.62	1.34	1.79	1.53	1.14	1.51
P	0.030	0.021	0.026	0.020	0.030	0.026	0.017
S	0.026	0.012	0.007	0.028	0.026	0.028	0.015
Si	0.58	0.55	0.51	0.44	0.35	0.49	0.75
Cu	0.25	0.24	0.16	0.06	0.29	0.34	0.26
Cr	18.96	18.00	17.66	17.69	17.40	18.65	18.43
Ni	10.38	9.57	10.30	12.44	13.39	9.09	8.94
Mo	0.36	0.20	0.19	2.86	2.10	0.61	0.28
Ti	–	0.58	0.42	–	–	–	–
Nb	0.81	–	–	–	–	–	–
$N^{(a)}$	(0.005)	(0.014)	(0.005)	–	–	–	–
$N^{(b)}$		(0.049)	(0.040)				

[a] Chemical analysis of the as-received material by private laboratory.
[b] Atomic absorption analysis of the solution-treated material by ARMCO.

(iii) AISI types 316 SS, 316L SS, 304 SS and 304L SS. The chemical compositions,
shown in Table IV, were provided by the suppliers according to earlier ASTM[1]
specifications. The as-received materials were cold drawn and annealed rods.

Five titanium-modified AISI type 316 stainless steels were obtained from
Oak Ridge National Laboratory. The chemical compositions are given in Table V.
The materials were received in the form of tensile test specimens, weighing about
5 grams each. The as-received specimens were re-annealed at 1150°C for 11 hours
in vacuum to eliminate the effect of previous aging at 650°C for 2000 hours.

3.2. Heat treatments

The commercial materials were heat treated in a sealed tube furnace under
an argon atmosphere. The solution treatments were carried out for 15 hours at

---

[1] American Society for Testing and Materials, Philadelphia, Pennsylvania.

TABLE V.   CHEMICAL COMPOSITIONS (wt%) OF TITANIUM-MODIFIED 316 SS

| Element | Alloy sample codes: | | | | |
	D0	R1	S1	T1	U1
C	0.050	0.06	0.05	0.05	0.04
Mn	1.92	0.5	0.4	0.3	0.2
P	0.013	0.010	0.009	0.009	0.008
S	0.016	0.012	0.010	0.009	0.004
Si	0.75	0.4	0.4	0.4	0.4
Cu	–	–	–	–	–
Cr	17.8	17	17	17	17
Ni	13.0	12	12	13	13
Mo	2.6	2.5	2.5	2.4	2.5
Ti	0.05	0.23	0.33	0.46	0.60
Nb	–	–	–	–	–
N	0.0560	0.0055	0.0060	0.0046	0.0048

1350°C for the stabilized and 1300°C for the standard stainless steels. The annealing treatments were carried out at 1150, 1050, 950 and 850°C for 25 hours. For the stabilized steels, one more annealing treatment was carried out at 1250°C for 20 hours. After heat treating, all specimens were water quenched to prevent precipitation reactions which might occur during the cooling process.

Each of the titanium-modified type 316 SS specimens were cut into two halves. One half of each specimen was studied in the as-received condition (annealed at 1150°C for 11 hours); the other half of each specimen was further annealed at 750°C for 25 hours in a Brew Furnace under vacuum. Helium gas was used to quench the specimens after annealing.

3.3. Phase extraction and identification

Three techniques were tested for the extraction of precipitates from the bulk specimens.

The first technique, anodic dissolution, was adopted from a summary compiled by Donachie and Kriege [14]. Since the anodic dissolution dissolves the matrix specimens fast enough and preserves the precipitates, it was used in the present experiments to extract precipitates from bulk specimens.

The second and third techniques, used by Krupowicz [15] for his precipitation studies, dissolve the specimens chemically. The second technique, which utilizes liquid bromine and methyl acetate, dissolves some of the sulphide and carbide. Hence, it was not suitable for this investigation. The third technique, which used a 60% HCl-40% H_2O solution, preserves the sulphide and carbide. However, it took more than a week to dissolve a specimen weighing about 5 grams.

Centrifuging was used to collect the suspended precipitates. The centrifuging was repeated five or more times with a methanol rinse, in order to wash and clean the precipitates. After cleaning, the precipitates were transferred into a pre-weighed 5 ml vial. The contents of the vial were allowed to set for several hours before being placed in a drying oven at 75°C.

Conventional X-ray diffraction analysis was used for phase identification. The amount of precipitates extracted was very small, thereby limiting the choice to the powder method using a Debye-Scherrer powder camera. The copper K_α radiation from a Norelco or a Picker unit was used in all X-ray diffraction experiments.

The identification of precipitates was accomplished primarily through comparison of the d-spacings obtained with those listed on ASTM X-ray Cards, supplemented by Edax analysis[2] The intensity values of the diffraction lines were determined by visual comparisons. The lattice parameters of TiN, TiC and Nb(CN) were calculated from the interplaner spacing equation and averaged from high-angle values.

4.   EXPERIMENTAL RESULTS AND DISCUSSION

4.1. Titanium-stabilized steels

The calculated precipitation starting temperature for TiN in liquid phase is approximately 1800°C, which is far above the solidification temperature; therefore, the formation of TiN should start in the liquid phase. The precipitation temperature is 200°C higher in the solid γ-phase; therefore any nitrogen left in the liquid should combine with titanium to form TiN during solidification. (The precipitation starting temperature for Ti_2S could not be obtained, because no thermodynamic data was available.) The detection of strong TiN and Ti_2S diffraction lines in solution-treated specimens and the observation of large TiN and smaller Ti_2S particles scattered randomly throughout the matrix supported the thermodynamic calculation for TiN and suggested that the free energy of formation of Ti_2S should be less negative than that of TiN, yet should still have a large negative value. The detection of TiN only in specimen[3] D0 of titanium-modified 316 SS supported the assumption.

---

[2] Energy dispersive analysis by X-rays.
[3] Alloy specimen codes (see Table V).

TABLE VI. WEIGHT PER CENT OF EXTRACTED PRECIPITATES FROM TITANIUM-STABILIZED AUSTENITIC SS (~ 5 g SAMPLES)

VIa. Commercial 321 SS

Specimens and heat treatment[a]	321(H)			321 (L)		
	Exp.		Calc.	Exp.		Calc.
	(mg)	(wt%)	(wt%)	(mg)	(wt%)	(wt%)
As received	27.7	0.383	–	20.2	0.293	–
Sol.T.: 1350°C; 15 h; WQ	35.9	0.354	0.265	21.1	0.354	0.205
+ Ann.: 1250°C; 20 h; WQ	10.0	0.247	0.391	4.6	0.131	0.268
+ Ann.: 1150°C; 25 h; WQ	11.8	0.266	0.472	8.9	0.308	0.359
+ Ann.: 1050°C; 25 h; WQ	13.8	0.277	0.514	12.1	0.340	0.415
+ Ann.: 950°C; 25 h; WQ	19.5	0.377	0.530	10.6	0.325	0.443
+ Ann.: 850°C; 25 h; WQ	34.0	0.485	0.534[b]	24.2	0.543	0.453[b]

[a] WQ: water quenched.
[b] Weight of $M_{23}C_6$ could not be calculated and was not included.

VIb. 321 (H) SS
Sol.T.: + Ann.; 950°C; WQ[a]

Ann. time (h)	Exp.	
	(mg)	(wt%)
0.25	11.6	0.274
2.5	10.3	0.275
50	11.0	0.255
250	11.5	0.272

[a] WQ: water quenched.

VIc. Ti-Modified 316 SS

Alloy sample codes	Ti (wt%)	As recd; 1150°C; 11 h		Calc. (wt%)	+750°C; 25 h	
		Exp.			Exp.	
		(mg)	(wt%)		(mg)	(wt%)
D0	0.05	0.8	0.038	0.065	11.3	0.496
R1	0.23	1.9	0.097	0.196	4.7	0.232
S1	0.33	2.2	0.094	0.228	4.6	0.212
T1	0.46	2.5	0.110	0.253	5.0	0.236
U1	0.60	3.1	0.139	0.207	6.4	0.298

The calculated precipitation starting temperatures for TiC in solid γ-phase are in the range of 1280 to 1350°C; they increase with increasing concentrations of titanium and carbon for 321 SS and titanium-modified 316 SS. The detection of TiC in 1150°C annealed specimens and the changes in intensities are in agreement with the calculated temperatures (Fig. 2).

The calculated precipitation temperature difference for TiN and TiC is in the range of 400 to 600°C, depending on the concentration of titanium, carbon and nitrogen. Because of the large temperature difference (almost all the TiN formation is completed above the precipitation starting temperature for TiC formation), therefore, relatively pure TiN and TiC (instead of Ti(CN)) are more likely to form in titanium-stabilized 18–8 stainless steels.

The weight per cent of the total precipitates was obtained from the weights of extracted precipitates ánd dissolved material. The theoretical weight per cent was calculated from the thermodynamic equations listed in Table II. The weights per cent of the extracted precipitates for titanium-stabilized 321 SS and titanium-modified 316 SS are given in Table VI. The amount of total precipitates increased with a lowering of the annealing temperatures, as expected. The experimental weights per cent were lower than the theoretical values. In the solution-treated specimens, the higher values were probably due to the presence of some silicon compounds, indicated by Edax analysis and X-ray diffraction analysis. The sudden increase in values for the specimens annealed at 850°C was caused by the formation of $M_{23}C_6$. Since the formation of $M_{23}C_6$ was due to precipitation kinetics, the amount formed could not be calculated; therefore, the theoretical weight per cent does not include the weight of $M_{23}C_6$. The effect of titanium addition on $M_{23}C_6$ formation was again clearly shown in the weight per cent analysis. The values increased by only 0.11% in the higher titanium content 321(H) SS, but by 0.22% in the lower titanium content 321(L) SS as the annealing temperatures were lowered from 950°C to 850°C. The same effect was observed in titanium-modified 316 SS after annealing at 750°C, especially in the case where a 0.46% increase in 0.05 wt% Ti commercial 316 SS (D0) was reduced to 0.14% in 0.23 wt% Ti titanium-modified 316 SS (R1).

The specimens heat treated at 950°C for different periods of time (0.25 to 250 hours) did not show any change in the amount of precipitates formed (Table VI). This indicates that the precipitation of TiC was very fast, and that equilibrium conditions were reached within 0.25 hour (15 minutes).

4.2. Niobium-stabilized steels

In niobium-stabilized 347 SS, a different situation exists. The precipitation starting temperatures for NbN and NbC in γ-phase are 1700°C and 1440°C, respectively. Since the temperature difference is about 250°C and the overlap in precipitation temperature range (Fig. 2) is much larger, coprecipitation could

TABLE VII. $Nb(C_{1-x}N_x)$ FORMATION IN 347 SS

Solving ΔG^0 equations for NbN and NbC simultaneously. (No interaction considered.)

$Nb = 0.81$ wt%; $C = 0.06$ wt%; $(N = 0.04$ wt% assumed); PPT starting temp. $= 1715°C$

Temp. (°C)	Theoretical values				Experimental result	
	Composition, x	Nb(CN), a_0 (Å)		(wt%)	a_0 (Å)	(wt%)
1350	0.54	4.416		0.52	4.437	0.97
1250	0.47	4.423		0.62	4.433	0.371
1150	0.42	4.428		0.71	4.435	0.63
1050	0.39	4.431		$0.77 + M_{23}C_6$	4.438	0.808
950	0.37	4.433		$0.81 + M_{23}C_6$	4.440	0.772
850	0.36	4.434		$0.83 + M_{23}C_6 + \sigma$	4.434	1.201

take place and a solid solution of Nb(CN) could form. Assuming that coprecipitation occurs, the composition and amount of the Nb(CN) at various annealing temperatures can be calculated using Eq. (4) and reaction (VII). To obtain the theoretical lattice parameter, it was assumed that it changes linearly with composition. The data obtained are given in Table VII. It must be borne in mind that precipitation starts in the liquid phase, and about one third of the precipitate is formed before solidification begins. The agreement between calculated and experimental data is very good.

4.3. $M_{23}C_6$ formation

Both TiC and $M_{23}C_6$ precipitates formed, as was shown by X-ray diffraction analysis of the specimens which were solution treated and annealed at 850°C for 25 hours. However, according to thermodynamic calculations, formation of $M_{23}C_6$ should not take place. The calculated carbon concentrations in equilibrium with TiC are 0.0007 wt% in 321(H) SS and 0.0017 wt% in 321(L) SS at 850°C, while the carbon concentrations in equilibrium with $Cr_{23}C_6$ are 0.0045 wt% and 0.0046 wt%, respectively. The amount of titanium in 321 SS is more than enough to prevent the $Cr_{23}C_6$ formation. Therefore, the formation of $M_{23}C_6$ in 321 SS is attributed to precipitation kinetics. The carbon atoms have a very high diffusion rate and are available for both TiC and $M_{23}C_6$ formation. However, the diffusion rates for the titanium and chromium atoms are much slower, so that they are the rate controlling factors. Since the titanium concentration becomes very low after some TiC precipitates have formed, the diffusion path to the precipitation sites becomes very long for the titanium atoms. Therefore, it becomes more and more difficult for TiC to form and grow. In the meantime, the chromium concentration remains very high (more than a factor of two higher than the titanium concentration), and so the chromium diffusion path is much shorter than that for titanium. So $M_{23}C_6$ can nucleate and grow faster within a smaller volume of space away from TiC precipitates. As the precipitate formation continues, the titanium and carbon concentrations decrease further, until the carbon concentration reaches the solubility limit (in equilibrium with $M_{23}C_6$); the formation of $M_{23}C_6$ then ceases. The equilibrium condition is not reached for TiC formation, and the carbon concentration will continue to decrease, with formation of TiC, but at an ever slower pace. Eventually, the $M_{23}C_6$ will redissolve as the carbon concentration is lowered below the equilibrium value and it will disappear as the equilibrium concentration with TiC is reached. Leitnaker and Bentley [16] examined a 321 SS after an in-service age of seventeen years at $\sim 600°C$ and found no $M_{23}C_6$. This finding confirmed the thermodynamic considerations for 321 SS, while the reported [17] formation of $M_{23}C_6$ away from NbC precipitates supported the diffusion concept.

The X-ray diffraction analysis of the extracted precipitates showed that $M_{23}C_6$ formed at a higher annealing temperature in niobium-stabilized 347 SS than in titanium-stabilized 321 SS. As the diffusion rate is lower for niobium than for titanium, the Nb(CN) complex would form even more slowly than the titanium compounds. For these reasons, formation of $M_{23}C_6$ in 347 SS could occur at higher temperatures than it can in 321 SS.

In the unstabilized 18–8 stainless steels, the $Cr_{23}C_6$ precipitation starting temperatures are 1050, 950, 1030, and 950°C for 316, 316L, 304, and 304L SS, respectively. If the actual formation of $M_{23}C_6$ instead of $Cr_{23}C_6$ is considered, the starting temperatures are raised. X-ray diffraction analysis showed reasonably good agreement with expectation for the various temperatures, except for the very low annealing temperature at which $M_{23}C_6$ was detected in 316L SS. The reason for this phenomenon appears to be the large amount of titanium and niobium impurities, which combine with carbon to form carbides and thus lower the carbon concentration in solution.

REFERENCES

[1] NORRIS, D.I.R., Radiat. Eff. **15** (1972) 1.

[2] BLOOM, E.E., STIEGLER, J.O., Trans. Am. Nucl. Soc. **15** (1972) 253.

[3] MININO, T., KINOSHITA, K., SHINODA, T., MINEGISHI, I., Trans. Iron Steel Inst. Jpn. **9** (1969) 472.

[4] CHEN, Y.S., Ph.D. Thesis, University of Cincinnati (1978).

[5] GERMAN, R.M., St.PIERRE, G.R., Metall. Trans. **3** (1972) 2819.

[6] VERNARDAKIS, T., personal communication.

[7] SPEICH, G.R., in American Society for Metals' Metals Handbook, 8th ed., Vol. 8, Source Book on Stainless Steels, ASM, Metals Park, Ohio (1976) 399.

[8] SPENCER, P., Technische Hochschule Aachen, Lehrstuhl für Metallurgie der Kernbrennstoffe und Theoretische Hüttenkunde, private communication.

[9] McGANNON, H.E., The Making, Shaping and Treating of Steel, 9th ed., US Steel Pittsburgh, Pennsylvania (1971) 326.

[10] ELLIOTT, J.F., GLEISER, M., RAMAKRISHNA, V., Thermochemistry for Steelmaking, Vol. 2, Addison-Wesley, Reading (USA) (1963) and Pergamon Press, Oxford (1964).

[11] ROSENQVIST, T., Principles of Extractive Metallurgy, McGraw-Hill, New York (1974).

[12] DUWEZ, P., ODELL, F., J. Electrochem. Soc. 97 10 (1950) 299.

[13] ROUBIN, M., PARIS, J.M., J. Less-Common Met. 24 (1971) 195 (in French).

[14] DONACHIE, Jr., M.J., KRIEGE, O.H., Phase Extraction and Analysis in Superalloys, Materials Engineering and Research Lab., Pratt and Whitney Aircraft, East Hartford, Connecticut (1971).

[15] KRUPOWICZ, J., Ph.D. thesis, University of Cincinnati (1974) 37.

[16] LEITNAKER, J.M., BENTLEY, J., Metall. Trans., A 8 10 (1977) 1605.

[17] SILCOCK, J.M., SIDDING, K.W., FRY, T.K., Mater. Sci. J. 4 (1970) 29.

Section L

MELT/CONCRETE INTERACTIONS

INFLUENCE OF GAS GENERATION
ON MELT/CONCRETE INTERACTIONS*

D.A. POWERS
Sandia Laboratories,
Albuquerque, New Mexico,
United States of America

Abstract

INFLUENCE OF GAS GENERATION ON MELT/CONCRETE INTERACTIONS.

Gases formed during the interaction of a high-temperature melt with concrete are shown to stem from the thermal dehydration and decarboxylation of the concrete. The kinetics of these decomposition reactions are described by rate equations of the form:

$$\frac{d\alpha}{dt} = K\{\exp[-E/RT]\}(1-\alpha)^n$$

where α is the fraction reacted, T is the absolute temperature, and E, K and n are rate constants. Values of the rate constants for the decomposition reactions in three varieties of concrete are presented. Dehydration reactions are satisfactorily described when $n = 1$. Decarboxylation reactions are best modelled when $n = 2/3$. Gases evolved from a horizontal concrete surface escape preferentially into the melt at localized sites on either side of the centre-line of the melt. Extensive gas evolution occurs at vertical concrete surfaces attacked by a melt. When gas evolution rates are sufficiently great vertical surfaces may be partially shielded from the melt. Gases within the melt cause an apparent swelling of the melt. The observed swelling is not easily correlated to the rate of gas evolution. Metallic melts cause carbon dioxide and water liberated from the melt to be reduced to carbon monoxide and hydrogen. When these gases escape from the melt they assist in aerosol formation. A linear correlation between aerosol concentration and gas evolution rate is suggested. As the gases cool they react along a pathway whose oxygen fugacity is apparently buffered by the iron-Wustite equilibrium. Methane is a product of the gas-phase reaction.

Accidents involving meltdown of reactor fuel are "design-basis" accidents for American fast breeder reactors. Such hypothetical accidents have been shown to contribute significantly to the risks associated with the use of light-water nuclear reactors [1]. Scenarios postulated for these accidents involve molten core materials penetrating the reactor pressure vessel and cascading into the concrete sump of the reactor.

---

* This work was supported by the Fuel Behavior Branch of the United States Nuclear Regulatory Commission under United States Department of Energy Contract AT(29-1)-789.

Interactions of the high-temperature molten material with concrete are believed to produce many of the phenomena responsible
for safety risks associated with core meltdown accidents. A
recent review, however, has shown that the data base concerning melt/concrete interactions is quite limited [2].

For the last three years the author and others at Sandia
Laboratories have been involved in an experimental program intended to provide data concerning melt/concrete interactions.
The program is an ongoing **exploratory** study to identify pertinent,
safety-related phenomena and provide evaluations of these
phenomena suitable for risk assessment purposes [3]. The
pertinent phenomena being studied may be broadly catergorized
as (a) concrete erosion, (b) gas generation, (c) heat transfer
from the molten materials, and (d) aerosol generation. The
second of these categories is the subject of this paper.

Gas generation is an important, safety-related phenomenon
in its own right. Sufficient gas generation may lead to failure
of the reactor containment by over-pressurization [1]. If the
gases are flammable they may fail containment by detonation.
Regardless of whether either of these drastic events occur,
gases generated during the melt/concrete interaction play a
central role in any hypothetical, meltdown-accident scenario.
The influence of gases on the other phenomena being studied in
Sandia research-concrete erosion, melt heat transfer, and
aerosol formation are discussed in this paper. The format of
the presentation is to describe generation of the gases,
their behavior as they pass through the melt and their behavior
above molten pools in contact with concrete. The paper brings
together the results of a large number of experiments. To
accommodate the space limitations here only brief descriptions
of the experimental techniques are given. More complete accounts
of these techniques are to be found in the referenced progress
reports for the Sandia research program.

A.   Generation of the Gases

Concrete is a heterogeneous material having no specific
composition. It consists of a cement paste binding together aggregate material. The aggregate is the source of the variability
in composition of concretes taken from various locales. The
compositions of three varieties of concrete used in this research
are listed in Table I. The first of these concretes is made of
basaltic aggregate and is rich in silica. It contains little
calcareous material and yields only 77 cm^3/g of volatile material

TABLE I.   CONCRETE COMPOSITIONS (wt%)

Species	Basaltic Concrete	Limestone- Common Sand Concrete	Generic Southeastern United States Concrete
Fe_2O_3	6.25	1.44	1.20
Cr_2O_3	nd	0.014	0.004
MnO	nd	0.03	0.01
TiO_2	1.05	0.18	0.12
K_2O	5.38	1.22	0.68
Na_2O	1.80	0.082	0.078
CaO	8.80	31.2	45.4
MgO	6.15	0.48	5.67
SiO_2	54.73	35.7	3.6
Al_2O_3	8.30	3.6	1.6
CO_2	1.5	22.0	35.7
SO_2	0.2	0.2	nil
H_2O	5.0	4.7	4.1

nd = not determined

upon firing to 1000°C. The second concrete contains both limestone and common sand aggregate. Firing it produces 187 cm^3/g volitiles. These first two varieties of concrete melt over the temperature range of 1100-1400°C [4]. The third variety of concrete has only limestone aggregate. Upon firing it yields 254 cm^3/g volatiles. This concrete melts over the temperature range of 1350-1600°C [5].

The concretes used in this study span a range of both gas generation potentials and melting ranges. It is hoped that the behaviors of other concretes can be found by interpolating results presented here.

Thermally initiated decomposition of concrete is responsible for gas generation during melt/concrete interactions. The natures of these decompositions were studied by thermogravimetric analysis with a Dupont Model 990 Thermal Analysis unit. Helium atmospheres flowing at 7-10 cubic centimetres at STP per minute were used for the analytic and kinetic determinations. Mixtures of helium and carbon dioxide flowing at 34 cm^3(STP)/min were used in studies of atmospheric composition effects.

Concrete samples for analysis were split from 5 to 7 kg lots of cast concrete crushed to a 250 μm   mean particle size. The samples were further crushed to pass a sieve with 100 μm openings. The samples were equilibrated for at least two weeks over a saturated magnesium nitrate solution (relative humidity = 56%) prior to analysis.

The thermal decomposition reactions of all three concrete varieties are qualitatively similar. Upon heating, water is lost over a temperature range of 40 to 250°C. This "evaporable" water loss comes from free water in the concrete and the loss of molecular water from cementitious species such as tobermorite ($Ca_3Si_6O_1$ $(OH)_2.nH_2O$), gypsum ($CaSO_4.2H_2O$), $Ca_3SiO_3OH.2H_2O$, and $Ca_{12}Al_{14}$ $(OH)_{24} (SO_4).50H_2O$. A second water loss occurs over the range of 380-520°C. This loss is due to decomposition of $Ca(OH)_2$ and other hydroxide-containing species such as those listed above. The magnitudes of these two weight losses are similar in all three concrete varieties.

Above 600°C another weight loss occurs in concrete due to the decomposition of calcareous species in the concrete. Carbonates occur in the cementitious phases of concrete due to the action of air on calcium salts. These are the only carbonates present in basaltic concrete. The other concretes utilize calcite and dolomite aggregates which decompose to yield carbon dioxide. Decomposition of these relatively crystalline carbonates in the aggregates occurs at a somewhat higher temperature and over a narrower temperature range than does decomposition of cementitious carbonates.

During melt/concrete interactions concrete is exposed to heating rates that may exceed 50°C/min. The kinetics of concrete decomposition must then be understood if predictions of gas generation rates are to be made. However, concrete is such a heterogeneous material that mechanistically correct kinetics are not easily obtained. It is necessary to adopt the empirical methods that have assumed wide popularity recently[6].

Each of the three weight loss steps described above may be assumed to be due to the decomposition of a single, hypothetical, species. The rates of decomposition of these species are assumed to obey the equation:

$$\frac{d\alpha}{dt} = K\left\{\exp\left(-E/RT\right)\right\} (1 - \alpha)^n \qquad (1)$$

where
α = fraction of the species decomposed
t = time
K = frequency factor
E = activation energy
R = universal gas constant
t = absolute temperature

and n is the order of reaction which can take on only prescribed values: 3/2, 1, 2/3, 1/2 and 1/3.

TABLE II.   KINETIC PARAMETERS FOR CONCRETE DECOMPOSITION[a,b]

Reaction	Basaltic Concrete	Limestone-Common Sand Concrete	Generic Southeastern United States Concrete
loss of evaporable water (n=1)	$E = 11.6$ $K = 4.4 \times 10^6$	$E = 11.0$ $K = 1.29 \times 10^6$	$E = 11.0$ $K = 1.29 \times 10^6$
loss of chemical water (n=1)	$E = 41.9$ $K = 2.8 \times 10^{12}$	$E = 40.8$ $K = 1.96 \times 10^{12}$	$E = 40.8$ $K = 1.96 \times 10^{12}$
decarboxylation (n=1)	$E = 42.6$ $K = 3.6 \times 10^9$	$E = 38.5$ $K = 1.98 \times 10^7$	$E = 45.8$ $K = 1.73 \times 10^9$
decarboxylation (n=2/3)	--- ---	$E = 37.0$ $K = 3.6 \times 10^7$	$E = 44.9$ $K = 1.94 \times 10^6$

[a] E in units of kcal/mole; K in units of minutes^{-1}

[b] Standard errors in E ~ 10%; standard errors in K ~ 30 to 50%

If the decomposing species is assumed to be exposed to linearly increasing temperatures and Equation 1 is differentiated with respect to temperature, the following expression for the temperature of maximum rate of reaction, T_M, is obtained:

$$\frac{\beta}{(1-\alpha_M)^{n-1} \, T_M^2} = \frac{n \, K \, R}{E} \, \exp\left(-E/RT_M\right) \qquad (2)$$

where
$\quad \beta$ = rate of temperature increase, and
$\quad \alpha_M$ = fraction of species decomposed at the temperature of maximum rate.

Thus by examining concrete at various heating rates, the kinetic parameters of decomposition may be determined. Results of such examinations for the three concrete varieties at heating rates of 0.5 to 100°C/minute are shown in Table II. First order kinetics (n = 1) were found to be appropriate for the dehydration reactions of concrete. No substantial statistical differences could be found between first order and 2/3 order kinetics for the decarboxylation reactions, and parameters for both orders are reported in Table II for the calcareous concretes. The 2/3 order model is consistent with previous studies of decarboxylation in polycrystalline and single-crystal calcium carbonate [7,8]. The 2/3 order model also yielded better predictions of the thermogravimetric results than the first-order model as shown in Figure 1.

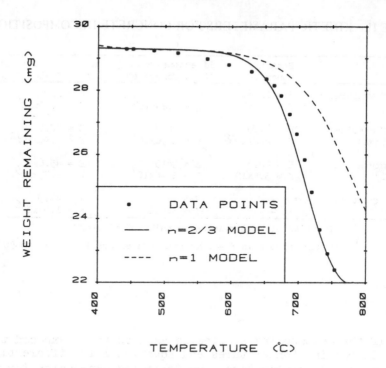

FIG.1. Comparison of first-order and 2/3-order kinetic models with the thermogravimetric data for the decarboxylation of limestone-common sand concrete.

The decarboxylation reactions are particularly susceptible to equilibrium effects. By varying the partial pressure of carbon dioxide in the gas stream it was found that the temperature of maximum reaction rate varied in the classic Clausius-Clapeyron fashion:

$$\frac{d \ln P_{CO_2}}{d T_M} = \frac{\Delta H}{RT_M^2}$$

where P_{CO_2} = partial pressure of carbon dioxide

             ΔH = latent heat of reaction

Least-squares procedures yielded a value of -48.5 kcal/mol for the latent heat of this reaction.

Cyclic variation of the heating cycle showed that only a small portion of the products of the decarboxylation could be recarboxylated. This fraction decreased with increases in the maximum temperature of the heating cycle and decreases in the heating rate. It is apparent that sintering of the solid reaction products inhibits the reverse reaction.

These results have shown that gases are generated during the melt/concrete interaction at rates influenced by both kinetic and thermodynamic factors. A computer model of the decomposition of bulk concrete based upon the above results is currently being developed by Knight and Beck [9].

B. Gas-Melt Interactions

It is when gases generated within the concrete emerge into the melt that they have the most pronounced influence on the behavior of the melt/concrete system. The interactions between the gases and the melt are both chemical and mechanical.

A test in which the contact between a high temperature melt and limestone-common sand concrete was followed by roentgenographic techniques has revealed the mechanical portion of the melt/gas interaction [10]. A melt was generated by the aluminothermic reduction of 1425 grams of magnetite within a concrete cavity 9.5 cm in diameter and 14.6 cm long. A Decalix image intensification system coupled with a Linatron Model 1500 pulsed x-ray source was used to monitor the melt behavior. The radiation source was operated at a potential of 7.5 MeV and a dose rate of 500 rads/minute. The real time image of the test melt was transferred at 24 frames per second to motion picture film.

The most obvious feature of the gas/melt interaction was that the gases vigorously agitated the melt. This agitation was of such magnitude that to a first approximation the melt could be considered an isothermal pool with sharp temperature gradients at its boundaries.

The agitation was not the result of random bubble or gas-jet formation. Rather, very soon after the start of the test, preferred sites of gas evolution could be identified at the melt/concrete interface. These sites are qualitatively depicted in Figure 2.

The limits of resolution with the x-ray technique are believed to be about 0.3 cm. Therefore, the controversy of whether or not a thin gas-film separates the melt from the

TABLE III.   PER CENT OF TIME MELT IS IN CONTACT WITH CONCRETE AT VARIOUS LOCATIONS AND DURING VARIOUS TIME INTERVALS

Location*	Mean % of Time Melt in Contact (Standard error of Mean) During the Time Interval:							
	0-9s	10-19s	20-29s	30-39s	40-49s	50-59s	60-67s	0-67s
A	31(6)	56(5)	46(4)	48(3)	51(5)	33(4)	4(7)	40(20)
B	59(9)	87(6)	92(1)	93(1)	89(2)	88(1)	91(4)	86(16)
C	66(10)	89(6)	97(1)	96(2)	84(3)	68(6)	42(9)	79(20)
D	19(5)	29(13)	33(6)	31(3)	12(2)	13(2)	3(3)	21(18)
E	59(5)	74(4)	89(4)	95(2)	94(3)	83(7)	66(4)	80(16)
F	64(15)	48(7)	29(4)	30(2)	17(3)	9(3)	7(1)	30(23)
G	40(3)	48(6)	79(14)	91(2)	79(5)	73(2)	62(1)	68(22)
H	84(3)	92(2)	94(1)	97(1)	95(1)	91(4)	90(2)	92(9)
I	25(5)	40(4)	44(4)	48(1)	46(3)	41(3)	68(16)	45(22)

*Locations depicted in Fig. 2.

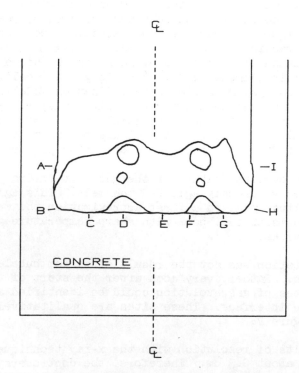

FIG.2.   *Schematic diagram of gas evolution at a high-temperature melt/concrete interface.*

concrete cannot be resolved by this test. Within the limitations of the x-ray technique, frame-by-frame analysis of the
motion picture record of the test was used to quantitatively
demonstrate the existence of localized gas evolution sites.
Determinations were made at 0.041 second intervals as to whether
melt was in "contact" with the concrete at the nine locations
marked in Figure 2 by letters. The observations were grouped
into 10 second intervals and a mean percent of time the melt was
in "contact" with the concrete calculated. A standard error in
the mean was determined not as a measure of the uncertainty in
the mean value, but as a measure of the variablility in the
contact during the time interval. Results of this analysis are
shown in Table III. Means and standard errors for the entire
test are also shown in this table.

        Contact between melt and concrete occurs less than 30% of
the time at points D and F on either side of the centerline.
Contact at the centerline occurred 80% of the time and contact
at points C and G on the bottom of the crucible cavity near the
cavity walls occurred about 70% of the time. The localized sites
at points D and F are separated by 2.3 cm. The points are about
3.5 cm from the cavity walls.

        The localized sites of gas evolution are located at posi-
tions very near to those that would be predicted based on
Taylor instability of a gas film between the melt and the concrete [11]. However, the test results do not definitively confirm this model, since other, qualitative, models may be invoked
to explain the positions of the sites.

        At the base of the cavity walls (points B and H) the melt
contacts the concrete 85 to 90% of the time. About 1/3 the pool
depth up the cavity walls melt/concrete contact occurs less than
45% of the time. At least a portion of contact at points A and
I was due to splashing of melt onto the cavity walls. Separation of the melt from the concrete walls is believed to be the
result of both local gas generation, and gases generated elsewhere escaping about the perimeter of the pool.

        Gases passing up the cavity walls between the melt and the
concrete tend to shield the concrete from attack by the melt.
This effect can be most clearly seen in comparing the erosion
of basaltic concrete, which contains the least amount of gas-
producing species, with erosion of the generic Southeastern
United States concrete, which has the greatest gas producing
potential. Comparisons of the post-test melt/concrete interfaces of these concrete types are shown in Figure 3. These concrete crucibles were exposed to about 200 kg of stainless steel
heated to 1700°C. The steel was maintained at temperature by in-
duction heating using coils embedded in the concrete [5].

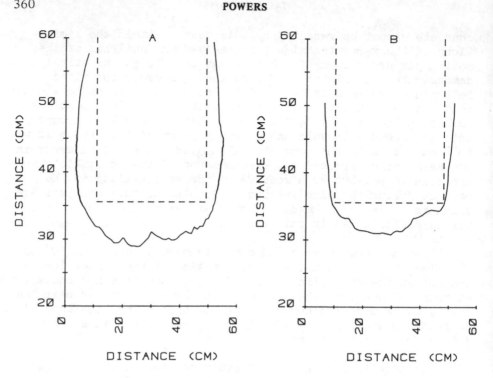

FIG.3. Post-test erosion profiles of (A) basaltic concrete and (B) generic Southeastern United States concrete crucibles after exposure to 1700°C stainless steel. The solid lines are the post-test profiles, and the broken lines are the original cavity shapes.

The ratio of maximum radial erosion to maximum axial erosion of basaltic concrete is 0.75 to 1.08. The same ratio for the calcareous concrete is 0.34.

As the gases expand into the melt they cause the melt to swell. Comparison of the melt depth observed in the test and that calculated based on the mass of melt present shows that the mean swelling can be as much as 250%. The swelling does not correlate simply with the rate of gas generation as shown in Figure 4. Complexity arises because the residence times of individual gas bubbles produced at low gas generation rates are much longer than those of gases in the form of jets that arise during high gas generation rates.

Gases emerging from the melt produce aerosols either by sparging modestly volatile constituents of the melt or by the well-known bubble breaking process [12]. The aerosol concentrations produced in the x-ray monitored test were determined

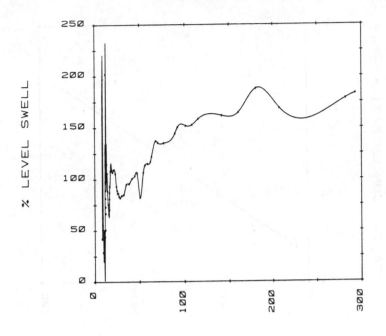

FIG.4. *Plot of the % of pool swell due to entrained gas versus the rate of gas flow for the x-ray monitored test.*

by optical and cascade impaction techniques [13]. The aerosol concentrations are plotted against the rate of gas generation in Figure 5. This plot shows that there is a linear correlation between the rate of gas generation and the concentration of aerosol in the gas.

C. Gas Chemistry In and Above the Melt

A core melt produced by a 3000 MW_{th} reactor would consist of 116 metric tons (t) of uranium dioxide, 24 t zirconium, 8.5 t zirconium dioxide, 53 t iron, 11 t chromium, and 6.4 t nickel [2]. Water and carbon dioxide released from the concrete would oxidize chromium, zirconium, and to some extent iron in such a melt. Heat liberated during this oxidation process has been shown to be comparable to the heat produced by the decay of fission products in the melt [14].

As the gases oxidize the metallic portions of the melt, they are themselves reduced. Compositions of gases liberated during

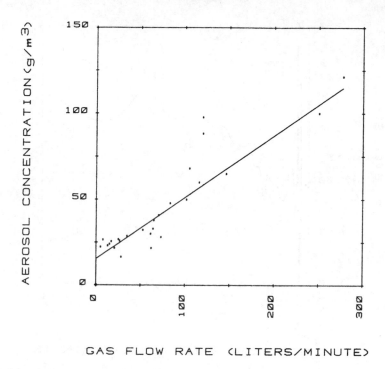

GAS FLOW RATE (LITERS/MINUTE)

FIG.5.   Plot of aerosol concentration versus the rate of gas flow in the x-ray monitored test. Solid line is the result of a least-squares fit of the data to a straight line.

TABLE IV.   COMPOSITIONS OF GAS SAMPLES

	Volume % in Sample			
	Q_{25} (as sampled)	Q_{34}	Q_{25} (last equilibrium)	Q_{34}
H_2	33.09±0.2	34.63±0.2	23.4	19.1
CO	45.62±0.3	29.03±0.3	45.6	29.0
CH_4	0.007±0.005	0.044±0.005	0.031	0.044
CO_2	20.66±0.2	35.49±0.2	20.7	35.5
H_2O	0.62±0.05	0.81±0.05	10.2	16.3

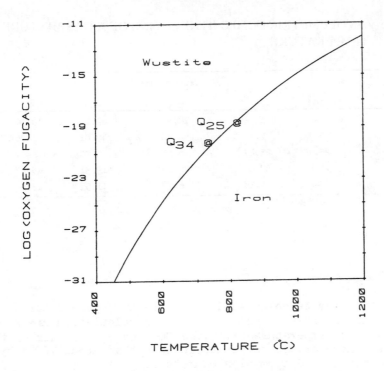

FIG.6.   *Oxygen fugacities of gas samples* Q_{34} *and* Q_{35} *compared to the temperature dependence of oxygen fugacity for the iron-Wustite equilibrium.*

the interaction of limestone-common sand concrete with a metallothermically generated melt of composition similar to that of a core melt are shown in Table IV.   These gases were obtained by quenching grab samples in stainless steel vessels.   Gas temperatures at the inlets to these vessels were found to be 475-500°C.

Gerlach (Sandia Laboratories) has analyzed these gas samples using the "correspondence" temperature technique [15].   Results of the analysis show that the gases were last at equilibrium at temperatures between 725 and 810°C.   The compositions at the last equilibrium temperature are also shown in Table IV.

Comparison of the oxygen fugacity in the samples with the oxygen fugacity produced by the equilibrium between iron and Wustite (FeO) is shown in Figure 6.   This comparison suggests that the oxygen fugacity of the gas above the melt is controlled by the iron-Wustite equilibrium.   If this is the case, and equilibrium is assumed, the dependence of the gas composition on

TABLE V. VOLUMETRIC COMPOSITION (vol.%) OF A GAS MIXTURE VERSUS TEMPERATURE

Species	Temperature (°C)					
	200	400	600	900	1300	1700
H_2	-	1. 35	18.52	22.15	18.12	16.01
CO	2.71	6.35	34.30	47.01	51.05	53.11
CH_4	25.40	23.49	6.25	0.003	-	-
CO_2	71.89	67.63	34.05	19.31	15.27	13.18
H_2O	-	1.18	6.88	11.53	15.56	17.64

temperature may be calculated since the hydrogen-to-carbon ratio in the gas is known. Results of such a calculation are shown in Table V. As temperature falls, the CO_2 and H_2 contents of the gas increase. However, at temperatures less than about 700°C hydrogen is consumed in the formation of methane.

This result suggests that detonation hazards posed by gases generated during melt/concrete interactions may come from methane rather than from H_2 or CO. The matter merits further considerations since detonation potentials within reactor containment calculated based on hydrogen may not be applicable to hazards posed by methane.

Acknowledgment

The author thanks T. M. Gerlach for the thermodynamic analyses of gas equilibria.

References

[1]   N. RASSMUSSEN, "The Reactor Safety Study", WASH-1400.
[2]   Core Meltdown Experimental Review, (W. B. MURFIN, Ed.), SAND74-0382 (Revision), Sandia Laboratories, Albuquerque, NM, March 1977.
[3]   D. A. POWERS, "Molten Core/Concrete Interactions Project Description", 189 No. A-1019. Available from the U.S. Nuclear Regulatory Commission Reading Room.

[4] Nuclear Fuel Cycle Safety Research Department, Light Water
Reactor Safety Research Program Quarterly Report - January-
March 1976, SAND76-0369, Sandia Laboratories, Albuquerque,
NM, Sept. 1976.

[5] D. A. POWERS, Sustained Molten Steel/Concrete Interactions
Tests", SAND77-1423, Sandia Laboratories, Albuquerque, NM,
June 1978.

[6] J. SESTAK, V. SETAVA, and W. WENDLANDT, Thermochimica Acta
1, 333 (1973).

[7] P. K. GALLAGHER and D. W. JOHNSON, JR., Thermochimica Acta,
14, 255 (1976)/

[8] J. M. THOMAS and G. D. RENSHAW, J. Chem. Soc. 4, 2058 (1967).

[9] R. KNIGHT and J. BECK, private communication.

[10] Nuclear Fuel Cycle Safety Research Department, Light Water
Reactor Safety Research Program Quarterly Report - June-
September, 1978, Sandia Laboratories, Albuquerque, NM, to
be published.

[11] V. K. DHIR, I. CATTON, and D. CHAO, Proceedings of the
Third Post-Accident Heat Removal Information Exchange, p. 141,
L. Baker, Jr. and J. D. Bingle, Eds. ANL-78-10, Argonne
National Laboratory, Argonne, Illinois.

[12] F. D. RICHARDSON, Physical Chemistry of Melts in Metallurgy
Academic Press, London (1974).

[13] Nuclear Fuel Cycle Safety Research Department, Light Water
Reactor Safety Research Program Quarterly Report - July-
September 1977, SAND78-0076, Sandia Laboratories, Albuquerque,
NM, April 1978.

[14] W. BAUKAL, J. NIXDORF, R. SKOUTAJAN, and H. WINTER, Inves-
tigation of the Relevancy and the Feasibility of Measure-
ments of Chemical Reactions During Core Meltdown on the
Integral Heat Content of Molten Cores, BMFT-RS-197, Battelle
Institut, e.v., Franfurt am Main, F. R. Germany, June 1977,
English translation NUREG/TR-0047, October 1978.

[15] Nuclear Fuel Cycle Safety Research Department, Light Water
Reactor Safety Research Program Quarterly Report - January-
March 1977, SAND77-1249, Sandia Laboratories, Albuquerque,
NM, October 1977.

DISCUSSION

M. TETENBAUM: From the known thermodynamic equilibria for the
gaseous species above the C—H system, one can predict that acetylene should be
present in the gas phase at temperatures of 2000 K and above. Did you detect
acetylene in the residual gas after reaction?

D.A. POWERS: No, we have never seen any acetylene in our gas samples. The predominant hydrocarbon which can be detected is methane. Ethene is also present at a concentration about half that of methane. Sometimes traces of ethane (C_2H_6) may be detected. Thermodynamic analysis of the gases has been confined to the inclusion of only those species which have been experimentally detected. I must also note that thermodynamic analysis may not be completely adequate for considering hydrocarbon production during melt-down situations. Often hydrocarbon-forming reactions are kinetically controlled. In the absence of suitable catalysts, thermodynamic equilibrium may not be attained in the gas phase.

INTERACTION OF MOLTEN 'CORIUM' WITH CONCRETE IN A HYPOTHETICAL LWR CORE-MELT-DOWN ACCIDENT*

Oxidation of core materials and hydrogen production

M. PEEHS, K. HASSMANN
Kraftwerk Union AG,
Erlangen,
Federal Republic of Germany

Abstract

INTERACTION OF MOLTEN 'CORIUM' WITH CONCRETE IN A HYPOTHETICAL LWR CORE-MELT-DOWN ACCIDENT: OXIDATION OF CORE MATERIALS AND HYDROGEN PRODUCTION.

Concrete degradation occurs only due to the thermal load applied. One of the main reasons for this is the oxidation by water released from the concrete of the metallic 'Corium' components, preventing possible reactions with the oxides from the concrete. (Corium is a term coined to describe the heterogeneous mass deriving from a reactor core after melt down) The water released from the heated concrete oxidizes the metals in the following sequence: $Zr \rightarrow ZrO_2$, $Cr \rightarrow Cr_2O_3$, $Fe \rightarrow FeO$. The oxygen potential of the Corium melt was determined to be about -300 kJ/mol. Only when all the iron is converted into FeO will a further increase of the oxygen potential occur, converting $FeO \rightarrow Fe_2O_3$ and $Ni \rightarrow NiO$. Corresponding amounts of H_2 are generated during the oxidation of the Corium components. Numerical calculations for a 1300 MW(e) PWR show that all metals in the Corium are oxidized in about 4.5 days. By 10 days some 2000 t of concrete are molten, the overall mass of the melt being 2700 t.

1. INTRODUCTION

The 'Kernschmelzen' (core melting) research project was started in the Federal Republic of Germany in 1971 to provide detailed information for light water reactor risk analyses. The objective of the research and development work is to investigate the consequences of a core-melt-down accident. Such an accident is considered to be hypothetical, but might occur if the emergency core cooling

* This work, part of the 'Kernschmelzen' (core melting) Project, is funded by the Bundesministerium für Forschung und Technologie.

systems fail completely after a loss-of-coolant accident. For the analysis, a core
melting accident was subdivided into four phases (Fig. 1):

1st phase   The core heats up until the core structure fails.

2nd phase   The 2nd phase is characterized by the evaporation of the water left
              in the lower plenum of the reactor pressure vessel (RPV).

3rd phase   The 3rd phase involves the heating up of the pressure vessel after a
              melt has been formed in the cavity of the RPV.

4th phase   Finally, after the pressure vessel has failed, the molten 'Corium'
              (footnote 1) will interact with the concrete structure beneath the
              pressure vessel.

It was recognized very early on that the concrete structure is an excellent
barrier for preventing the propagation of molten Corium [1–4]. Detailed
investigations of the interaction of molten Corium with concrete [2] have shown
that the concrete degradation is due only to the thermal load applied. One of
the main reasons for this result is that oxidation of the metallic components of
the Corium by the steam released from the heat affected zone of the concrete
prevents possible reactions between metals and oxides. Owing to the oxidation
of the metallic components of the core melt, a large amount of H_2 is generated.

The objective of the work reported here was to provide detailed information
concerning:

• the oxidation of molten LWR core material in contact with concrete;
• the analysis of hydrogen generation during the time interval during which the
  molten Corium penetrates into the concrete foundation;
• thermodynamic data describing the complex multi-component system.

To achieve the research and development objectives, experiments on the
interaction between concrete and Corium or the individual components of Corium
have been undertaken. These experiments were amply instrumented so as to be
able to measure all parameters. In addition, detailed post-experiment investigations
of the melts were carried out. Based on the data obtained, a theoretical
thermodynamic model was completed on the basis of which the oxidation
and the hydrogen production during the 4th phase of a hypothetical core-melt-
down accident could be calculated.

---

1 Corium is a term that has been coined to describe the heterogeneous mass comprising
the various elements deriving from a reactor core after melt down (see §2).

2.   DEFINITION OF CORIUM

A common definition of core melts (Corium) is needed in order to limit
the number of experiments necessary and to ensure comparability of results
from different laboratories. An additional objective is to find a definition of
Corium which would be generally valid for all LWRs. Constituents present in
the melt at only a low concentration have a negligible influence on the overall
materials behaviour and can therefore be neglected. Thus the Corium components
of interest are UO_2, Zircaloy and stainless steel. Based on these considerations,
two Corium states were defined [4–6]:
(a)   Corium A (A ≏ Anfang, i.e. beginning) contains all the material of the core
      and that of the support structure beneath the core;
(b)   Corium E (E ≏ Ende, i.e. finish) contains all the material in the interior of
      the reactor pressure vessel.

The melt interaction with the concrete structure contains, in addition, the molten
steel from the part of the reactor pressure vessel affected by the melt. Those
melts are called Corium (A + R) and Corium (E + R), respectively (R ≏ reactor
pressure vessel part). Table I summarizes the different Corium compositions.

3.   EXPERIMENTAL EQUIPMENT

Figure 2 shows, schematically, the experimental equipment which has been
used to investigate oxidation and hydrogen generation when molten core materials
are penetrating the concrete. The steel reinforced concrete crucible is situated
in a double walled, water cooled, stainless steel chamber. The power is supplied
by arc melting. The total pressure in the test chamber is monitored by an
inductive pressure transducer. The gas analysis is performed using a vacuum
partial-pressure monitor in combination with a pressure reducing system operating
up to a system pressure limit of 2 bars. The calibration of this on-line gas
analysing system was performed using test-gases of known composition. The
water content in the gas phase is recorded by a commercial dew-point measuring
device. The gas temperature above the melt is measured by thermocouples
protected against direct heat radiation from the melt by shadow shields.

4.   EXPERIMENTAL RESULTS

4.1. The water release from heated concrete

Figures 3 and 4 show the results of the temperature-dependent water release
from concrete. The water is released in three steps, at 100–150°C, at 550°C

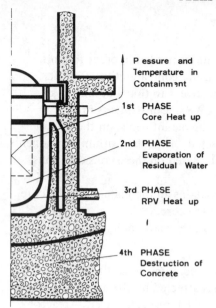

Pressure and
Temperature in
Containment

1st PHASE
Core Heat up

2nd PHASE
Evaporation of
Residual Water

3rd PHASE
RPV Heat up

4th PHASE
Destruction of
Concrete

FIG.1.   The four phases of a hypothetical core-melt-down accident (RPV: reactor pressure vessel).

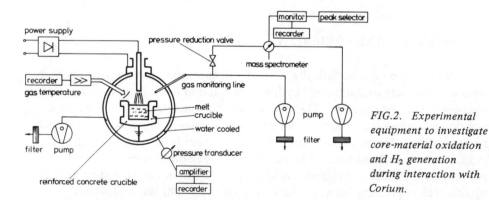

FIG.2.   Experimental equipment to investigate core-material oxidation and H_2 generation during interaction with Corium.

TABLE I.   CORIUM COMPOSITIONS (wt%)

	Corium A	Corium (A + R)	Corium E	Corium (E + R)
UO_2	65	59	35	30
Zirconium	18	16	10	9
1.4550 SS	17	16	55	49
1.6751 SS	—	9	—	12

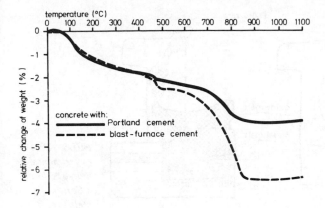

FIG.3. Water release from concrete — relative change of weight.

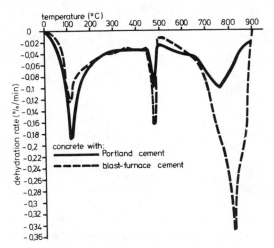

FIG.4. Water release from concrete — dehydration rate.

and at 800°C. The first water-release step, at the lower temperatures, can be attributed to the release of absorbed water. The steps at 500°C and at 800°C are related to dehydration of the calcium silicate phases in the cement [7]. Variations in the type of cement do not influence the release temperatures of the water bound in the concrete, though the distribution of the release fractions will be affected (see Figs 3 and 4). Due to the thermal conductivity and the thermal diffusivity of concrete, the hot zone between the concrete disintegration temperature isotherm at 1300°C [4, 5] and the 100°C isotherm increases as a function of the heat flux with time. With a power input of 100 W/cm², the heat affected zone (100°C isotherm) moves at an average speed of 1 cm/min [3] at the beginning of the interaction experiment.

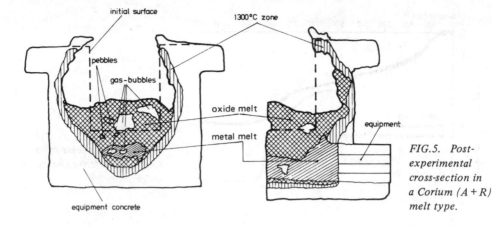

FIG.5. Post-experimental cross-section in a Corium (A + R) melt type.

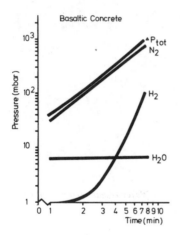

FIG.6. H_2 generation from Corium components.

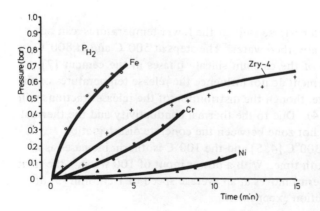

FIG.7. Hydrogen generation during interaction of Zircaloy-4, iron, chromium and nickel with concrete.

TABLE II. H_2 AND H_2O PARTIAL PRESSURES AT THE END OF THE
EXPERIMENTS

Melt	Concrete	Gas composition at the end of the experiments:						
		P_{H_2} (mbar)	P_{H_2O} (mbar)	P_{CO_2} (mbar)	P_{CO} (mbar)	P_{CH_4} (mbar)	P_{H_2}/P_{H_2O}	P_{CO}/P_{CO_2} (footnote a)
Corium (A + R)	basaltic	150	8	–	–	–	1.9×10^1	–
Corium (E + R)	basaltic	200	8	–	–	–	2.5×10^1	–
Corium (A + R)	limestone	180	8	500	b	b	2.3×10^1	1 – 10
Corium (E + R)	limestone	180	8	1000	b	b	2.3×10^1	1 – 10

[a] Calculated.   [b] Identified by gas chromatography.

4.2. Interaction of molten core materials with concrete

Figure 5 shows a typical result from an interaction experiment with Corium
(A + R) and concrete. The experiment was made with a Corium mass of 1 kg.
The power input was 20 kW for 8 minutes. The gas generated during the
interaction penetrates the melt in bubbles 1–2 cm in diameter. The gas
temperature above the melt is about 1200°C. Figure 6 shows the partial
pressures of N_2, H_2 and H_2O and the total pressure recorded. The water vapour
pressure stays nearly constant during the experiments, indicating a complete
reduction of the water released from concrete to give hydrogen and metal oxides.
This fact is also verified by the steeply increasing H_2 partial pressure. The
increasing N_2 partial pressure reflects the heating up of the atmosphere in the
experimental device. If air is used as cover gas, the results show an oxygen-
consuming process, within the first few minutes, due to oxidation of the metallic
components of the melt.

In addition to the interaction experiments with Corium, experiments have
been performed with the individual metallic components of Corium. The hydrogen
generation decreases in the sequence iron, Zircaloy-4, chromium, nickel (Fig. 7).
After taking into account the oxidation of the steel reinforcement of the concrete
entering the melt, the hydrogen generation from nickel melts is negligible.

5.   THERMODYNAMIC EVALUATION OF THE EXPERIMENTAL DATA

Table II shows the H_2 and H_2O partial pressures measured at the end of the
experiments. In addition, CO and CO_2 partial pressures at the end of these
experiments when limestone concrete was tested are given in Table II, also.

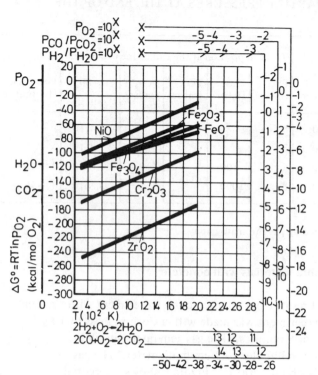

FIG.8.   Richardson diagram for zirconium, chromium, iron and nickel oxides.

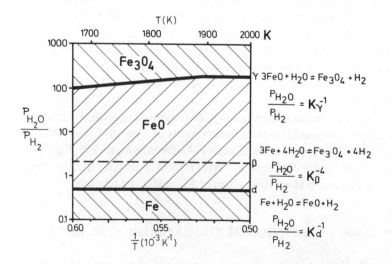

FIG.9.   Equilibrium composition in the Fe–H₂O system.

TABLE III. CORIUM ANALYSIS

	U (wt%)	Zr (wt%)	Si + Ca (wt%)	Fe (wt%)	Cr (wt%)	Ni (wt%)	Other (wt%)	Type of lattice
Metallic phase	1	5	1	40–70	5–10	5–10	remaining	α-Fe, and austenite
Oxidic phase	55–65	14–18	6–10	1–1.5	1–1.5	0.1	remaining	$U_{1-x}Zr_xO_2$

Taking into account the measured partial pressure ratios and the gas temperature, a gas phase oxygen potential of −330 to −550 kJ/mol can be estimated.

Table III contains results from the chemical analyses of a Corium (A + R) melt interacting for up to 30 minutes with basaltic concrete. The metallic phase mainly consists of iron, chromium and nickel. X-ray analysis shows the structures α-Fe and austenite. The oxide phase mainly consists of uranium and zirconium oxides, silica and calcium oxides. X-ray investigations indicate structures typical for U−Zr−O system crystals.

Both chemical and X-ray analyses show clearly that nickel is only in the metallic phase. From the free energies in the Richardson diagram (Fig. 8), it can be concluded that the oxygen potential in the melt is lower than that in the Ni−NiO system. The elements uranium and zirconium are almost completely in the oxide phase. The oxygen potential for the melt is therefore always higher than that for the $Zr−ZrO_2$ and $U−UO_2$ systems. Iron and chromium are found in the metallic and the oxide phases. Therefore the oxygen potential of the core melt has to be near the oxygen potentials of chromium and iron precipitating as oxide phases. For a melting temperature of 1300−1400°C, this assumption results in an oxygen potential for the core melt in the range of −330 to −550 kJ/mol, in accord with the value derived from the partial pressure measurements of the gas phase.

The oxygen potential can be better determined by taking into account the results of the single-component experiments. The experiments resulted in a complete oxidation of zirconium, chromium and iron. Only nickel remained in the metallic state. Considering the well known Fe−O system [8], from Fig. 9 and the results from the H/H_2O partial pressure ratio measurements it can be seen that iron oxidized only to FeO, which has a free energy of −330 kJ/mol at 1400°C. The formation of Fe_3O_4 needs a very high H_2O/H_2 partial pressure ratio, i.e. about 500, not observed in the experiments up to now. The oxygen potential during the interactions of Corium melts with concrete investigated so far was, therefore, about −300 kJ/mol.

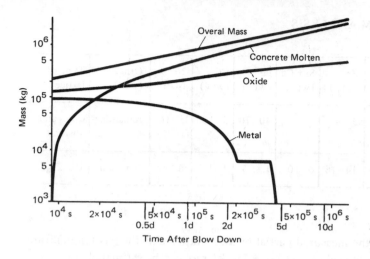

FIG.10. Masses in the melt during concrete destruction.

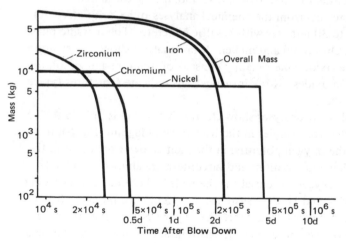

FIG.11. Metals in the melt during concrete destruction.

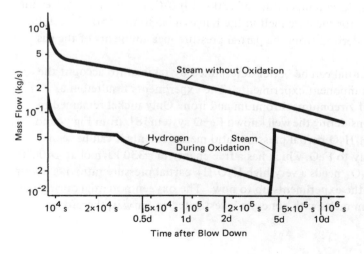

FIG.12. Mass flow into containment atmosphere during concrete destruction.

6.   DEVELOPMENT OF A THERMODYNAMIC MODEL TO COMPUTE
     THE OXIDATION AND THE H_2 GENERATION RATE DURING
     CORIUM/CONCRETE INTERACTION

Based upon the metallurgical phenomena determined experimentally,
analytical models have been developed to describe the oxidation of the metallic
constituents and the release of hydrogen to the containment atmosphere during
the 4th phase, which is the melt/concrete interaction phase [8]. The analytical
models developed were integrated using the computer code BETZ. BETZ analyses
the penetration shape, the concrete destruction rate, the mass and energy release
to the containment and the duration of the 4th phase [9, 10].

The calculation reported in this paper was made for a 1300 MW(e) PWR,
starting with the first accident phase, the heat-up of the core, a 2nd phase
blowdown, followed by a total reflooding of the reactor pressure vessel (RPV)
1150 s after blowdown. The bottom of the RPV will fail and the melt will come
into contact with the concrete floor of the cavity after approximately 9000 s
($\sim 2\frac{1}{2}$ h).

In Fig. 10 the amount of melt, the increase of oxidic melt and the decrease
of metallic melt are shown as functions of time. All results are influenced by
the following two assumptions:

(a)   Based on the experimental investigations reported above, the *sequence* of
      oxidation of the metallic constituents in the melt was assumed to be:
      $Zr \rightarrow ZrO_2$; $Cr \rightarrow Cr_2O_3$; $Fe \rightarrow FeO$. (In order to show the maximum
      H_2 generation rates only (cf. the results reported in §5), oxidation of
      $FeO \rightarrow Fe_3O_4$ and $Ni \rightarrow NiO$ were taken into account also.)
(b)   The oxidation of one element is completed before the oxidation of the
      next element starts.

After approximately 400 000 s (4.6 d), all the metal, including the molten
iron from the steel reinforcement of the concrete, is completely oxidized. After
about 250 000 s, for a period of some 150 000 s, the 6400 kg of nickel remain
unoxidized in the melt because iron and FeO are being oxidized to Fe_3O_4.

At about 900 000 s (10 days) after blowdown, more than 2×10^6 kg of
concrete are molten. The overall mass of the melt at this time amounts to
2.7×10^6 kg.

The progress of oxidation can be seen in Fig. 11 in more detail. It shows
the decrease of the metallic constituents in the above-mentioned sequence:
zirconium, chromium, iron and nickel. After about 17 000 s (~ 5 h) the oxidation
of zirconium is completed and the $Cr \rightarrow Cr_2O_3$ reaction starts. After about
11 hours, all chromium has been oxidized, and the oxidation of iron to FeO starts.
(Up to around 40 000 s the iron mass increases owing to melting of the reinforcing
in the concrete.) Some 200 000 s (2.3 days) after blowdown, the $Fe/FeO \rightarrow Fe_3O_4$

reaction starts. It takes about another two days before all the metals originally contained in the melt are oxidized. During this interval of time large amounts of hydrogen enter the containment atmosphere (see Fig. 12). After those two days, a mixture of some hydrogen deriving from the concrete reinforcing and steam will be leaving the upper surface of the melt. To show the influence of the model involving oxidation, the results derived therewith can be compared with the uppermost line in Fig. 12, which shows the steam flowing off into the containment calculated without allowing for any oxidation of the metallic constituents.

7.   CONCLUSION

In the event of a hypothetical core-melt-down accident, the molten pool inter-acting with the concrete foundation contains large amounts of metals. The large quantities of hydrogen generated by their oxidation will be given off into the containment atmosphere. Based upon experimental results, analytical models have been developed in the 'Kernschmelzen' research project in order to calculate the oxidation of the metallic constituents and the corresponding hydrogen generation rates. The experimental results have shown that zirconium, chromium and iron (to FeO) will be oxidized by the steam released from the heated concrete. Oxidation of FeO to Fe_3O_4 and nickel to NiO has not been experimentally observed, because for these two reactions very high H_2O/H_2 partial pressure ratios would be needed. The numerical calculations show that oxidation of the metals in the Corium melt is completed after approximately 4.6 days from the beginning of an accident. As reported in Ref. [3], overpressure failure of the containment of the PWR would not occur, neither due to the hydrogen generated nor due to subsequent reaction of the hydrogen.

REFERENCES

[1] PEEHS, M., SKOKAN, A., REIMANN, M., Investigations in Germany of the barrier effect of reactor concrete against propagating molten Corium in the case of a hypothetical core melt-down accident of a LWR", ENS/ANS Int. Topical Mtg on Nuclear Power Reactor Safety (Brussels, October 1978), Vol. 1, American Nuclear Society, Mol, Belgium (1979) 215.

[2] SKOKAN, A., HOLLECK, W., PEEHS, M., "Chemical reactions between core melt and concrete and their effects on the course and consequences of a hypothetical core melt-down accident", ENS/ANS Int. Topical Mtg on Nuclear Power Reactor Safety (Brussels, October 1978), Vol. 1, American Nuclear Society, Mol, Belgium (1979) 641.

[3] HASSMANN, K., HERKOMMER, E., HOSEMANN, J.P., KÖRBER, W., REINEKE, H.,
     "Computer codes developed in Germany to analyse hypothetical core melt-down
     accidents", ENS/ANS Int. Topical Mtg on Nuclear Power Reactor Safety (Brussels,
     October 1978), Vol.1, American Nuclear Society, Mol, Belgium (1979) 227.
[4] PEEHS, M., ARTNIK, J., DÖRR, W., LÖSCHER, H., MOLLWITZ, K., PETRI, W.,
     Untersuchung der Wechselwirkung von Kernschmelze und Reaktorbeton, Kraftwerk
     Union final report to Bundesministerium für Forschung und Technologie, BMFT/RS 154
     (May 1978).
[5] PEEHS, M., HASSMANN, K., HAGEN, S., Analysis of a Hypothetical Core Melt-down
     Accident of a PWR, Siemens Forschungs- und Entwicklungsberichte (Feb. 1979).
[6] PEEHS, M., "Investigations of molten 'Corium' phases", Thermodynamics of Nuclear
     Materials, 1974 (Proc. Symp. Vienna, 1974) Vol. 1, IAEA, Vienna (1975) 355.
[7] PEEHS, M., DÖRR, W., "Physical properties and behaviour of heated basaltic concrete",
     Post-Accident Heat Removal Information Exchange, Argonne National Laboratory,
     Illinois, Nov. 1977.
[8] HASSMANN, K., JACOBSEN, X., PEEHS, M., HOSEMANN, J.P., SKOGAN, A.,
     REIMANN, M., DORNER, S., "Abschätzung der H_2-Entwicklung aus der mit Beton
     wechselwirkenden Kernschmelze", Kraftwerk Union final report to the Bundesministerium
     für Forschung und Technologie, BMFT/RS 237 (1978).
[9] ARTNIK, J., HASSMANN, K., JACOBSEN, X., Arbeiten zu den Energiebilanzen nach
     hypothetischem RDB-Versagen unter Berücksichtigung der Betonzerstörung, Kraftwerk
     Union, 1. Technischer Fachbericht zu BMFT/RS 183 (Mai 1977).
[10] ARTNIK, J., HASSMANN, K., PEEHS, M., "The significance of concrete destruction
     during a hypothetical melt-down accident for PWRs", PAHR-Information Exchange
     Meeting, Ispra, Oct. 1978.

Section M

SUMMARY

A SUMMARY OF THE SYMPOSIUM

P.A.G. O'HARE
Argonne National Laboratory,
Argonne, Illinois,
United States of America

At the invitation of the Government of the Federal Republic of Germany, the fifth IAEA International Symposium on Thermodynamics of Nuclear Materials was held at the Jülich Nuclear Research Centre during the week of 29 January to 2 February, 1979. Previous Symposia in this series were held in 1962, 1965, 1967 and 1974. The first three meetings concentrated on fundamental studies of nuclear materials. In 1974 and 1979, however, the application of thermodynamics to practical problems in nuclear energy has come to the fore, although a significant number of basic studies were also described: it is quite clear that the interaction between 'science' and 'technology' has increased significantly.

The 1979 Symposium was attended by 103 participants, representing 21 Member States and two international organizations. A total of 55 scientific papers were presented in nine technical sessions.

An excellent illustration of the interplay between basic science and technology was provided by the session devoted to equation-of-state studies. Here, fundamental thermodynamics and thermophysical properties at very high temperatures were discussed. This information has particular relevance in the area of safety analysis.

One such investigation dealt with vapour pressures produced by very rapid ($\sim 10^{-3}$ to 10^{-6} s) heating of UO_2. In such experiments, account has to be taken of sample stoichiometry, chemical interactions with the sample container and non-equilibrium evaporation brought about by rapid heating, with the consequential large thermal gradients. The problem of non-equilibrium conditions was also dealt with in connection with laser pulse heating of uranium monocarbide. Using this technique, vapour pressures were measured between 6000 and 8000 K. A vapour pressure equation valid between the melting point and 7000 K was derived, together with the enthalpy and entropy of vaporization.

To what extent are the thermophysical and heat transport properties of vaporized reactor core components, specifically UO_2, sodium and caesium, affected by thermal ionization? To help answer this question, calculations were presented which dealt with, inter alia, the degree of thermal ionization and the effect of the plasma state on convective heat transfer at high temperatures. In a related study, thermophysical properties, including the normal reflectivity and emissivity of solid UO_2, UC, ThO_2 and Nd_2O_3, and of liquid Nd_2O_3 up to

383

4800 K, were measured. The chemistry of reactions of molten 'Corium' with concrete was dealt with in two other reports.[1]

Two non-experimental papers related to equations-of-state were also presented. One paper employed a modified version of Eyring's Significant Structure Theory to predict the equation-of-state of hypostoichiometric urania, which is used as a stand-in for other materials that evaporate non-congruently. This model was used to extrapolate partial pressures into the liquid region; as yet, there are no experimental data with which to compare the theoretical predictions. The other theoretical paper in this session dwelt on the contribution of electronic transitions to the heat capacity of UO_2 in the condensed state. It was concluded that, because of the complexity of these transitions, extrapolation of existing data beyond the experimental temperature region would yield unreliable heat capacity values. Evidently, only experimental measurements at the higher temperatures of interest will solve this problem.

The calculation of thermodynamic properties of actinide gases, with specific reference to the effect of unobserved energy levels, was also dealt with. These missing energy levels can have a very significant effect on, for example, the atomic heat capacities at temperatures that are of interest in reactor safety calculations. It was pointed out that, with the exception of thorium, existing compilations of thermodynamic data for the monatomic actinide gases are unreliable above approximately 2000 K.

A significant portion of the Symposium programme was devoted to oxide fuels. Thermodynamic and chemical properties of thoria-urania and thoria-plutonia solid solutions were reviewed, with particular emphasis being given to fast breeder reactor applications. Other topics discussed included fuel/cladding and fuel/coolant interactions and chemical reactions with fission products.

The oxygen potential plays a very important role in determining the chemical and physical behaviour of oxide fuels. One technique was described for measuring the oxygen potentials in urania-plutonia solid solutions by means of thermogravimetry. Another investigation in which the oxygen potential was an important factor was concerned with the use of oxygen buffer/getter materials for limiting fuel/cladding chemical interactions. The optimum placement of these materials is governed to a large extent by the axial diffusion of oxygen through the fuel.

Theoretical studies of oxides were also described. A statistical thermodynamic model incorporating a tetrahedral defect (one oxygen vacancy bonded to two reduced cations) was used to calculate thermodynamic and structural properties of PuO_{2-x} and CeO_{2-x}. Another model was discussed which combined

---

[1] Corium is a term that has been coined to describe the mass comprising various elements deriving from a reactor core after melt down.

theoretical (point-charge model) considerations with experimental (photoelectron spectroscopic) measurements to describe the partial pressure of oxygen in equilibrium with hyperstoichiometric urania.

Diffusion of oxygen in oxide fuels, per se, was the primary subject matter of three papers, two of which dealt with the use of galvanic cells for measuring the change of O/M ratios due to diffusion. A third paper dealt with the Seebeck effect. Diffusion in other systems was discussed. Interdiffusion of krypton and xenon in high-pressure helium was monitored by using ^{85}Kr and ^{127}Xe, hydrogen diffusion in group IV metal hydrides was described, and the only paper in the Symposium dealing with waste management gave results for the diffusion of ^{233}U and ^{238}Pu in waste glasses.

Thermodynamic aspects of the behaviour of fission products were dealt with in five papers. The U-C-(X) and Pu-C-(Y) phase diagrams were updated; new ternary compounds were reported in the U-C-Tc and Pu-C-Cr systems. Phase diagrams were calculated for the Mo-Tc, Tc-Rh and Tc-Pd binaries and for the Mo-Tc-Rh, Mo-Tc-Pd and Mo-Ru-Pd ternaries. Solubilities of a large number of fission product and transition elements in the carbides and nitrides of uranium and plutonium were determined. Empirical rules, devised to represent these results, were used to predict hitherto unmeasured solubilities.

Out-of-pile studies of reactions of caesium, a major fission product, with UO_2, PuO_2, and $(U,Pu)O_2$ to form a variety of uranates were described in two papers. A U-Cs-O phase diagram was constructed. In order to explain some of the experimental observations, formation of compounds such as Cs_3UO_4 and $Cs_{2-x}UO_{4-y}$ (neither of which has yet been isolated) was postulated. Various Cs-Nb-O, Cs-V-O, and Cs-Ti-O phases were proposed to account for the interactions between caesium and oxidized buffer/getter metals within irradiated fuel pins. An examination of the behaviour of solid fission products in irradiated coated particles showed that in UO_2 kernels molybdenum, technetium, ruthenium and palladium form metallic inclusions, whereas in UC_2 and UO_2/UC_2 kernels the fission products form complex carbides containing uranium. Rare-earth fission products form oxides in UO_2 and UO_2/UC_2 kernels, whereas the alkaline earths form carbides even in UO_2/UC_2 kernels.

Four studies specifically related to clad performance were described. A thermodynamic evaluation of the compatibility of thorium carbides with the Fe-Cr-Ni content of clad alloys indicated that solid solutions of $ThNi_5$ and $ThFe_5$ and of Cr_7C_3 and Fe_7C_3 would be the main products of reaction. Of the seven stainless steels considered in this study, the Fe-Cr-Ni content of type 316 was significantly the least reactive. The stability of Fe-Cr-Ni was also the subject of a theoretical study in which the effects of adding niobium, titanium, aluminium, molybdenum, cobalt and carbon to the alloys were assessed. Confirmation was obtained for the ferritizing effects of niobium, titanium and aluminium and the austenitizing effect of carbon. Limited agreement with experimental observations was obtained.

Precipitation in stainless steels was considered. In one investigation, precipitates such as TiN, $Ti_4C_2S_2$ and $M_{23}C_6$ were identified, and thermodynamic calculations yielded precipitation temperatures in good agreement with experimental values. A theoretical study of the radiation-induced precipitation of Ni_3Si could not be extended to complex systems because information on the dose rate/temperature/composition relationships is lacking.

Although the subject of phase diagrams permeates the entire topic of thermodynamics of nuclear materials, one session of the Symposium was set aside specifically for consideration of this topic. This session opened with a review paper entitled "Advanced fuels for fast breeder reactors: A critical assessment of some phase equilibria". The phases considered were U-C-O, Pu-C-O, U-C-N, Pu-C-N, U-N-O and Pu-N-O. There is complete miscibility in the U-C-N and Pu-C-N systems, with the uranium solid solution likely to be ideal. The hypothetical monoxides have significant solubility in the monocarbides, with PuC-PuO being close to ideal, while the UN-UO system may also be close to ideal. This paper also delineated many of the basic data that are still required in order to satisfactorily describe the above systems. The UC-UN and U-Cr-C systems were the subjects of another paper, and the session concluded with a theoretical presentation that dealt with multiple chemical interactions and the application of this approach to the calculation of phase diagrams.

The EMF method for obtaining Gibbs energies of reaction continues to be used extensively in thermodynamic studies of refractory materials. This technique was employed in four investigations described at the present Symposium: thermodynamics of formation of solid solutions of γ-uranium with d-transition metals; determination of the activities of molybdenum in Mo-Pd, Mo-Rh and Mo-Ru-Pd alloys at elevated temperatures; thermodynamic properties of Mg-In-Sn liquid solutions; and the determination of the activities of oxygen and hydrogen in a sodium test loop.

Vaporization data for nuclear materials, most of which operate at high temperatures, are of importance not only, as already discussed, in reactor safety analysis, but also in many other areas of nuclear technology. Five papers which dealt with essentially basic aspects of vaporization thermodynamics were presented. A reassessment of the results for UO_2 was reported; based on several reliable investigations, an "international" average value for the vapour pressure at 2150 K was selected. The enthalpy of sublimation at 298.15 K was given together with recalculated thermodynamic functions for $UO_2(g)$. The vaporization of oxygen-deficient thoria was studied by means of isopiestic and effusion techniques. For ThO_{2-x}, partial molar enthalpies and entropies of solution of oxygen were determined together with the Gibbs energy of formation as a function of temperature. Transpiration and boiling temperature techniques were used with the thorium and uranium tetrahalides; results were obtained for the behaviour of the vapour pressure with change in temperature and for the enthalpies of fusion and vaporization. Although these compounds have been extensively investigated, there are still significant areas of disagreement. High-temperature mass spectrometry was used to study the vaporization of certain alloys which have been proposed for use in HTGCR systems. Chemical activities and activity coefficients of chromium, nickel, iron and cobalt in these alloys were measured. Equilibrium pressure/temperature/composition investigations in the Ti-B-N system yielded a result for the enthalpy of formation of titanium diboride, TiB_2, which does not agree either with the selected value or with more recent calorimetric studies.

Although the preliminary announcement of the Symposium encouraged contributions in the area of fusion, only two papers directly related to this topic were presented. A plenary lecture entitled "The current status of fusion reactor blanket thermodynamics" dealt essentially with three categories of materials that may find use in fusion reactors, namely, liquid lithium, solid lithium alloys and lithium-containing ceramics. Highlights of the paper included a discussion of recent hydrogen isotope sorption studies related to the Li-LiH, Li-LiD, and Li-LiT systems, solubilities in lithium of Li_3N, Li_2O and Li_2C_2 and a summary of thermodynamic properties of lithium-containing materials. Future research work in specific areas was also recommended. In the only other fusion-related paper, thermodynamic data were presented for a number of lithium-containing materials. Among the species of interest considered were the deuterides, oxides and aluminates of lithium. Ionization potentials were also given for gaseous Li, Li_2 and Li_3.

About twenty per cent of the total number of papers presented at the Symposium were placed in the 'Basic Thermodynamic Studies' category. Most of those papers reported new experimental measurements of thermodynamic properties.

One of the compounds suggested as a possible product of the reaction of fission-product caesium with urania is $Cs_2U_4O_{12}$, new results were presented for its low-temperature heat capacity, enthalpy of formation at 298 K, and high-temperature enthalpy increments. These results were used to predict the stability of $Cs_2U_4O_{12}$ as a function of caesium and oxygen pressure. Other solution-calorimetric work yielded enthalpies of formation for non-stoichiometric yttrium hydrides, from which values for the dihydride and trihydride, YH_2 and YH_3, can be deduced. Calorimetric techniques were also used to investigate the thermodynamic instability of solid solutions of aluminium in delta-plutonium at 298 K; enthalpies of formation of two such solutions were reported. The aqueous solubility of magnetite, Fe_3O_4, was determined as a function of temperature, and the Gibbs energy of formation of $HFeO_2^-$(aq.) was derived.

Formation constants that have particular relevance for fuel reprocessing were obtained for complexes of lanthanide and actinide ions with orthophenanthroline and dibutylphosphate, and for chromium with dibutylphosphate.

Formation of complexes was also the subject matter of two other papers, one
of which described a current IAEA-sponsored project concerning the compilation
and critical assessment of thermochemical data for complexes of the actinides;
the other described an empirical procedure for predicting overall stability
constants of metal—ligand complexes.

Calculations of oxide solubility, such as those carried out in the nuclear
industry, are often hindered by the lack of high-temperature Gibbs energy
values for mononuclear ions. A method was described by which such data can
be estimated from the more readily available room temperature Gibbs energies.

Low and high-temperature calorimetric techniques were combined to
investigate the thermodynamics of the following alkaline earth uranates: $MgUO_4$,
$CaUO_4$, $SrUO_4$, $BaUO_4$ and MgU_4O_{12}.

In the area of solid-state chemistry at high temperatures, results were
presented for interactions of metals with uranium monocarbide, monosulphide
and carbosulphide, for the silicothermic reduction of uranium fluorides, and
for the synthesis of thorium monocarbide by carbothermic reduction of the dioxide.

Obituaries

In the course of the Symposium, obituaries for two outstanding scientists
in the field of experimental thermodynamics were presented. Dr. R.J. Ackermann
(Argonne National Laboratory, Argonne, USA) and Prof. O.S. Ivanov
(A.A. Bajkov Institute of Metallurgy, Moscow, USSR) had both been enthusiastic
supporters of Agency activities in thermodynamics for many years.

CONCLUSION

The IAEA Symposium on Thermodynamics of Nuclear Materials continues
to provide a unique international forum for the exchange and publication of
information in this area of research. A perusal of the Proceedings of the five
Symposia held to date gives a vivid picture of the change in emphasis of and
progress in this research over the past two decades. As mentioned earlier, the
interplay between basic and applied thermodynamics appears to have increased
dramatically over the past few years.

Be that as it may, a recurrent theme in this Symposium concerned the lack
of *basic* thermodynamic information for application to the solution of practical
problems or for use in checking theoretical models in such areas as fuel performance
and safety studies. It appears that the rate at which new data become available
is being outstripped by the amount of additional data needed to keep abreast
of current problems.

To help experimentalists, a short list follows of some of the requirements noted by Symposium participants:

- More thermodynamic data are needed at temperatures in excess of 2500–3000 K for use in areas such as reactor safety studies. Included here are items such as heat capacities of condensed state materials and electronic levels for gases;
- Basic thermodynamic data are needed for thorium carbide, for the Cs-U-O system, for caesium titanates and vanadates, for the effect of oxygen dissolution in $UC_{1.9}$ and for the Ti-B system;
- Experimental results are needed for alloys, with particular emphasis on dose rate/composition relationships and on the effects on ternary phases of the addition of a fourth element;
- In fusion thermodynamics, new information is needed concerning the properties of lithium-containing systems such as alloys; sorption and thermodynamic data are lacking for lithium-containing ceramics which are candidate materials for use in reactor blankets.

In this reviewer's opinion, the high scientific standards of the earlier IAEA Symposia were maintained at the Jülich meeting, and it would appear appropriate to hold a sixth symposium in mid or late 1983 to ensure that the experimentalists and the users of the thermodynamic data are kept aware of the current problems in each other's fields.

CHAIRMEN OF SESSIONS

General Chairman	P.A.G. O'HARE	United States of America
Session I	H.R. IHLE	Federal Republic of Germany
Session II	P.E. POTTER	United Kingdom
Session III	A. PATTORET	France
Session IV	D.D. SOOD	India
Session V	H. BLANK	Commission of the European Communities
Session VI	V.V. AKHACHINSKIJ	Union of Soviet Socialist Republics
Session VII	Z. MOSER	Poland
Session VIII	J. FUGER	Belgium
Session IX	M.G. ADAMSON	United States of America

SECRETARIAT OF THE SYMPOSIUM

Scientific
  Secretary: J.D. NAVRATIL   Division of Research and
                                                  Laboratories,
                                                  IAEA

Administrative
  Secretary: G. SEILER   Division of External Relations,
                                                  IAEA

Editor: E.R.A. BECK   Division of Publications,
                                                  IAEA

Records Officer: S.K. DATTA   Division of Languages,
                                                  IAEA

Liaison Officers:
  Government of the   G. HERRMANN   Bundesministerium für
  Federal Republic   Forschung und Technologie,
  of Germany   Bonn

  Local   O. RENN   Kernforschungsanlage
                                                  Jülich GmbH

LIST OF PARTICIPANTS

ARGENTINA

Marajofsky, A.
    Comisión Nacional de Energía Atómica,
Avenida del Libertador 8250, Buenos Aires 1429

AUSTRIA

Holub, F.
    Österreichische Studiengesellschaft für Atomenergie,
Lenaugasse 10, A-1082 Vienna

BELGIUM

Delbrassine, A.
    Centre d'étude de l'énergie nucléaire (SCK/CEN),
Boeretang 200, B-2400 Mol

Drowart, J.D.
    Laboratorium Fysische Chemie, Vrije Universiteit Brussel,
Pleinlaan 2, B-1050 Brussels

Fuger, J.
    Institut de Radiochimie – B16, Université de Liège,
Sart Tilman, B-4000 Liège

Gilissen, R.
    Centre d'étude de l'énergie nucléaire (SCK/CEN),
Boeretang 200, B-2400 Mol

BRAZIL

Cardoso, P.E.
    NUCLEBRAS, Centro de Desenvolvimento
da Tecnologia Nuclear,
3° Andar, Rua da Alfandega 80, Rio de Janeiro

Sette-Camara, A.
    NUCLEBRAS, Centro de Desenvolvimento
da Tecnologia Nuclear,
3° Andar, Rua da Alfandega 80, Rio de Janeiro

BULGARIA

Strezov, A.S.
    Institute of Nuclear Research and Nuclear Energy,
Bulgarian Academy of Sciences,
Boulevard Lenin 72, Sofia 1113

CANADA

Tremaine, P.R.

Atomic Energy of Canada Ltd.,
Whiteshell Nuclear Research Establishment,
Pinawa, Manitoba, R0E 1L0

DENMARK

Sørensen, O. Toft

Risø National Laboratory,
DK-4000 Roskilde

FRANCE

Darras, R.L.

Département de recherche et analyse,
CEA, Centre d'études nucléaires de Saclay,
B.P. 2, F-91190 Gif-sur-Yvette

Ducroux, R.

Département d'études des combustibles
à base de plutonium,
CEA, Centre d'études nucléaires de Fontenay-aux-Roses,
B.P. 6, F-92260 Fontenay-aux-Roses

Dulon, M.

Commissariat à l'énergie atomique,
29-33 Rue de la Fédération,
B.P. 510, F-75752 Paris Cedex 15

Fromont, M.

Département d'études des combustibles
à base de plutonium,
CEA, Centre d'études nucléaires de Fontenay-aux-Roses,
B.P. 6, F-92260 Fontenay-aux-Roses

Geleznikoff, F.

CEA, Centre d'études de Vaujours,
B.P. 7, F-93270 Sevran

Lambert, I.

Département de recherche et analyse,
CEA, Centre d'études nucléaires de Saclay,
B.P. 2, F-91190 Gif-sur-Yvette

Le Marois, G.

Division de chimie,
CEA, Centre d'études nucléaires de Fontenay-aux-Roses,
B.P. 6, F-92260 Fontenay-aux-Roses

Lorenzelli, R.

Département d'études des combustibles
à base de plutonium,
CEA, Centre d'études nucléaires de Fontenay-aux-Roses,
B.P. 6, F-92260 Fontenay-aux-Roses

FRANCE (cont.)

Marcon, J.-P. — Section du plutonium irradié,
CEA, Centre d'études nucléaires de Fontenay-aux-Roses,
B.P. 6, F-92260 Fontenay-aux-Roses

Marcucci, S. — CEA, Centre d'études de Bruyères-le-Châtel,
F-92542 Montrouge Cedex

Musikas, C. — Division de chimie,
CEA, Centre d'études nucléaires de Fontenay-aux-Roses,
B.P. 6, F-92260 Fontenay-aux-Roses

Noe, M.C. — CEA, Centre d'études nucléaires de Cadarache,
B.P. 1, F-13115 Saint-Paul-lez-Durance

Pattoret, A. — Département d'études des combustibles
à base de plutonium,
CEA, Centre d'études nucléaires de Fontenay-aux-Roses,
B.P. 6, F-92260 Fontenay-aux-Roses

GERMAN DEMOCRATIC REPUBLIC

Kreis, E.M. — Ministerium für Elektrotechnik und Elektronik,
Alexanderplatz 6, DDR-1026 Berlin

Reetz, T. — Zentralinstitut für Kernforschung,
Rossendorf, Postfach 19, DDR-8051 Dresden

Ullmann, H. — Zentralinstitut für Kernforschung,
Rossendorf, Postfach 19, DDR-8051 Dresden

GERMANY, FEDERAL REPUBLIC OF

Benz, R. — Institut für Reaktorwerkstoffe,
Kernforschungsanlage Jülich GmbH,
Postfach 1913, D-5170 Jülich 1

Bober, M. — Kernforschungszentrum Karlsruhe GmbH,
Postfach 3640, D-7500 Karlsruhe 1

Brandt, W. — Abt. R-441, B.51, Kraftwerk Union AG,
Hammerbacherstr. 12, Postfach 3220, D-8520 Erlangen

Fenyi, S. — Institut für Datenverarbeitung in der Technik,
Kernforschungszentrum Karlsruhe GmbH,
Postfach 3640, D-7500 Karlsruhe 1

GERMANY, FEDERAL REPUBLIC OF (cont.)

Fischer, E.A.

Institut für Neutronenphysik und Reaktortechnik,
Kernforschungszentrum Karlsruhe GmbH,
Postfach 3640, D-7500 Karlsruhe 1

Förthmann, R.

Institut für Reaktorwerkstoffe,
Kernforschungsanlage Jülich GmbH,
Postfach 1913, D-5170 Jülich 1

Grübmeier, H.

Institut für Reaktorwerkstoffe,
Kernforschungsanlage Jülich GmbH,
Postfach 1913, D-5170 Jülich 1

Guggi, D.J.

Institut für Chemie, 1. Nuklearchemie,
Kernforschungsanlage Jülich GmbH,
Postfach 1913, D-5170 Jülich 1

Heuvel, H.J.

Internationale Atomreaktorbau GmbH (INTERATOM),
Friedrich-Ebert-Strasse, D-5060 Bergisch Gladbach 1

Hilpert, K.

Institut für Chemie, 4. Angewandte Physikalische Chemie,
Kernforschungsanlage Jülich GmbH,
Postfach 1913, D-5170 Jülich 1

Joos, V.E.

Institut für Physikalische Chemie,
Universität Stuttgart,
Pfaffenwaldring 55, D-7000 Stuttgart 80

Kania, M.J.

Institut für Reaktorwerkstoffe,
Kernforschungsanlage Jülich GmbH,
Postfach 1913, D-5170 Jülich 1

Karow, H.U.

Kernforschungszentrum Karlsruhe GmbH,
Postfach 3640, D-7500 Karlsruhe 1

Naoumidis, A.

Institut für Reaktorwerkstoffe,
Kernforschungsanlage Jülich GmbH,
Postfach 1913, D-5170 Jülich 1

Neubert, A.

Institut für Chemie, 1. Nuklearchemie,
Kernforschungsanlage Jülich GmbH,
Postfach 1913, D-5170 Jülich 1

Nickel, H.

Kernforschungsanlage Jülich GmbH,
Postfach 1913, D-5170 Jülich 1

Odoj, R.

Institut für Chemische Technologie,
Kernforschungsanlage Jülich GmbH,
Postfach 1913, D-5170 Jülich 1

GERMANY, FEDERAL REPUBLIC OF (cont.)

Peehs, M.
Kraftwerk Union AG,
Hammerbacherstr. 12, Postfach 3220, D-8520 Erlangen

Scharfer, U.
Kernforschungsanlage Jülich GmbH,
Postfach 1913, D-5170 Jülich 1

Schumacher, G.
Institut für Neutronenphysik und Reaktortechnik,
Kernforschungszentrum Karlsruhe GmbH,
Postfach 3640, D-7500 Karlsruhe 1

Schuster, H.
Kernforschungsanlage Jülich GmbH,
Postfach 1913, D-5170 Jülich 1

Steinwandel, J.
Institut für Physikalische Chemie,
Universität Stuttgart,
Pfaffenwaldring 55, D-7000 Stuttgart 80

Wenzl, H.F.
Kernforschungsanlage Jülich GmbH,
Postfach 1913, D-5170 Jülich 1

Wu, C.H.
Institut für Chemie, 1. Nuklearchemie,
Kernforschungsanlage Jülich GmbH,
Postfach 1913, D-5170 Jülich 1

INDIA

Sood, D.D.
Radiochemistry Division,
Bhabha Atomic Research Centre,
Trombay, Bombay 400 085

IRAQ

Zainel, H.A.
Department of Chemistry, College of Science,
University of Baghdad,
Baghdad

ITALY

De Maria, R.
Dipartimento Reattori Veloci,
Comitato Nazionale per l'Energia Nucleare,
Via dell'Arcoveggio 56/23, I-40129 Bologna

JAPAN

Yamawaki, M.

Research Center for Nuclear Science and Technology,
University of Tokyo,
2-11-16 Yayoi, Bunkyo-ku, Tokyo 113

THE NETHERLANDS

Cordfunke, E.H.P.

Netherlands Energy Research Foundation (ECN),
Westerduinweg 3, PO Box 1, Petten (N.H.)

Nagel, W.

N.V. KEMA,
Utrechtseweg 310, Arnhem

Prins, G.

Netherlands Energy Research Foundation (ECN),
Westerduinweg 3, PO Box 1, Petten (N.H.)

POLAND

Moser, Z.

Institute for Metal Research,
Polish Academy of Sciences,
25 Reymonta Street, PL-30059 Cracow

SOUTH AFRICA

Lessing, J.G.V.

South African Atomic Energy Board,
National Nuclear Research Centre,
Pelindaba, Private Bag X256, Pretoria 0001

SWITZERLAND

Groner, P.K.

Eidgenössisches Institut für Reaktorforschung (EIR),
CH-5303 Würenlingen

UNION OF SOVIET SOCIALIST REPUBLICS

Akhachinskij, V.V.

USSR State Committee on the Utilization of Atomic Energy,
Moscow

Panov, A.S.

A.A. Bajkov Institute of Metallurgy,
Academy of Sciences of the USSR,
Leninskij Prospect 49, Moscow

UNITED KINGDOM

Bussey, P.R.

Health and Safety Executive,
Nuclear Installations Inspectorate,
Thames House North, Millbank, London SW19 4QL

Chilton, G.R.

United Kingdom Atomic Energy Authority,
Windscale Nuclear Power Development Laboratories,
Windscale Works, Sellafield, Seascale, Cumbria CA20 1PF

Ewart, F.T.

Atomic Energy Research Establishment,
Harwell, Didcot, Oxfordshire OX11 0RA

Findlay, J.R.

Atomic Energy Research Establishment,
Harwell, Didcot, Oxfordshire OX11 0RA

Hesketh, R.

Central Electricity Generating Board,
Berkeley Nuclear Laboratories,
Berkeley, Gloucestershire GL13 9PB

MacInnes, D.A.

Safety and Reliability Directorate,
United Kingdom Atomic Energy Authority,
Wigshaw Lane, Culcheth, Warrington WA3 4NE

Marschall, E.M.

School of Molecular Sciences, University of Sussex,
Falmer, Brighton, Sussex BN1 9Q5

Pedley, J.B.

School of Molecular Sciences, University of Sussex,
Falmer, Brighton, Sussex BN1 9Q5

Potter, P.E.

Chemistry Division,
Atomic Energy Research Establishment,
Harwell, Didcot, Oxfordshire OX11 0RA

Rand, M.H.

Atomic Energy Research Establishment,
Harwell, Didcot, Oxfordshire OX11 0RA

Watkin, J.S.

Springfields Nuclear Power Laboratories (B328),
United Kingdom Atomic Energy Authority,
Springfields Works, Springfields, Salwick, Preston PR4 0RR

UNITED STATES OF AMERICA

Adamson, M.G.

General Electric Company,
Vallecitos Nuclear Center,
Pleasanton, CA 94566

UNITED STATES OF AMERICA (cont.)

Benson, D.A.   Thermomechanical and Physical Research Division 5534,
                                      Sandia Laboratories,
                                      Albuquerque, NM 87185

Besmann, T.M.   Chemical Technology Division, Bldg 4501,
                                      Oak Ridge National Laboratory,
                                      PO Box X, Oak Ridge, TN 37830

Campana, R.J.   General Atomic Company,
                                      PO Box 81608, San Diego, CA 92138

Gilles, P.W.   Department of Chemistry, University of Kansas,
                                      Lawrence, KS 66044

Hoch, M.   Department of Materials Science and
                                           Metallurgical Engineering,
                                      University of Cincinnati,
                                      Cincinnati, OH 45221

Morss, L.R.   Rutgers University,
                                      New Brunswick, NJ 08903

Murch, G.E.   Chemistry Division, Argonne National Laboratory,
                                      9700 South Cass Avenue, Argonne, IL 60439

O'Hare, P.A.G.   Chemistry Division, Argonne National Laboratory,
                                      9700 South Cass Avenue, Argonne, IL 60439

Powers, D.A.   Division 5831, Sandia Laboratories,
                                      Albuquerque, NM 87185

Spear, K.E.   270 Materials Research Laboratory,
                                      Pennsylvania State University,
                                      University Park, PA 16802

Szasz, G.J.   General Electric Co. (USA),
                                      Pelikanstrasse 37, Zurich, Switzerland

Tetenbaum, M.   Chemical Engineering Division,
                                      Argonne National Laboratory,
                                      9700 South Cass Avenue, Argonne, IL 60439

Veleckis, E.   Chemical Engineering Division,
                                      Argonne National Laboratory,
                                      9700 South Cass Avenue, Argonne, IL 60439

Westrum Jr., E.F.   Department of Chemistry, University of Michigan,
                                      Ann Arbor, MI 48109

ORGANIZATIONS

COMMISSION OF THE EUROPEAN COMMUNITIES (CEC)

Babelot, J.F.	CEC, European Institute for Transuranium Metals, Forschungsanstalt Karlsruhe, Postfach 2266, D-7500 Karlsruhe, Federal Republic of Germany
Benedict, U.	CEC, European Institute for Transuranium Metals, Forschungsanstalt Karlsruhe, Postfach 2266, D-7500 Karlsruhe, Federal Republic of Germany
Blank, H.	CEC, European Institute for Transuranium Metals, Forschungsanstalt Karlsruhe, Postfach 2266, D-7500 Karlsruhe, Federal Republic of Germany
Lindner, R.	CEC, European Institute for Transuranium Metals, Forschungsanstalt Karlsruhe, Postfach 2266, D-7500 Karlsruhe, Federal Republic of Germany
Long, K.A.	CEC, European Institute for Transuranium Metals, Forschungsanstalt Karlsruhe, Postfach 2266, D-7500 Karlsruhe, Federal Republic of Germany
Magill, J.	CEC, European Institute for Transuranium Metals, Forschungsanstalt Karlsruhe, Postfach 2266, D-7500 Karlsruhe, Federal Republic of Germany
Malein, M.A.	200, rue de la Loi, Brussels B-1049, Belgium
Manes, L.	CEC, European Institute for Transuranium Metals, Forschungsanstalt Karlsruhe, Postfach 2266, D-7500 Karlsruhe, Federal Republic of Germany
Matzke, Hj.	CEC, European Institute for Transuranium Metals, Forschungsanstalt Karlsruhe, Postfach 2266, D-7500 Karlsruhe, Federal Republic of Germany
Ohse, R.W.	CEC, European Institute for Transuranium Metals, Forschungsanstalt Karlsruhe, Postfach 2266, D-7500 Karlsruhe, Federal Republic of Germany
Pierini, G.	Euratom Joint Research Centre, I-21020 Ispra (VA), Italy
Scotti, A.	Euratom Joint Research Centre, I-21020 Ispra (VA), Italy

INTERNATIONAL ATOMIC ENERGY AGENCY (IAEA)

Kakihana, H.   Deputy Director General,
                                Department of Research and Isotopes,
                                IAEA, PO Box 100, A-1400 Vienna, Austria

Sundermann, H.   IAEA, PO Box 100, A-1400 Vienna, Austria

AUTHOR INDEX

Roman numerals are volume numbers.
Italic numerals refer to the first page of a paper by the author concerned.
Upright numerals denote comments and questions in discussions.
Literature references are not indexed.

TRANSLITERATION INDEX

PRE-PRINT SYMBOL INDEX

Symbol IAEA-SM-	Session	Vol.	Page	Lang.	Symbol IAEA-SM-	Session	Vol.	Page	Lang.
236/01	VII	2	59	E	236/48	–	withdrawn		
236/02	VII	2	75	E	236/50	IX	2	331	E
236/03	II	withdrawn			236/51	–	withdrawn		
236/04	II	1	171	E	236/52	–	withdrawn		
236/05	VI	1	311	E	236/53	I	1	73	E
236/06	IV	1	405	E	236/54	VIII	2	143	E
236/07	IV	1	369	E	236/55	IX	withdrawn		
236/09	V	1	453	E	236/56	VI	2	3	E
236/11	VII	2	47	E	236/57	I	1	29	E
236/12	VII	2	89	F	236/58	II	2	351	E
236/13	–	withdrawn			236/59	I	1	11	E
236/14	–	withdrawn			236/60	II	1	93	E
236/15	VI	2	31	E	236/61	IX	2	277	E
236/17	II	1	115	E	236/62	IV	1	333	E
236/19	V	1	565	E	236/63	V	1	503	E
236/21	II	1	141	E	236/65	V	1	289	E
236/22	II	1	155	E	236/66	–	withdrawn		
236/23	VI	withdrawn, *but see:*			236/68	III	1	219	E
		1	325	E	236/69	IV	1	427	E
236/24	IX	2	271	E	236/71	VI	withdrawn		R
236/25	IV	1	439	E	236/72	III	withdrawn		R
236/26	II	2	367	E	236/75	VIII	2	155	R
236/27	I	1	61	E	236/78	VII	2	105	F
236/28	III	1	249	E	236/80	IV	1	273	E
236/29	IX	2	185	E	236/85	VIII	withdrawn		R
236/31	I	1	45	E	236/87	V	1	539	F
236/34	VIII	2	125	E	236/89	IV	1	385	F
236/35	III	1	263	E	236/91	VIII	withdrawn		
236/36	IV	withdrawn			236/92	VIII	2	229	R
236/37	II	1	129	E	236/93	III	1	235	R
236/39	III	1	197	E	236/94	IX	2	247	R
236/40	VIII	2	195	E	236/95	–	withdrawn		R
236/42	V	1	471	E	236/96	VIII	2	161	R
236/43	IV	1	357	E	236/97	VI	1	303	R
236/44	IX	2	297	E	236/98	III	2	171	R
236/45	IX	2	315	E					
236/46	III	withdrawn			Summary	–	2	383	E

SUBJECT INDEX

This index of Papers and Discussions is divided into two parts. Part I indexes key phrases and Part II indexes elements, alloy systems and compounds by chemical symbol. Although the two parts do overlap, they are intended to be complementary and there is no guaranteed correspondence between them.

Volume numbers are shown in Roman numerals, page numbers followed by an asterisk indicate an extended reference, and F and T stand for figures and tables, respectively. Semi-colons indicate the end of a reference to a particular paper.[1]

In the alphabetical ordering, prepositions etc. and words in parentheses are ignored, and general references come before the specific.

PART I

---

[1] The following examples are given from Parts I and II:

"entropy of fusion of UO_2: I 131*, T1". This means that there is an extended reference starting on page 131 of Vol.I, and Table I of the paper should also be consulted.

"Cr activity in carbides: I T3(66)". This means that Table III on page 66 should be consulted.

PART II

FACTORS FOR CONVERTING SOME OT THE MORE COMMON UNITS TO INTERNATIONAL SYSTEM OF UNITS (SI) EQUIVALENTS

NOTES:

(1) SI base units are the metre (m), kilogram (kg), second (s), ampere (A), kelvin (K), candela (cd) and mole (mol).

(2) ▶ indicates SI derived units and those accepted for use with SI;
▷ indicates additional units accepted for use with SI for a limited time.
[For further information see The International System of Units (SI), 1977 ed., published in English by HMSO, London, and National Bureau of Standards, Washington, DC, and International Standards ISO-1000 and the several parts of ISO-31 published by ISO, Geneva.]

(3) The correct abbreviation for the unit in column 1 is given in column 2.

(4) ✳ indicates conversion factors given exactly; other factors are given rounded, mostly to 4 significant figures.
≡ indicates a definition of an SI derived unit: [ ] in column 3+4 enclose factors given for the sake of completeness.

Column 1 Multiply data given in:	Column 2	Column 3 by:	Column 4 to obtain data in:	
Radiation units				
▶ becquerel	1 Bq	(has dimensions of s^{-1})		
disintegrations per second (= dis/s)	$1\ s^{-1}$	$\equiv 1.00 \times 10^0$	Bq	✳
▷ curie	1 Ci	$= 3.70 \times 10^{10}$	Bq	✳
▷ roentgen	1 R	$[= 2.58 \times 10^{-4}$	C/kg]	✳
▶ gray	1 Gy	$[\equiv 1.00 \times 10^0$	J/kg]	✳
▷ rad	1 rad	$= 1.00 \times 10^{-2}$	Gy	✳
sievert *(radiation protection only)*	1 Sv	$[= 1.00 \times 10^0$	J/kg]	✳
rem *(radiation protection only)*	1 rem	$[= 1.00 \times 10^{-2}$	J/kg]	✳
Mass				
▶ unified atomic mass unit ($\frac{1}{12}$ of the mass of ^{12}C)	1 u	$[= 1.660\,57 \times 10^{-27}$	kg, approx.]	
▶ tonne (= metric ton)	1 t	$[= 1.00 \times 10^3$	kg]	✳
pound mass (avoirdupois)	1 lbm	$= 4.536 \times 10^{-1}$	kg	
ounce mass (avoirdupois)	1 ozm	$= 2.835 \times 10^1$	g	
ton (long) (= 2240 lbm)	1 ton	$= 1.016 \times 10^3$	kg	
ton (short) (= 2000 lbm)	1 short ton	$= 9.072 \times 10^2$	kg	
Length				
statute mile	1 mile	$= 1.609 \times 10^0$	km	
nautical mile (international)	1 n mile	$= 1.852 \times 10^0$	km	✳
yard	1 yd	$= 9.144 \times 10^{-1}$	m	✳
foot	1 ft	$= 3.048 \times 10^{-1}$	m	✳
inch	1 in	$= 2.54 \times 10^1$	mm	✳
mil (= 10^{-3} in)	1 mil	$= 2.54 \times 10^{-2}$	mm	✳
Area				
▷ hectare	1 ha	$[= 1.00 \times 10^4$	m^2]	✳
▷ barn *(effective cross-section, nuclear physics)*	1 b	$[= 1.00 \times 10^{-28}$	m^2]	✳
square mile, (statute mile)^2	1 mile^2	$= 2.590 \times 10^0$	km^2	
acre	1 acre	$= 4.047 \times 10^3$	m^2	
square yard	1 yd^2	$= 8.361 \times 10^{-1}$	m^2	
square foot	1 ft^2	$= 9.290 \times 10^{-2}$	m^2	
square inch	1 in^2	$= 6.452 \times 10^2$	mm^2	
Volume				
▶ litre	1 l *or* 1 ltr	$[= 1.00 \times 10^{-3}$	m^3]	✳
cubic yard	1 yd^3	$= 7.646 \times 10^{-1}$	m^3	
cubic foot	1 ft^3	$= 2.832 \times 10^{-2}$	m^3	
cubic inch	1 in^3	$= 1.639 \times 10^4$	mm^3	
gallon (imperial)	1 gal (UK)	$= 4.546 \times 10^{-3}$	m^3	
gallon (US liquid)	1 gal (US)	$= 3.785 \times 10^{-3}$	m^3	
Velocity, acceleration				
foot per second (= fps)	1 ft/s	$= 3.048 \times 10^{-1}$	m/s	✳
foot per minute	1 ft/min	$= 5.08 \times 10^{-3}$	m/s	✳
mile per hour (= mph)	1 mile/h	$=\begin{cases}4.470 \times 10^{-1} \\ 1.609 \times 10^0\end{cases}$	m/s km/h	
▷ knot (international)	1 knot	$= 1.852 \times 10^0$	km/h	✳
free fall, standard, g		$= 9.807 \times 10^0$	m/s^2	
foot per second squared	1 ft/s^2	$= 3.048 \times 10^{-1}$	m/s^2	✳

Column 1 Multiply data given in:	Column 2	Column 3 by:	Column 4 to obtain data in:
Density, volumetric rate			
pound mass per cubic inch	$1\ lbm/in^3$	$= 2.768 \times 10^4$	kg/m^3
pound mass per cubic foot	$1\ lbm/ft^3$	$= 1.602 \times 10^1$	kg/m^3
cubic feet per second	$1\ ft^3/s$	$= 2.832 \times 10^{-2}$	m^3/s
cubic feet per minute	$1\ ft^3/min$	$= 4.719 \times 10^{-4}$	m^3/s
Force			
▶ newton	$1\ N$	$[\equiv 1.00 \times 10^0$	$m \cdot kg \cdot s^{-2}]$ ✻
dyne	$1\ dyn$	$= 1.00 \times 10^{-5}$	N ✻
kilogram force (= kilopond (kp))	$1\ kgf$	$= 9.807 \times 10^0$	N
poundal	$1\ pdl$	$= 1.383 \times 10^{-1}$	N
pound force (avoirdupois)	$1\ lbf$	$= 4.448 \times 10^0$	N
ounce force (avoirdupois)	$1\ ozf$	$= 2.780 \times 10^{-1}$	N
Pressure, stress			
▶ pascal	$1\ Pa$	$[\equiv 1.00 \times 10^0$	$N/m^2]$ ✻
▷ atmosphere[a], standard	$1\ atm$	$= 1.013\ 25 \times 10^5$	Pa ✻
▷ bar	$1\ bar$	$= 1.00 \times 10^5$	Pa ✻
centimetres of mercury ($0°C$)	$1\ cmHg$	$= 1.333 \times 10^3$	Pa
dyne per square centimetre	$1\ dyn/cm^2$	$= 1.00 \times 10^{-1}$	Pa ✻
feet of water ($4°C$)	$1\ ftH_2O$	$= 2.989 \times 10^3$	Pa
inches of mercury ($0°C$)	$1\ inHg$	$= 3.386 \times 10^3$	Pa
inches of water ($4°C$)	$1\ inH_2O$	$= 2.491 \times 10^2$	Pa
kilogram force per square centimetre	$1\ kgf/cm^2$	$= 9.807 \times 10^4$	Pa
pound force per square foot	$1\ lbf/ft^2$	$= 4.788 \times 10^1$	Pa
pound force per square inch (= psi)[b]	$1\ lbf/in^2$	$= 6.895 \times 10^3$	Pa
torr ($0°C$) (= mmHg)	$1\ torr$	$= 1.333 \times 10^2$	Pa
Energy, work, quantity of heat			
▶ joule ($\equiv W \cdot s$)	$1\ J$	$[\equiv 1.00 \times 10^0$	$N \cdot m]$ ✻
▶ electronvolt	$1\ eV$	$[= 1.602\ 19 \times 10^{-19}$	$J, approx.]$
British thermal unit (International Table)	$1\ Btu$	$= 1.055 \times 10^3$	J
calorie (thermochemical)	$1\ cal$	$= 4.184 \times 10^0$	J ✻
calorie (International Table)	$1\ cal_{IT}$	$= 4.187 \times 10^0$	J
erg	$1\ erg$	$= 1.00 \times 10^{-7}$	J ✻
foot-pound force	$1\ ft \cdot lbf$	$= 1.356 \times 10^0$	J
kilowatt-hour	$1\ kW \cdot h$	$= 3.60 \times 10^6$	J ✻
kiloton explosive yield (PNE) ($\equiv 10^{12}$ g-cal)	$1\ kt\ yield$	$\simeq 4.2 \times 10^{12}$	J
Power, radiant flux			
▶ watt	$1\ W$	$[\equiv 1.00 \times 10^0$	$J/s]$ ✻
British thermal unit (International Table) per second	$1\ Btu/s$	$= 1.055 \times 10^3$	W
calorie (International Table) per second	$1\ cal_{IT}/s$	$= 4.187 \times 10^0$	W
foot-pound force/second	$1\ ft \cdot lbf/s$	$= 1.356 \times 10^0$	W
horsepower (electric)	$1\ hp$	$= 7.46 \times 10^2$	W ✻
horsepower (metric) (= ps)	$1\ ps$	$= 7.355 \times 10^2$	W
horsepower (550 ft·lbf/s)	$1\ hp$	$= 7.457 \times 10^2$	W

Temperature

▶ temperature in degrees Celsius, t
    where T is the thermodynamic temperature in kelvin
    and T_0 is defined as 273.15 K

$$t = T - T_0$$

degree Fahrenheit	$t_{°F} - 32$	t (in degrees Celsius) ✻
degree Rankine	$T_{°R}$	$\times \left(\dfrac{5}{9}\right)$ gives T (in kelvin) ✻
degrees of temperature difference[c]	$\Delta T_{°R}\ (= \Delta t_{°F})$	$\Delta T\ (= \Delta t)$ ✻

Thermal conductivity[c]

$1\ Btu \cdot in/(ft^2 \cdot s \cdot °F)$	(International Table Btu)	$= 5.192 \times 10^2$	$W \cdot m^{-1} \cdot K^{-1}$
$1\ Btu/(ft \cdot s \cdot °F)$	(International Table Btu)	$= 6.231 \times 10^3$	$W \cdot m^{-1} \cdot K^{-1}$
$1\ cal_{IT}/(cm \cdot s \cdot °C)$		$= 4.187 \times 10^2$	$W \cdot m^{-1} \cdot K^{-1}$

[a] atm abs, ata: atmospheres absolute;
    atm (g), atü: atmospheres gauge.

[b] lbf/in^2 (g)   (= psig): gauge pressure;
    lbf/in^2 abs (= psia): absolute pressure.

[c] The abbreviation for temperature difference, deg (= degK = degC), is no longer acceptable as an SI unit.

HOW TO ORDER IAEA PUBLICATIONS

An exclusive sales agent for IAEA publications, to whom all orders
and inquiries should be addressed, has been appointed
in the following country:

UNITED STATES OF AMERICA   UNIPUB, 345 Park Avenue South, New York, NY 10010

In the following countries IAEA publications may be purchased from the
sales agents or booksellers listed or through your
major local booksellers. Payment can be made in local
currency or with UNESCO coupons.

ARGENTINA	Comisión Nacional de Energía Atomica, Avenida del Libertador 8250, RA-1429 Buenos Aires
AUSTRALIA	Hunter Publications, 58 A Gipps Street, Collingwood, Victoria 3066
BELGIUM	Service Courrier UNESCO, 202, Avenue du Roi, B-1060 Brussels
CZECHOSLOVAKIA	S.N.T.L., Spálená 51, CS-113 02 Prague 1
	Alfa, Publishers, Hurbanovo námestie 6, CS-893 31 Bratislava
FRANCE	Office International de Documentation et Librairie, 48, rue Gay-Lussac, F-75240 Paris Cedex 05
HUNGARY	Kultura, Hungarian Foreign Trading Company P.O. Box 149, H-1389 Budapest 62
INDIA	Oxford Book and Stationery Co., 17, Park Street, Calcutta-700 016
	Oxford Book and Stationery Co., Scindia House, New Delhi-110 001
ISRAEL	Heiliger and Co., Ltd., Scientific and Medical Books, 3, Nathan Strauss Street, Jerusalem 94227
ITALY	Libreria Scientifica, Dott. Lucio de Biasio "aeiou", Via Meravigli 16, I-20123 Milan
JAPAN	Maruzen Company, Ltd., P.O. Box 5050, 100-31 Tokyo International
NETHERLANDS	Martinus Nijhoff B.V., Booksellers, Lange Voorhout 9-11, P.O. Box 269, NL-2501 The Hague
PAKISTAN	Mirza Book Agency, 65, Shahrah Quaid-e-Azam, P.O. Box 729, Lahore 3
POLAND	Ars Polona-Ruch, Centrala Handlu Zagranicznego, Krakowskie Przedmiescie 7, PL-00-068 Warsaw
ROMANIA	Ilexim, P.O. Box 136-137, Bucarest
SOUTH AFRICA	Van Schaik's Bookstore (Pty) Ltd., Libri Building, Church Street, P.O. Box 724, Pretoria 0001
SPAIN	Diaz de Santos, Lagasca 95, Madrid-6
	Diaz de Santos, Balmes 417, Barcelona-6
SWEDEN	AB C.E. Fritzes Kungl. Hovbokhandel, Fredsgatan 2, P.O. Box 16356, S-103 27 Stockholm
UNITED KINGDOM	Her Majesty's Stationery Office, Agency Section PDIB, P.O. Box 569, London SE1 9NH
U.S.S.R.	Mezhdunarodnaya Kniga, Smolenskaya-Sennaya 32-34, Moscow G-200
YUGOSLAVIA	Jugoslovenska Knjiga, Terazije 27, P.O. Box 36, YU-11001 Belgrade

---

Orders from countries where sales agents have not yet been appointed and
requests for information should be addressed directly to:

Division of Publications
International Atomic Energy Agency
Wagramerstrasse 5, P.O. Box 100, A-1400 Vienna, Austria

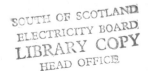